The Definitive Guide to Azure Data Engineering

Modern ELT, DevOps, and Analytics on the Azure Cloud Platform

Ron C. L'Esteve

Apress®

The Definitive Guide to Azure Data Engineering: Modern ELT, DevOps, and Analytics on the Azure Cloud Platform

Ron C. L'Esteve
Chicago, IL, USA

ISBN-13 (pbk): 978-1-4842-7181-0   ISBN-13 (electronic): 978-1-4842-7182-7
https://doi.org/10.1007/978-1-4842-7182-7

Managing Director, Apress Media LLC: Welmoed Spahr
Acquisitions Editor: Jonathan Gennick
Development Editor: Laura Berendson
Coordinating Editor: Jill Balzano

Cover image designed by Freepik (www.freepik.com)

Distributed to the book trade worldwide by Springer Science+Business Media LLC, 1 New York Plaza, Suite 4600, New York, NY 10004. Phone 1-800-SPRINGER, fax (201) 348-4505, e-mail orders-ny@springer-sbm.com, or visit www.springeronline.com. Apress Media, LLC is a California LLC and the sole member (owner) is Springer Science + Business Media Finance Inc (SSBM Finance Inc). SSBM Finance Inc is a **Delaware** corporation.

For information on translations, please e-mail booktranslations@springernature.com; for reprint, paperback, or audio rights, please e-mail bookpermissions@springernature.com.

Apress titles may be purchased in bulk for academic, corporate, or promotional use. eBook versions and licenses are also available for most titles. For more information, reference our Print and eBook Bulk Sales web page at http://www.apress.com/bulk-sales.

Any source code or other supplementary material referenced by the author in this book is available to readers on GitHub via the book's product page, located at www.apress.com/9781484271810. For more detailed information, please visit http://www.apress.com/source-code.

Printed on acid-free paper

For Mom and Dad.

Table of Contents

About the Author

Ron C. L'Esteve is a professional author residing in Chicago, IL, USA. His passion for Azure Data Engineering originates from his deep experience with designing, implementing, and delivering modern Azure data projects for numerous clients. Ron is a trusted technology leader and digital innovation strategist, responsible for scaling key data architectures, defining the road map and strategy for the future of data and business intelligence (BI) needs, and challenging customers to grow by thoroughly understanding the fluid business opportunities and enabling change by translating them into high-quality and sustainable technical solutions that solve the most complex challenges and promote digital innovation and transformation. He applies a practical and business-oriented approach of taking transformational ideas from concept to scale. Ron is an advocate for data excellence across industries and consulting practices and empowers self-service data, BI, and AI through his contributions to the Microsoft technical community.

About the Technical Reviewer

Greg Low is one of the better-known database consultants in the world. In addition to deep technical skills, Greg has experience with business and project management and is known for his pragmatic approach to solving issues. His skill levels at dealing with complex situations and his intricate knowledge of the industry have seen him cut through difficult problems.

Microsoft has specifically recognized his capabilities and appointed him to the Regional Director program. They describe it as consisting of "150 of the world's top technology visionaries chosen specifically for their proven cross-platform expertise, community leadership, and commitment to business results."

Greg leads a boutique data consultancy firm called SQL Down Under. His clients range from large tier 1 organizations to start-ups.

Greg is a long-term Data Platform MVP and considered one of the foremost consultants in the world on SQL Server and Microsoft data-related technologies. He has provided architectural guidance for some of the largest SQL Server implementations in the world and helped them to resolve complex issues. Greg was one of the two people first appointed as SQL Server Masters worldwide. Microsoft use him to train their own staff.

For several years, Greg served on the global board for the Professional Association for SQL Server. He is particularly proud of having helped it triple the size of its community and, more importantly to him, taken it from being 90% US based to being a truly global community with 60% of chapters outside the United States.

A talented trainer and presenter, Greg is known for his ability to explain complex concepts with great clarity to people of all skill levels. He is regularly invited to present at top-level tier 1 conferences around the world. Greg's SQL Down Under podcast has a regular audience of over 40,000 listeners.

Outside of work and family, Greg's current main passion is learning Mandarin Chinese, and he is determined to learn to read, write, speak, and understand it clearly.

Acknowledgments

Writing this book has been both a solitary and accompanied journey with sacrifices and victories along the way. Thank you to all who have supported me on the path to completing this book.

Introduction

With the numerous cloud computing technologies being at the forefront of the modern-day data architectural and engineering platforms, Microsoft Azure's cloud platform has contributed over 200 products and services that have been specifically designed to solve complex data challenges, empower self-service data engineering, and pave the way for the future of data and AI.

Navigating through these many offerings in the Azure Data Platform can become daunting for aspiring Azure Data Engineers, architects, consultants, and organizations that are seeking to build scalable, performant, and production-ready data solutions. This book is intended to uncover many of the complexities within the Azure data ecosystem with ease through structured end-to-end scenario-based demonstrations, exercises, and reusable architectural patterns for working with data in Azure and building highly performant data ingestion and ELT pipelines.

As Azure continues to introduce numerous data services to their ever-growing and evolving platform, this book will demystify many of the complexities of Azure Data Engineering with ease and introduce you to tried, tested, and production-ready patterns and pipelines for data of all different volumes, varieties, and velocities.

Additionally, you will be introduced to the many capabilities of bringing value and insights to your data through real-time and advanced analytics, continuously integrating and deploying your data ingestion pipelines, and getting started with many Azure data services to help you progress through your journey within the Azure Data Engineering ecosystem.

PART I

Getting Started

CHAPTER 1

The Tools and Prerequisites

This chapter will cover a few key tips for getting started with Azure Data Engineering by empowering you to master the traditional Microsoft Business Intelligence (BI) Stack. This chapter will help you to understand Azure's Modern Enterprise Data and Analytics Platform, the various big data options and performance tuning recommendations, along with fundamental requirements of the Azure Data Engineer Associate certification. Finally, this chapter will cover the value of expanding your knowledge across other Azure specialties to address the business value of an Azure data solution and introduce you options on getting started with Azure Data Engineering through Azure Portal. The chapter will conclude with an introduction to the Azure services that will be covered in this book along with the Azure services that will not be covered.

Master the Traditional Microsoft Business Intelligence Stack

Many of Azure's Data Platform tools have their roots in the traditional Microsoft SQL Server BI Platform. For example, Azure Data Factory's (ADF) Mapping Data Flows (MDF) is much like SQL Server Integration Services (SSIS), while Azure Analysis Services has its roots in SQL Server Analysis Services (SSAS). While Azure has many complexities that are not prevalent in Microsoft's traditional BI Stack, having a strong understanding of SSIS, SSRS, SSAS, T-SQL, C#, Data Warehousing, and more will help with gaining a stronger understanding of Azure's data services. Many organizations have a very similar history of working with the traditional Microsoft BI Stack for many years and maybe searching for Azure Data Engineers to help with pioneering their journey into the Modern Enterprise Data and Analytics Platform. By having knowledge of and experience

3

© Ron C. L'Esteve 2021
R. C. L'Esteve, *The Definitive Guide to Azure Data Engineering*, https://doi.org/10.1007/978-1-4842-7182-7_1

in the traditional tooling that those organizations currently have on-premises, you will be able to relate well to the environment and demonstrate a mastery of the traditional Microsoft BI Stack along with data warehousing concepts. For example, understanding how to design and implement a process to incrementally sync data from an on-premises SQL database to Azure Data Lake Storage Gen2 (ADLS Gen2) is a skill that many organizations are seeking and a topic that this book covers in future chapters.

There are many online resources and books to help with mastering the traditional Microsoft BI skillset. For example, *The Data Warehouse Toolkit* by Ralph Kimball presents an authoritative guide to dimensional modeling. There are many other online resources for learning Microsoft BI, ranging from paid video courses to free YouTube tutorials such as from WiseOwlTutorials, TechBrothersIT, and more. Figure 1-1 depicts the typical end-to-end flow of a traditional Microsoft BI data architecture.

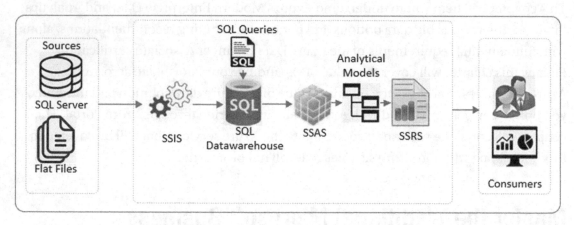

Figure 1-1. *Use of SSIS, SSAS, and SSRS in Microsoft BI Platform*

Understand Azure's Modern Enterprise Data and Analytics Platform

While Microsoft Azure has a vast collection of resources, the most common components within the Modern Enterprise Data and Analytics Platform are listed in Figure 1-2. As an Azure Data Engineer, it will be critical to be able to design and implement an end-to-end solution that follows this architectural process or custom variations of it while accounting for security, high availability, and more. It will also be critical to understand the differences and similarities between multiple data storage and data integration options.

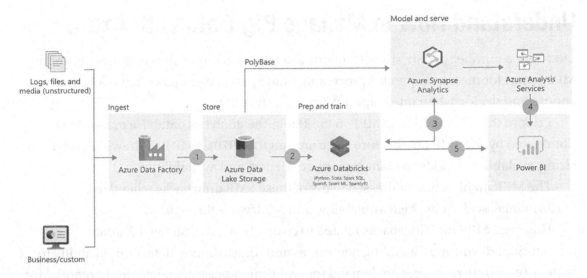

Figure 1-2. *Modern Azure Data Architecture Platform*

In the next chapter, one such topic will be covered on choosing between Azure Data Factory, SQL Server Integration Services (SSIS), and Azure Databricks to learn more about how to choose the right ETL (Extract-Load-Transform) tool for the job based on certain use cases and scenarios and to enable you to pick the best tool for the job. It is also good practice to know when to pick Azure Synapse Analytics Data Warehouse (DW) vs. Azure SQL Database (ASQLDB). The following article by Melissa Coates discusses these various options in detail:

"Is Azure SQL Data Warehouse a Good Fit?" (`www.bluegranite.com/blog/is-azure-sql-data-warehouse-a-good-fit-updated`)

Finally, having a good understanding of recent trends, feature updates, availability releases, and more for new and existing Azure data resources will only help you make more educated, experienced, and well-thought-out decisions as you build more Azure Data Engineering solutions. Azure updates *(https://azure.microsoft.com/en-us/updates/)* is a great place to find many of these updates, and you could filter the product categories to the data engineering–specific resources. There are many other free learning resources available, ranging from articles to video tutorials to help with keeping up to date on the Azure Data Platform.

Understand How to Manage Big Data with Azure

There are a number of data storage options available, and given the similarity between MPP (short for massively parallel processing) and Spark, customers often ask how to choose and decide when one is appropriate over the other.

Prior to the introduction of MPP in the 1990s, the analytics database market was dominated by the SMP architecture since around the 1970s. SMP had drawbacks around sizing, scalability, workload management, resilience, and availability.

The MPP architecture addressed many of these SMP drawbacks related to performance, scalability, high availability, and read/write throughput.

However, MPP had drawbacks related to cost; it had a critical need for data distribution, downtime for adding new nodes and redistributing data, limited ability to scale up compute resources on demand for real-time processing needs, and potential for overcapacity given the limitations to isolate storage from compute.

An RDD in Spark is similar to a distributed table in MPP in that many of the RDD operations have an equivalent MPP operation. RDD does, however, offer better options for real-time processing needs and the ability to scale up nodes for batch processing while also scaling storage (data lake) independently and cost-effectively from compute. Also, it is recommended over MPP for highly unstructured data processing (text, images, videos, and more). Additionally, it offers the capability for large-scale advanced analytics (AI, ML [machine learning], text/sentiment analysis, and more).

Both Spark and MPP complement each other well, and the choice between the two depends on the use case. Spark is well suited for big data engineering, ETL, AI/ML, real-time data processing, and non-reporting/non-querying use cases, while MPP is great for big data reporting and BI analytics platforms that require heavy querying and performance capabilities since it is based on the traditional RDBMS and brings with it the best features of SQL Server such as automated query tuning, data shuffling, ease of analytics platform management, even data distribution based on a primary key, and much more.

While Spark is great for large data processing, small files can cause slow query performance. Also, the lack of data shuffling in Spark can cause performance issues. Since Spark is designed for big data batch processing, it can also have poor performance when querying the data from a reporting dashboard.

Apache Spark in Azure Synapse Analytics, which you will learn about in subsequent chapters of this book, is attempting to bridge the divide between MPP and Spark through combining the benefits of both Spark and MPP into one unified analytics platform.

With all the hype around big data, along with the multiple big data products and resources available within Azure, the topic of big data is becoming increasingly important to many organizations. As an Azure Data Engineer, these organizations will be looking to you as their resident expert in the big data realm. Having a good understanding around the following will be key:

- Big data architectures

- Tuning performance and scalability of ADF's Copy activity

- Creating and configuring Databricks Spark clusters

- Performance tuning ADF's Mapping Data Flows

Understand the Fundamental Requirements for the Data Engineer Associate

With all the numerous Microsoft Azure certifications available for aspiring Azure Experts, it is clear that there is quite a lot to learn about Azure. The Data Engineer Associate certification path is most relevant for the Azure Data Engineer and originally consisted of passing two exams, which are depicted in Figure 1-3: DP-200 (Implementing an Azure Data Solution) and DP-201 (Designing an Azure Data Solution).

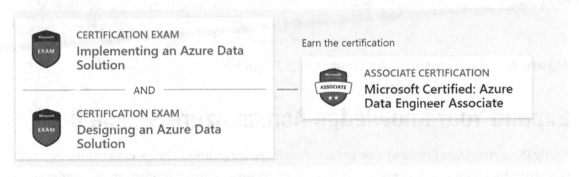

Figure 1-3. *Original Azure Data Engineer Associate certification exam requirements*

Committed earners of the Data Engineer Associate certification typically spend approximately 80 hours to prepare for the exams while having 3 hours to complete each exam. Microsoft offers both online (free) and instructor-led training programs to prepare for these exams. The exams can then be scheduled online and taken at a remote or on-site test center. Completing this certification will help with acquiring the fundamental foundations to the Azure Data Engineering landscape. Additionally, for those individuals who work for Microsoft partners, the cost can be fully expensed and may even include a bonus payout for passing the exams, along with the glory of becoming a Microsoft Certified Azure Data Engineer Associate.

Note   Microsoft recently planned to replace exams DP-200 and DP-201 with one exam, DP-203, as they work to retire the original exams.

Figure 1-4 shows Microsoft's learning path for the Azure Data Engineer, which covers designing and implementing the management, monitoring, security, and privacy of data using Azure data resources.

Figure 1-4. *Learning path for Azure Data Engineer*

Expand Your Knowledge Across Azure Specialties

While the Azure Data Engineer role covers a lot, there are many other specializations within Azure including DevOps, security, AI, data science, Azure administration, database administration, and more. While it would be nice to have specializations across these areas, it may take time to acquire this knowledge base, and it will help to have a basic understanding of DevOps CI/CD, security, infrastructure as code best practices, subscription and billing management, and more along with how it relates to data engineering in Azure to embrace Azure from a holistic view beyond the fundamentals of the role. Figure 1-5 lists the additional relevant Azure certifications that are available, along with their corresponding exams.

1. Microsoft Certified: Azure – Fundamentals Exam AZ-900
2. Microsoft Certified: Azure AI – Fundamentals Exam AI-900
3. Microsoft Certified: Azure Data – Fundamentals Exam DP-900
4. Microsoft Certified: Azure Administrator – Associate Exam AZ-104
5. Microsoft Certified: Azure Developer – Associate Exam AZ-204
6. Microsoft Certified: Azure Security Engineer – Associate Exam AZ-500
7. Microsoft Certified: Azure AI Engineer – Associate Exam AI-100
8. Microsoft Certified: Azure Data Scientist – Associate Exam DP-100
9. Microsoft Certified: Azure Data Engineer – Associate Exams DP-200 & DP-201
10. Microsoft Certified: Azure Database Administrator – Associate Exam DP-300
11. Microsoft Certified Solutions Architect – Expert Exams AZ-303 and AZ-304
12. Microsoft Certified: Azure DevOps Engineer – Expert Exam AZ-400

Figure 1-5. *List of additional and relevant Azure certifications*

With free online video tutorials, along with Microsoft's vast knowledge base of documentation that's easily accessible, understanding the end-to-end architectural process and how it relates to connectivity, security, infrastructure as code, Azure administration, DevOps CI/CD, and billing and cost management will instill confidence in your holistic understanding of Azure as you help your organization and team evangelize Azure Data Engineering and pioneer their journey into the cloud. Figure 1-6 presents a diagram with multiple components, along with how it all ties together from an architectural standpoint. In future chapters of this book, we will cover additional topics around DevOps CI/CD.

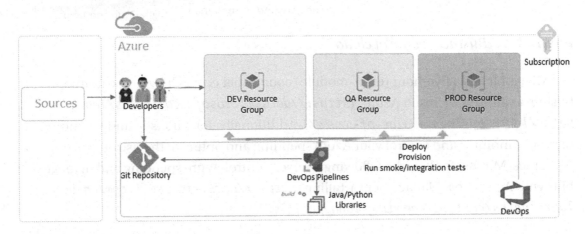

Figure 1-6. *High-level diagram of Azure data architecture with DevOps CI/CD*

Be Able to Address the Business Value of the Azure Data Platform

Azure Data Engineers empower and challenge organizations to embrace digital transformation at scale. Frequently, Azure Data Engineers are involved in conversations with C-level executives at the organization and may be asked to contribute to Business Requirement Documents that cover cost, security, and how an Azure solution brings true business value to the organization. Being able to speak about the business value of an Azure data and AI digital transformation solution is a valuable asset for many organizations. Figure 1-7 illustrates a few factors that impact the business values of the cloud.

Business Value of Cloud

Cost	**Agility**	**Quality**	**New Opportunities**
Transparency	DevOps CI/CD	Performance	Big Data
Cost Savings	Time to Market	Scalability	IoT
		Reliability	AI and ML
		Security and Compliance	Digital Transformation

Figure 1-7. Business value of cloud

Microsoft has a five-hour (eight-module) course that covers how to "Learn the business value of Microsoft Azure" *(https://docs.microsoft.com/en-us/learn/ paths/learn-business-value-of-azure/).* Additionally, for a holistic understanding of how to monitor and control your Azure spending and optimize the use of Azure resources, Microsoft offers the following course: Control Azure spending and manage bills with Azure Cost Management + Billing *(https://docs.microsoft.com/en-us/ learn/paths/control-spending-manage-bills/).*

Get Hands-On with Azure Data Engineering Through Azure Portal

One of the best ways to get hands-on with Azure Data Engineering is to explore and create these resources through Azure Portal. To quickly get started with Azure, you could create a free trial account *(https://azure.microsoft.com/en-us/free/)* in Azure Portal as shown in Figure 1-8. The demonstrations, exercises, and tutorials that follow in the subsequent chapters of this book will be hands-on, so if you are able to follow along with your own account, this may supplement your learning experience. If you do not have access to an Azure Portal account, then this book will still help by serving as a definitive guide to Azure Data Engineering.

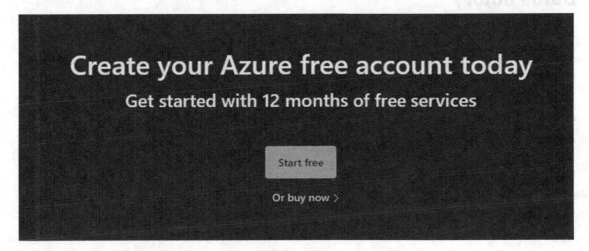

Figure 1-8. *Create your Azure free account today*

Azure Services Covered in This Book

Microsoft's Azure cloud platform has over 6,000 service offerings and counting. This book is intended to explore and deep dive into the capabilities of some of these resources that are specific to data extraction, transformation and ingestion, DevOps, advanced analytics, and governance. This section will describe the Azure data services that are covered in this book along with some context on how these services are used in this book.

Data Lake Storage Gen2

Azure Data Lake Storage Gen2 is a set of capabilities dedicated to big data analytics, built on Azure Blob Storage. A fundamental part of Data Lake Storage Gen2 is the addition of a hierarchical namespace to Blob Storage. The hierarchical namespace organizes objects/files into a hierarchy of directories for efficient data access. In Chapter 3, I will discuss in detail how to effectively design an Azure Data Lake Storage Gen2 account (ADLS Gen2). In many of the subsequent chapters, you will get more familiar with ADLS Gen2 from the ETL demonstrations and exercises on how it can be used as both source and sink in data ingestion and transformation pipelines.

Data Factory

Azure Data Factory (ADF) is Azure's cloud ETL service for scaled-out serverless data integration and data transformation. It offers a code-free UI for intuitive authoring and single-pane-of-glass monitoring and management. You can also lift and shift existing SSIS packages to Azure and run them with full compatibility in ADF. The majority of chapters in this book, ranging from Chapters 4 through 15, cover in-depth topics around Azure Data Factory and its vast capabilities.

Ingest and Load

Data Factory supports over 90 sources and sinks as part of its ingestion and load process. In the subsequent chapters, you will learn how to create pipelines, datasets, linked services, and activities in ADF:

- **Pipelines:** A logical grouping of activities that together perform a task.

- **Activities:** Defined actions to perform on your data (e.g., Copy data activity, ForEach loop activity, etc.)

- **Datasets:** Named view of data that simply points or references the data you want to use in your activities within the pipeline

- **Linked services:** Much like connection strings, which define the connection information needed for Data Factory to connect to external resources

Within ADF, integration runtimes (IRs) are the compute infrastructure used to provide data integration capabilities such as data flows and data movement. ADF has the following three IR types:

1. **Azure integration runtime:** All patching, scaling, and maintenance of the underlying infrastructure are managed by Microsoft, and the IR can only access data stores and services in public networks.

2. **Self-hosted integration runtimes:** The infrastructure and hardware are managed by you, and you will need to address all the patching, scaling, and maintenance. The IR can access resources in both public and private networks.

3. **Azure-SSIS integration runtimes:** VMs (virtual machines) running the SSIS engine allow you to natively execute SSIS packages. All the patching, scaling, and maintenance are managed by Microsoft. The IR can access resources in both public and private networks.

Mapping Data Flows for Transformation and Aggregation

Mapping Data Flows are visually designed data transformations in Azure Data Factory. Data flows allow data engineers to develop data transformation logic without writing code. The resulting data flows are executed as activities within Azure Data Factory pipelines that use scaled-out Apache Spark clusters. We will explore some of the capabilities of Mapping Data Flows in Chapters 11–15 for data warehouse ETL using SCD Type I, big data lake aggregations, incremental upserts, and Delta Lake. There are many other use cases around the capabilities of Mapping Data Flows that will not be covered in this book (e.g., dynamically splitting big files into multiple small files, managing small file problems, Parse transformations, and more), and I would encourage you to research many of these other capabilities of Mapping Data Flows.

Additionally, data can be transformed through Stored Procedure activities in the regular Copy data activity of ADF.

There are three different cluster types available in Mapping Data Flows – general purpose, memory optimized, and compute optimized:

- **General purpose:** Use the default general-purpose cluster when you intend to balance performance and cost. This cluster will be ideal for most data flow workloads.

- **Memory optimized:** Use the more costly per-core memory-optimized clusters if your data flow has many joins and lookups since they can store more data in memory and will minimize any out-of-memory errors you may get. If you experience any out-of-memory errors when executing data flows, switch to a memory-optimized Azure IR configuration.

- **Compute optimized:** Use the cheaper per-core-priced compute-optimized clusters for non-memory-intensive data transformations such as filtering data or adding derived columns.

Wrangling Data Flows

Data wrangling in Data Factory allows you to build interactive Power Query mashups natively and then execute those at scale inside of an ADF pipeline. This book will not cover topics on wrangling data flows outside of this introductory section.

Schedule and Monitor

You can schedule and monitor all of your pipeline runs natively in the Azure Data Factory user experience for triggered pipelines. Additionally, you can create alerts and receive texts and/or emails related to failed, succeeded, or custom pipeline execution statuses.

ADF Limitations

Currently, there are a few limitations with ADF. Some of these limitations include the following:

1. Inability to add a ForEach activity or Switch activity to an If activity.

2. Inability to nest ForEach loop, If, and Switch activities.

3. Lookup activity has a maximum of 5,000 rows and a maximum size of 4 MB.

4. Inability to add CRON functions for modular scheduling.

Some of these limitations are either on the ADF product team's road map for future enhancement, or there are custom solutions and workarounds. For example, the lack of modular scheduling within a single pipeline can be offset by leveraging tools such as Azure Functions, Logic Apps, Apache Airflow, and more. Ensure that you have a modular design with many pipelines to work around other limitations such as the 40 activities per pipeline limit. Additional limitations include 100 queued runs per pipeline and 1,000 concurrent pipeline activity runs per subscription per Azure integration runtime region.

Databricks

Azure Databricks is a data analytics platform optimized for the Microsoft Azure cloud services platform. Through the course of this book, we will explore Databricks as both a data ingestion and transformation tool in addition to its advanced analytics capabilities around use cases such as graph analytics (Chapter 21) and machine learning (Chapter 23). Additionally, Chapter 17 will discuss how to use Databricks for real-time IoT (Internet of Things) analytics by using Spark compute.

Synapse Analytics

Azure Synapse Analytics, much like Databricks, is an enterprise service that accelerates time to insight across data warehouses and big data systems. Azure Synapse brings together the best of SQL technologies used in enterprise data warehousing, Spark technologies used for big data, pipelines for data integration and ETL/ELT, and deep integration with other Azure services such as Power BI, Cosmos DB, and Azure ML. In the subsequent chapters on ELT using Azure Data Factory, we will explore how to dynamically load Azure Synapse Analytics DW. Additionally, in Chapter 22, we will explore how to get started with Azure Synapse Analytics workspace samples to promote self-service and advanced analytics capabilities.

Choosing between Synapse Analytics and Databricks might be a challenge for customers. Although Databricks has reached a level of maturity and respect as the industry standard for Apache Spark–based distributed computing in Azure, Synapse is slowly but surely catching up to be a unified data and analytics platform that may have tighter integration with other Azure services than Databricks. As Synapse Analytics continues to mature in its capabilities and offerings, many customers that are beginning new projects in Azure are very interested in exploring its capabilities.

DevOps CI/CD

A CI/CD pipeline is used to automate the process of continuous integration and continuous deployment. The pipeline facilitates the software delivery process via stages like Build, Test, Merge, and Deploy. Azure DevOps Server is a Microsoft product that provides version control, reporting, requirements management, project management, automated builds, testing, and release management capabilities. It covers the entire application lifecycle and enables DevOps capabilities such as CI/CD. In Chapter 19, we will explore how to deploy a Data Factory to multiple environments using DevOps CI/CD, and in Chapter 20, we will explore how to deploy an Azure SQL Database to multiple environments using DevOps CI/CD pipelines.

IoT Hub

IoT Hub is a Platform-as-a-Service (PaaS) managed service, hosted in the cloud, that acts as a central message hub for bidirectional communication between an IoT application and the devices it manages. In Chapter 16, we will set up an IoT Hub that will collect events from a Device Simulator to facilitate real-time analytics and insights on the incoming events.

Prior to selecting an IoT Hub, understand the differences in capabilities between the various IoT Hub options and compare them to Event Hubs. Table 1-1 lists out the capabilities of Event Hubs and IoT Hub options.

Table 1-1. *IoT Hub vs. Event Hubs*

IoT Capability	IoT Hub Standard Tier	IoT Hub Basic Tier	Event Hubs
Device-to-cloud messaging	Yes	Yes	Yes
Protocols: HTTPS, AMQP, AMQP over webSockets	Yes	Yes	Yes
Protocols: MQTT, MQTT over webSockets	Yes	Yes	
Per-device identity	Yes	Yes	
File upload from devices	Yes	Yes	
Device provisioning service	Yes	Yes	
Cloud-to-device messaging	Yes		
Device twin and device management	Yes		
Device streams (preview)	Yes		
IoT Edge	Yes		

Stream Analytics

Azure Stream Analytics is a real-time analytics and complex event-processing engine that is designed to analyze and process high volumes of fast-streaming data from multiple sources simultaneously. Patterns and relationships can be identified in information extracted from a number of input sources including devices, sensors, clickstreams, social media feeds, and applications. In Chapter 16, we will explore a practical use case for Stream Analytics that is specific to anomaly detection, which can be used to detect fraud as an example.

In time-streaming scenarios, performing operations on the data contained in temporal windows is a common pattern. Stream Analytics has native support for windowing functions, enabling developers to author complex stream processing jobs with minimal effort.

The following windowing functions are available within Stream Analytics:

- **Tumbling window** allows you to segment data into distinct time segments (count of tweets per time zone every 10 seconds).

- **Session window** allows group streaming events that arrive at a similar time and filter out no data (count of tweets that occur within 5 minutes of each other).

- **Hopping window** looks backward to determine when an event occurs (every 5 seconds, count of tweets in the last 10 seconds).

- **Sliding window** produces output when an event occurs (count of tweets for a single topic over the last 10 seconds).

- **Snapshot window** groups events that have the same timestamp. You can apply a snapshot window by adding System.Timestamp() to the GROUP BY clause.

Power BI

Power BI, which is a part of Microsoft's Power Platform, is a reporting platform that provides interactive visualizations and business intelligence capabilities, and it empowers end users to create their own reports and dashboards. In Chapter 16, we will explore how to build a Power BI dashboard for real-time monitoring and analytics of incoming IoT events from a Device Simulator.

There are a few options to choose from when considering Power BI. These options include Premium and Pro, which come with their own capabilities and price points. Table 1-2 shows a side-by-side comparison between Power BI Pro and Premium to help with choosing the right feature for your reporting needs.

Table 1-2. *Power BI Pro vs. Premium*

Feature	Power BI Pro	Power BI Premium Per User	Power BI Premium Per Capacity
Collaboration and analytics			
Mobile app access	Yes	Yes	Yes
Publish reports to share and collaborate	Yes	Yes	
Paginated (RDL) reports		Yes	Yes
Consume content without a per-user license			Yes
On-premises reporting with Power BI Report Server			Yes
Data prep, modeling, and visualization			
Model size limit	1 GB	100 GB	400 GB
Refresh rate	8/day	48/day	48/day
Connect to 100+ data sources	Yes	Yes	Yes
Create reports and visualizations with Power BI Desktop	Yes	Yes	Yes
Embed APIs and controls	Yes	Yes	Yes
AI visuals	Yes	Yes	Yes
Advanced AI (text analytics, image detection, automated machine learning)		Yes	Yes
XMLA endpoint read/write connectivity		Yes	Yes
Data flows (direct query, linked and computed entities, enhanced compute engine)		Yes	Yes
Analyze data stored in Azure Data Lake Storage		Yes	Yes

(*continued*)

Table 1-2. (*continued*)

Feature	Power BI Pro	Power BI Premium Per User	Power BI Premium Per Capacity
Governance and administration			
Data security and encryption	Yes	Yes	Yes
Metrics for content creation, consumption, and publishing	Yes	Yes	Yes
Application lifecycle management		Yes	Yes
Multi-geo deployment management			Yes
Bring your own key (BYOK)			Yes
Auto-scale add-on availability (preview)			Yes
Maximum storage	10 GB/user	100 TB	100 TB
Continuous integration and deployment			
Deployment pipelines (including paginated reports management)		Yes	Yes

Purview

Azure Purview is a unified data governance service that helps you manage and govern your data easily by creating a holistic, up-to-date map of your data landscape with automated data discovery, sensitive data classification, and end-to-end data lineage. It empowers data consumers to find valuable, trustworthy data and can be integrated with many of the services that are covered in this book ranging from Synapse Analytics, SQL Database, Data Factory, and more. In Chapter 24, we will explore a practical example of implementing Purview through a detailed exercise.

Snowflake

Snowflake is a cloud provider that enables data storage, processing, and analytics solutions that are faster, easier to use, and far more flexible than traditional offerings. The Snowflake data platform is not built on any existing database technology or "big data" software platforms such as Hadoop. Instead, Snowflake combines a completely new SQL query engine with an innovative architecture natively designed for the cloud. Snowflake offers some pretty neat capabilities around data sharing, data warehousing, and data applications. In Chapter 10, you will learn more about how to dynamically load Snowflake DW using Data Factory and Databricks.

SQL Database

Azure SQL Database is a cloud computing database service (Database as a Service) that is offered by Microsoft Azure platform and helps to host and use a relational SQL database in the cloud without requiring any hardware or software installation. We will be using the standard Azure SQL Database in many of the chapters in this book as a source, sink, and metadata-driven control database.

Purchasing Models (SQL DTU vs. vCore Database)

There are a vast number of SQL database options in Azure including DTU and vCore. DTU and vCore are two different purchasing models for Azure SQL that include variations in computation, memory, storage, and I/O. Azure Pricing Calculator can help with aligning cost and capability with the appropriate SQL database solution.

A DTU unit is a combination of CPU, memory, and read and write operations and can be increased when more power is needed. It is a great solution if you have a preconfigured resource configuration where the consumption of resources is balanced across CPU, memory, and I/O. You can increase the number of DTUs reserved once you reach a limit on the allocated resources and experience throttling, which translates into slower performance or timeouts.

The disadvantage of DTUs is that you don't have the flexibility to scale only a specific resource type, like the memory or the CPU. Because of this, you can end up paying for additional resources without needing or using them.

The vCore model allows you to scale each resource (CPU, memory, IO) independently. You could scale the storage space up and down for a database based on how many GB of storage is needed, and you can also scale the number of cores (vCores). The disadvantage is that you can't control the size of memory independently. Also, it is important to note that vCore serverless compute resources are twice the price of provisioned compute resources, so a constant high load would cost more in serverless than it would in provisioned. vCore can use the SQL Server licenses that you have from your on-premises environment.

Deployment Models

There are two available deployment models for Azure SQL Database:

- **Single database** represents a fully managed, isolated database. You might use this option if you have modern cloud applications and microservices that need a single reliable data source. A single database is similar to a contained database in the SQL Server database engine.

- **Elastic pool** is a collection of single databases with a shared set of resources, such as CPU or memory, and single databases can be moved into and out of an elastic pool.

Service Tiers

There are three available service tiers for Azure SQL Database:

- **General Purpose/Standard** service tier is designed for common workloads. It offers budget-oriented balanced compute and storage options.

- **Business Critical/Premium** service tier is designed for OLTP applications with a high transaction rate and lowest-latency I/O. It offers the highest resilience to failures by using several isolated replicas.

- **Hyperscale** service tier is designed for big data OLTP databases and the ability to scale storage and compute gracefully.

Cosmos DB

Azure Cosmos DB is a fully managed NoSQL database for modern app development. Single-digit millisecond response times and automatic and instant scalability guarantee speed at any scale. Business continuity is assured with SLA-backed availability and enterprise-grade security. In Chapter 18, we will explore Cosmos DB in the context of establishing a real-time Synapse Analytics Link for Cosmos DB to learn about the capabilities of real-time analytics on transactional data in Cosmos DB. Azure Synapse Link for Azure Cosmos DB is a cloud-native hybrid transactional and analytical processing (HTAP) capability that enables you to run near-real-time analytics over operational data in Azure Cosmos DB. Azure Synapse Link creates a tight seamless integration between Azure Cosmos DB and Azure Synapse Analytics.

When deciding when to choose between a SQL and NoSQL database for your data solution, ensure that you take the following comparison factors into consideration, as shown in Table 1-3.

Table 1-3. *SQL vs. NoSQL*

	SQL	NoSQL
Definition	SQL databases are primarily called RDBMSs or relational databases.	NoSQL databases are primarily called as non-relational or distributed databases.
Designed for	Traditional RDBMS uses SQL syntax and queries to analyze and get the data for further insights. They are used for OLAP systems.	NoSQL database system consists of various kinds of database technologies. These databases were developed in response to the demands presented for the development of the modern application.
Query language	Structured Query Language (SQL)	No declarative query language
Type	SQL databases are table-based databases.	NoSQL databases can be document-based, key-value pair, graph databases.
Schema	SQL databases have a predefined schema.	NoSQL databases use a dynamic schema for unstructured data.

(*continued*)

Table 1-3. (*continued*)

	SQL	NoSQL
Ability to scale	SQL databases are vertically scalable.	NoSQL databases are horizontally scalable.
Examples	Oracle, Postgres, and MS SQL.	MongoDB, Redis, Neo4j, Cassandra, HBase.
Best suited for	An ideal choice for the complex query-intensive environment.	It is not a good fit for complex queries.
Hierarchical data storage	SQL databases are not suitable for hierarchical data storage.	More suitable for the hierarchical data store as it supports key-value pair methods.
Variations	One type with minor variations.	Many different types that include key-value stores, document databases, and graph databases.
Development year	It was developed in the 1970s to deal with issues with flat file storage.	Developed in the late 2000s to overcome issues and limitations of SQL databases.
Consistency	It should be configured for strong consistency.	It depends on DBMS as some offer strong consistency like MongoDB, whereas others offer only eventual consistency, like Cassandra.
Best used for	RDBMS is the right option for solving ACID problems.	NoSQL is best used for solving data availability problems.
Importance	It should be used when data validity is super important.	Use when it's more important to have fast data than correct data.
Best option	When you need to support dynamic queries.	Use when you need to scale based on changing requirements.
ACID vs. BASE model	ACID (Atomicity, Consistency, Isolation, and Durability) is a standard for RDBMS.	BASE (Basically Available, Soft state, Eventually consistent) is a model of many NoSQL systems.

The following APIs are available in Cosmos DB:

- **SQL:** Provides capabilities for data users who are comfortable with SQL queries. Even though the data is stored in JSON format, it can easily be queried by using SQL-like queries.

- **MongoDB:** Existing instances of MongoDB can be migrated to Azure Cosmos DB without major effort.

- **Gremlin:** Can be used for storing and performing operations on graph data and supports native capabilities for graph modeling and traversing.

- **Casandra:** Dedicated data store for applications created for Apache Cassandra where users have the ability to interact with data via CQL (Cassandra Query Language).

- **Table:** Can be used by applications prepared natively for close working with Azure storage tables.

Relevant Azure Services Not Covered

There are a vast array of relevant Azure services that will not be covered in this book. I would encourage you to have a basic understanding of these services since they are frequently part of the modern Azure Data Platform architecture stack. This section will provide a basic introduction to some of these services.

Analysis Services

Azure Analysis Services is a fully managed Platform-as-a-Service that provides enterprise-grade data models in the cloud. Use advanced mashup and modeling features to combine data from multiple data sources, define metrics, and secure your data in a single, trusted tabular semantic data model. Analysis Services integrates well with Logic Apps and can also be integrated with Azure Data Factory and Azure Functions for incremental processing of models.

When considering Analysis Services, it can be compared to Power BI Premium since Power BI Premium is expected to provide a super-set of the capabilities.

Currently, the following feature considerations in Table 1-4 can be used to compare pros and cons when choosing between Power BI Premium and Analysis Services.

Table 1-4. *Power BI Premium vs. Azure Analysis Services*

	Power BI Premium	Azure Analysis Services
Unlimited Power BI content viewing	Yes	No
Paginated reports	Yes	No
Data flows	Yes	No
AI workload	Yes	No
Multi-model memory management	Yes	No
Pre-aggregated tables	Yes	No
Composite models	Yes	No
Automated incremental refresh	Yes	No
Large datasets	Yes	Yes
Third-party application support	Yes	Yes
Bring your own key	Yes	No
Scale-out	Not yet	Yes
Metadata translations	Not yet	Yes
Object-level security	Not yet	Yes
Perspectives	Not yet	Yes

Cognitive Services

Cognitive Services brings AI to developers through APIs and offers a variety of services to provide the ability for AI to see, hear, speak, search, understand, and accelerate decision-making into apps. Developers of all skill levels and those that do not have an expertise in machine learning can easily add AI capabilities to their apps.

The currently available Cognitive Services include the following.

Decision

- **Anomaly Detector:** Identify potential problems early on.
- **Content Moderator:** Detect potentially offensive or unwanted content.
- **Metrics Advisor:** Monitor metrics and diagnose issues.
- **Personalizer:** Create rich, personalized experiences for every user.

Language

- **Immersive Reader:** Help comprehend text using audio and visual cues.
- **Language Understanding:** Build natural language understanding in apps.
- **QnA Maker:** Create a conversational question-and-answer layer over data.
- **Text Analytics:** Detect sentiment, key phrases, and named entities.
- **Translator:** Detect and translate more than 90 supported languages.

Speech

- **Speech to Text:** Transcribe audio speech into readable text.
- **Text to Speech:** Convert text to lifelike speech for natural interfaces.
- **Speech Translation:** Integrate real-time speech translation into apps.
- **Speaker Recognition:** Identify and verify speaking based on audio.

Vision

- **Computer Vision:** Analyze content in images and videos.
- **Custom Vision:** Customize image recognition to fit your business needs.
- **Face:** Detect and identify people and emotions in images.

- **Form Recognizer:** Extract text, key-value pairs, and tables from documents.

- **Video Indexer:** Analyze visuals and audio of video and index content.

Search

- **Search:** Azure Cognitive Search is a cloud search service that gives developers APIs and tools for building a rich search experience over private, heterogeneous content in web, mobile, and enterprise applications.

Azure Machine Learning

Azure Machine Learning is a service that delivers a complete data science platform. It supports both code-first and low-code experiences. Azure Machine Learning Studio is a web portal in Azure Machine Learning that contains low-code and no-code options for project authoring and asset management. Azure Machine Learning integrates well with other Azure services such as Databricks and Data Factory.

The three main machine learning techniques include the following.

Supervised Learning

Algorithms make predictions based on a set of labeled examples that you provide. This technique is useful when you know what the outcome should look like.

Unsupervised Learning

Algorithms label the datasets for you by organizing the data or describing its structure. This technique is useful when you don't know what the outcome should look like.

Reinforcement Learning

Algorithms learn from outcomes and decide which action to take next. After each action, the algorithm receives feedback that helps it determine whether the choice it made was correct, neutral, or incorrect. It's a good technique to use for automated systems that have to make a lot of small decisions without human guidance.

Monitor

Azure Monitor helps you maximize performance and availability of your applications and proactively identify problems in seconds by allowing you to monitor your infrastructure and network.

Log Analytics

With Log Analytics, you can edit and run log queries from data collected by Azure Monitor Logs and interactively analyze their results. You can use Log Analytics queries to retrieve records matching particular criteria, identify trends, analyze patterns, and provide a variety of insights into your data.

Event Hubs

Event Hubs is a big data streaming platform and event ingestion service. It can receive and process millions of events per second. Data sent to an Event Hub can be transformed and stored by using any real-time analytics provider or batching/storage adapters.

Data Share

Azure Data Share enables organizations to simply and securely share data with multiple customers and partners. In just a few clicks, you can provision a new Data Share account, add datasets, and invite your customers and partners to your data share. Data providers are always in control of the data that they have shared. It is important to note that many other multi-cloud platforms such as Snowflake also offer robust data sharing mechanisms and platforms that contribute to the modern architectural pattern known as data mesh. Data meshes federate data ownership among domain data owners who are held accountable for providing their data as products while also facilitating communication between distributed data across different locations.

Logic Apps

Logic Apps is a cloud service that helps you schedule, automate, and orchestrate tasks, business processes, and workflows when you need to integrate apps, data, systems, and services across enterprises or organizations.

Power Apps

Power Apps, which is a part of Microsoft's Power Platform, is a suite of apps, services, connectors, and data platform that provides a rapid application development environment to build custom apps for your business needs. There are two styles of these apps: canvas apps and model-driven apps.

Canvas apps provide you with a blank canvas onto which you can drag and drop components in any formation to design a user interface. Model-driven apps are based on underlying data stored in Common Data Service (CDS), which is a secure, cloud-based storage space that organizations can use to store business application data. Canvas apps are ideal for building task-based or role-based applications. Model-driven apps, on the other hand, are better for creating end-to-end solutions.

App Service

Azure App Service is a fully managed web hosting service for building web apps, mobile back ends, and RESTful APIs.

SQL Managed Instance

SQL Managed Instance is a deployment option of Azure SQL Database, providing near-100% compatibility with the latest SQL Server on-premises (Enterprise Edition) database engine, a native virtual network (VNet) implementation that addresses common security concerns, and a business model favorable for on-premises SQL Server customers.

Data Box

Azure Data Box lets you send terabytes of data into and out of Azure in a quick, inexpensive, and reliable way. The secure data transfer is accelerated by shipping you a proprietary Data Box storage device. Each storage device has a maximum usable storage capacity of 80 TB and is transported to your data center through a regional carrier. The device has a rugged casing to protect and secure data during the transit.

Data Sync

Azure SQL Data Sync is a service that is used to replicate the tables in Azure SQL Database to another Azure SQL Database or on-premises database. Data can be replicated one way or bidirectionally.

Data Gateway

The gateway works as a bridge that provides quick data transfer and encryption between data sources on-premises and your Logic Apps, Power BI, Power Apps, Microsoft Flow, and Analysis Services.

Cost Management + Billing

Azure Cost Management + Billing helps you understand your Azure bill, manage your billing account and subscriptions, monitor and control Azure spending, and optimize resource use.

Digital Twins

Azure Digital Twins is an Internet of Things (IoT) platform that enables you to create a digital representation of real-world things, places, business processes, and people. Gain insights that help you drive better products, optimize operations and costs, and create breakthrough customer experiences.

Mobile

Azure Mobile Services provides a scalable cloud back end for building Windows Store, Windows Phone, Apple iOS, Android, and HTML/JavaScript applications.

Networking

Azure Networking provides the connectivity and scale you need without requiring you to build or manage down to the fiber. Additionally, it allows you to manage traffic for applications using Azure App Gateway, protect using Azure WAF, define and monitor global routing with Azure Front Door, and get turnkey firewall capabilities with Azure Firewall.

Security

Azure Security Center is a unified infrastructure security management system that provides advanced threat protection across your hybrid workloads in the cloud and strengthens the security stance of your data centers.

Identity

Azure Active Directory (Azure AD) is Microsoft's cloud-based identity and access management service, which helps your employees sign in and access resources such as apps on your corporate network and intranet, along with any cloud apps developed by your own organization.

Kubernetes

Azure Kubernetes Service (AKS) simplifies deploying a managed Kubernetes cluster in Azure by offloading the operational overhead to Azure. As a hosted Kubernetes service, Azure handles critical tasks, like health monitoring and maintenance.

Functions

Azure Functions is an on-demand service that provides serverless compute for Azure and can be used to build web APIs, respond to database changes, process IoT streams, manage message queues, and more. Additionally, Functions can be called from and integrated within Azure Data Factory pipelines.

HVR Real-Time Data Replication

HVR for Azure can be used for integrating Azure data with on-premises systems and inter-cloud integration and migration and zero-downtime migrations to and from Azure, as it offers real-time heterogeneous data replication in Azure and allows you to move large volumes of data faster and experience continuous data flow between on-premises and in the cloud with low latency.

Summary

In this chapter, I have covered tips for getting started with Azure Data Engineering by highlighting the importance of the traditional Microsoft BI Stack and its influential role in Azure's Modern Enterprise Data and Analytics Platform. I also introduced concepts around big data options for performance tuning recommendations, along with the fundamental requirements of the Azure Data Engineer Associate certification. Additionally, this chapter covered the value of expanding your knowledge across other Azure specialties to address the business value of an Azure data solution and introduce you options on getting started with Azure Data Engineering through Azure Portal. Finally, this chapter introduced you to some of the Azure resources that will be covered in this book and also discussed the Azure resources that will not be covered in this book.

## CHAPTER 2

Data Factory vs. SSIS vs. Databricks

Choosing the right ELT tool can be difficult based on the many data integration offerings from Microsoft's ever-growing Azure data engineering and integration ecosystem. Technology professionals ranging from data engineers to data analysts are interested in choosing the right ELT tool for the job and often need guidance when determining when to choose between Azure Data Factory (ADF), SQL Server Integration Services (SSIS), and Azure Databricks for their data integration projects.

Both SSIS and ADF are robust GUI-driven data integration tools designed for ELT and ETL workflows, pipelines, and operations with connectors to multiple sources and sinks. SSIS development is hosted in SQL Server Data Tools, while ADF development is a browser-based experience; both have robust scheduling and monitoring features. With ADF's Mapping Data Flows, transforming data through aggregations, derived columns, fuzzy lookups, and other visually designed data transformations similar to SSIS is a capability that allows data engineers to build ELT in a code-free manner. Both ADF's Mapping Data Flows and Databricks utilize Spark clusters to transform and process big data and analytics workloads in Azure. This chapter aims to cover the similarities and differences between ADF, SSIS, and Databricks, in addition to providing some guidance to help you determine how to choose between these various data integration services.

Choosing the Right Data Integration Tool

When choosing between Azure Data Factory (ADF) and SQL Server Integration Services (SSIS) for a new project, it is critical to understand whether your organization has an Azure footprint, and if so, can your data integration project be hosted in Azure? If the answer is yes, then ADF is the perfect tool for the job. On the other hand, if the new project must be completed on-premises either for security reasons or because

© Ron C. L'Esteve 2021
R. C. L'Esteve, *The Definitive Guide to Azure Data Engineering*, https://doi.org/10.1007/978-1-4842-7182-7_2

there is already an existing SSIS ecosystem, then SSIS is the tool of choice. Oftentimes organizations reap the benefits of combining SSIS with ADF through lift and shift scenarios in which they leverage ADF's cloud-based computing services to schedule, run, and execute SSIS packages.

SSIS is a part of SQL Server's several editions, ranging in price from free (Express and Developer editions) to ~$14K per core (Enterprise), and SSIS integration runtime nodes start at $0.84 per hour on Azure. That said, data volume can become a concern from both a price and performance standpoint when running big data workloads using SSIS since hardware will need to be purchased and frequently maintained.

Azure Data Factory V2's pay-as-you-go plan starts at $1 per 1,000 orchestrated runs and $1.5 per 1,000 self-hosted IR runs. ADF would be a great resource for organizations that have hundreds of SSIS packages that they would not want to rewrite in ADF but would like to reduce operational costs, increase high availability, and increase scalability by leveraging Azure. For this scenario, a hybrid lift and shift of SSIS workloads to the cloud would be ideal.

From a data velocity perspective, ADF natively supports event-based and Tumbling window triggers in addition to scheduled batch triggers, whereas SSIS only supports batching natively with the capability of potentially building custom triggers for near-real-time data streams. For example, developing a file watcher task for SQL Server Integration Services would automate the process of continuously checking a directory for incoming files before processing them.

From a data variety perspective, ADF can natively connect to over 90 sources ranging from REST APIs to CRM systems to complex JSON structures, while SSIS is better suited for structured data sources but can integrate well to either third-party or custom C# connectors for JSON, REST APIs, and more.

From a programmability perspective, Azure Data Factory does not have a native programming SDK but does support automation through PowerShell without any third-party components, whereas SSIS has a programming SDK, along with automation through BIML and a variety of other third-party components. Figure 2-1 lists the various similarities and differences between SSIS and ADF.

		SSIS	ADF
Volume			
	Big Data Volume		X
	Medium Data Volume	X	X
Velocity			
	Batch	X	X
	Streaming		X
Variety			
	Structured	X	X
	Unstructured		X
Platform			
	On-Premise Data Platform	X	
	Cloud Data Platform		X
	Hybrid Data Platform	X	X
	Managed		X
	Purchase Hardware	X	
Load Pattern			
	Extract-Load-Transform		X
	Extract-Transform-Load	X	
Dev Tools			
	SSDT	X	
	Browser		X
Dev Interface			
	GUI	X	X
Pricing			
	License	X	
	Pay as you go		X

Figure 2-1. *Capabilities of SSIS and ADF*

When to Use Azure Data Factory, Azure Databricks, or Both

For big data projects, both Data Factory and Databricks are Azure cloud-based data integration tools that are available within Microsoft Azure's data ecosystem and can handle big data, batch/streaming data, and structured/unstructured data. Both have browser-based interfaces along with pay-as-you-go pricing plans.

ADF's Mapping Data Flows uses scaled-out Apache Spark clusters, which is similar to Databricks' underlying architecture, and performs similarly for big data aggregations and transformations. It is important to note that Mapping Data Flows currently does not support connectivity to on-premises data sources. Also, ADF's original Copy activity does not use Spark clusters but rather self-hosted integration runtimes and does allow connectivity to on-premises SQL Servers. Based on these options to connect to on-premises SQL Servers, Databricks does have capabilities to connect to on-premises data sources and may outperform ADF on big data workloads since it utilizes Spark clusters.

From a velocity perspective, both ADF and Databricks support batch and streaming options. ADF does not natively support real-time streaming capabilities, and Azure Stream Analytics would be needed for this. Databricks supports Structured Streaming, which is an Apache Spark API that can handle real-time streaming analytics workloads.

From a development interface perspective, ADF's drag-and-drop GUI is very similar to that of SSIS, which fosters a low learning curve and ease of use for developers that are familiar with the code-free interface of SSIS. Additionally, cluster types, cores, and nodes in the Spark compute environment can be managed through the ADF activity GUI to provide more processing power to read, write, and transform your data.

Databricks does require the commitment to learn either Spark, Scala, Java, R, or Python for data engineering– and data science–related activities. This can equate to a higher learning curve for traditional MS SQL BI developers that have been ingrained in the SSIS ETL process for over a decade. For data engineers and scientists that are familiar and comfortable with the Databricks programming languages, Databricks offers a neat and organized method of writing and managing code through notebooks.

The last and most notable difference between ADF and Databricks is related to their primary purposes. ADF, which resembles SSIS in many aspects, is mainly used for ETL/ELT, data movement, and orchestration, whereas Databricks can be used for real-time data streaming and collaboration across data engineers, data scientists, and more, along with supporting the design and development of AI and machine learning models by data scientists.

For example, MLflow from Databricks simplifies the machine learning lifecycle by tracking experiment runs between multiple users within a reproducible environment and manages the deployment of models to production. Additionally, Databricks supports a variety of third-party machine learning tools. In the subsequent chapters, we will cover more details on the capabilities of MLflow and a variety of other advanced features of Databricks.

Once these Databricks models have been developed, they can be integrated within ADF's Databricks activity and chained into complex ADF ETL/ELT pipelines, coupled with a seamless experience for passing parameters from ADF to Databricks. Additionally, the Databricks models can be scheduled and monitored via ADF. We will explore these various combined pipelines and parameter passing capabilities in future chapters. Figure 2-2 lists the various similarities and differences between Databricks and ADF.

		Databricks	ADF
Volume			
	Big Data	X	X
	Medium Data	X	X
Velocity			
	Batch	X	X
	Streaming	X	X
	Real-Time	X	
Variety			
	Structured	X	X
	Unstructured	X	X
Dev Tools			
	Browser	X	X
Dev Interface			
	GUI		X
	Code	X	
Pricing			
	Pay as you go	X	X
Primary Purpose			
	Movement		X
	Orchestration		X
	ETL/ELT		X
	Preparation	X	
	Collaboration	X	
	AI/ML	X	

Figure 2-2. *Capabilities of Databricks and ADF*

Summary

In this chapter, I explored the differences and similarities between ADF, SSIS, and Databricks and made recommendations on when to choose one over the other and when to use them together. The solution truly depends on a number of different factors such as performance, cost, preference, security, feature capability, and more. In the upcoming chapters, we will use a combination of ADF and Databricks to demonstrate real-world end-to-end workflows and pipelines.

CHAPTER 3

Design a Data Lake Storage Gen2 Account

There are a variety of considerations to account for while designing and architecting an Azure Data Lake Storage Gen2 account. Some of these considerations include security, zones, folder and file structures, data lake layers, and more.

This chapter will explain the various considerations to account for while designing an Azure Data Lake Storage Gen2 account. Topics that will be covered include

- Data lake layers along with some of their properties

- Design considerations for zones, directories, and files

- Security options and considerations at the various levels

Data Lake Layers

Azure Data Lake Storage Gen2 offers the flexibility to divide repositories into multiple layers. These layers help with organizing, securing, and managing the data lake with ease. Figure 3-1 outlines the various layers in a data lake that can be accounted for when designing a data lake. These layers include

- Environments

- Storage accounts

- File systems

- Zones

41

- Directories

- Files

The subsequent sections will discuss these layers in greater detail.

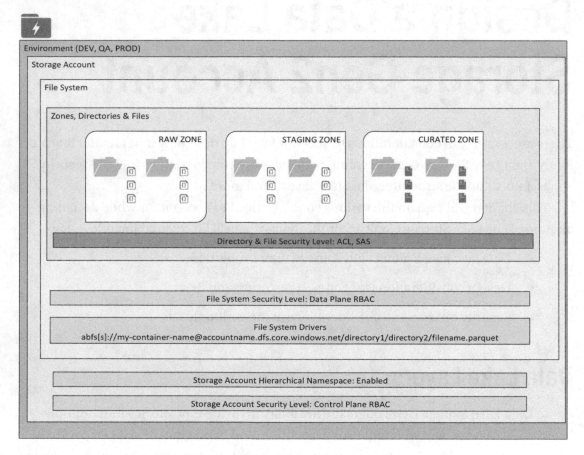

Figure 3-1. *Various layers in a data lake*

Environments

The environments define the top-level layer that needs to be accounted for when designing a data lake. For example, if a Dev, QA, and Prod environment is needed, then these environments must also include one or many ADLS Gen2 storage accounts. Figure 3-2 depicts how this multi-environment can be governed and orchestrated through Azure DevOps pipelines.

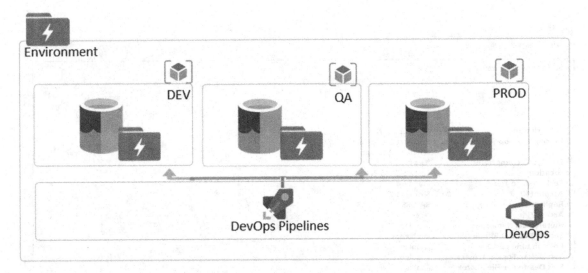

Figure 3-2. *Dev, QA, Prod Environments orchestrated by DevOps Pipelines*

Storage Accounts

There are several properties that need to be configured when creating an Azure Data Lake Storage account, which are listed in Figure 3-3. Additionally, when designing a storage account, considering the limits and capacity of the storage account will be critical to determine whether to have multiple storage accounts. Security at the storage account level will be defined by the Control Plane RBAC (Role-Based Access Control), and more details will be covered in the "Security" section.

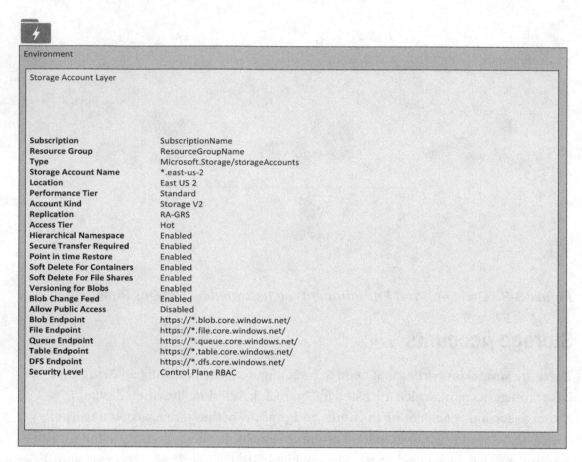

Environment	
Storage Account Layer	
Subscription	SubscriptionName
Resource Group	ResourceGroupName
Type	Microsoft.Storage/storageAccounts
Storage Account Name	*.east-us-2
Location	East US 2
Performance Tier	Standard
Account Kind	Storage V2
Replication	RA-GRS
Access Tier	Hot
Hierarchical Namespace	Enabled
Secure Transfer Required	Enabled
Point in time Restore	Enabled
Soft Delete For Containers	Enabled
Soft Delete For File Shares	Enabled
Versioning for Blobs	Enabled
Blob Change Feed	Enabled
Allow Public Access	Disabled
Blob Endpoint	https://*.blob.core.windows.net/
File Endpoint	https://*.file.core.windows.net/
Queue Endpoint	https://*.queue.core.windows.net/
Table Endpoint	https://*.table.core.windows.net/
DFS Endpoint	https://*.dfs.core.windows.net/
Security Level	Control Plane RBAC

Figure 3-3. *Storage account layer properties*

The following are the various properties that are configurable at the storage account level:

- **Performance Tier**: Standard storage accounts are backed by magnetic drives and provide the lowest cost per GB. They're best for applications that require bulk storage or where data is accessed infrequently. Premium storage accounts are backed by solid-state drives and offer consistent, low-latency performance. They are best for I/O-intensive applications, like databases. Additionally, virtual machines that use premium storage for all disks qualify for a 99.9% SLA, even when running outside of an availability set. This setting can't be changed after the storage account is created.

- **Account Kind**: General-purpose storage accounts provide storage for blobs, files, tables, and queues in a unified account. Blob Storage accounts are specialized for storing blob data and support choosing an access tier, which allows you to specify how frequently data in the account is accessed. Choose an access tier that matches your storage needs and optimizes costs.

- **Replication**: The data in your Azure storage account is always replicated to ensure durability and high availability. Choose a replication strategy that matches your durability requirements. Some settings can't be changed after the storage account is created.

- **Point in time Restore**: Use point-in-time restore to restore one or more containers to an earlier state. If point-in-time restore is enabled, then versioning, change feed, and blob soft delete must also be enabled.

- **Soft Delete for Containers**: Soft delete enables you to recover containers that were previously marked for deletion.

- **Soft Delete for File Shares**: Soft delete enables you to recover file shares that were previously marked for deletion.

- **Versioning for Blobs**: Use versioning to automatically maintain previous versions of your blobs for recovery and restoration.

- **Blob Change Feed**: Keep track of create and delete changes or modifications to blobs in your account.

- **Connectivity Method**: You can connect to your storage account either publicly, via public IP addresses or service endpoints, or privately, using a private endpoint.

- **Routing Preferences**: Microsoft network routing will direct your traffic to enter the Microsoft cloud as quickly as possible from its source. Internet routing will direct your traffic to enter the Microsoft cloud closer to the Azure endpoint.

- **Secure Transfer Required**: The secure transfer option enhances the security of your storage account by only allowing requests to the storage account by secure connection. For example, when calling REST APIs to access your storage accounts, you must connect using HTTPS. Any requests using HTTP will be rejected when "Secure Transfer Required" is enabled. When you are using the Azure file service, connection without encryption will fail, including scenarios using SMB 2.1, SMB 3.0 without encryption, and some flavors of the Linux SMB client. Because Azure storage doesn't support HTTPS for custom domain names, this option is not applied when using a custom domain name.

- **Allow Public Access**: When "Allow Public Access" is enabled, one is permitted to configure container ACLs (access control lists) to allow anonymous access to blobs within the storage account. When disabled, no anonymous access to blobs within the storage account is permitted, regardless of underlying ACL configurations.

- **Hierarchical Namespace**: The ADLS Gen2 hierarchical namespace accelerates big data analytics workloads and enables file-level access control lists (ACLs).

File Systems

A file system, also known as a container, holds hierarchical file systems for logs and data. Figure 3-4 shows the container-level properties that can be configured. The Data Plane RBAC security level will be discussed in detail in the "Security" section.

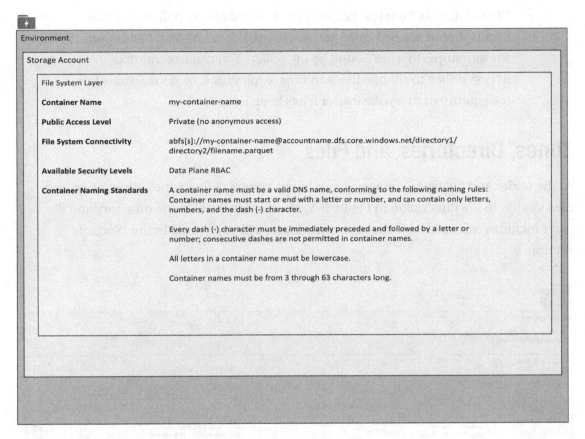

Figure 3-4. File system properties

The following properties can be configured at the container level:

- **Public Access Level**: Specifies whether data in the container may be accessed publicly. By default, container data is private to the account owner. Use "Blob" to allow public read access for blobs. Use "Container" to allow public read and list access to the entire container.

- **Immutable Policies**: Immutable storage provides the capability to store data in a write once, read many (WORM) state. Once data is written, the data becomes non-erasable and non-modifiable, and you can set a retention period so that files can't be deleted until after that period has elapsed. Additionally, a legal hold can be placed on data to make that data non-erasable and non-modifiable until the hold is removed.

- **Stored Access Policies**: Establishing a stored access policy serves to group shared access signatures and to provide additional restrictions for signatures that are bound by the policy. You can use a stored access policy to change the start time, expiry time, or permissions for a signature or to revoke it after it has been issued.

Zones, Directories, and Files

At the folder and file layer, storage accounts' containers define zones, directories, and files similar to the illustration in Figure 3-5. The security level at the directory and file layer includes ACL and SAS. These security levels will be covered in the "Security" section.

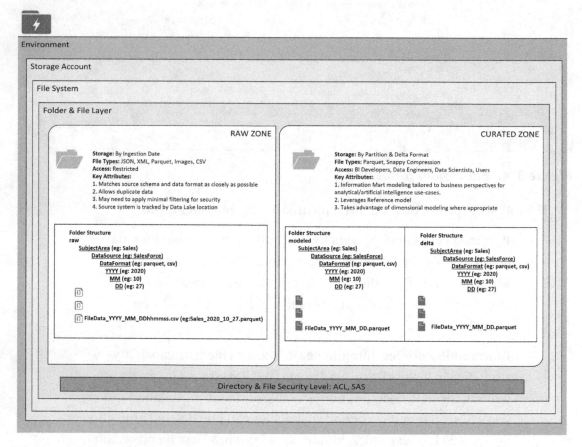

Figure 3-5. *Folder and file–level zones, directories, and structures*

Zones

Zones define the root-level folder hierarchies in a data lake container. Zones can be defined by multiple containers in a storage account or multiple folders in a container. The following sample zones in Table 3-1 describe their purpose and typical user base.

Zones do not need to always reside in the same physical data lake and could also reside as separate file systems or different storage accounts, even in different subscriptions. If large throughput requirements are expected in a single zone exceeding a request rate of 20,000 per second, then multiple storage accounts in different subscriptions may be a good idea.

Table 3-1. *Data Lake Zones, Access, and Descriptions*

Zones	Access	Description
Raw	Service accounts (read-only)	No transformations; Original format; stored by ingestion date
Staging	Scientists, engineers	General staging; Delta preparation
Curated	Analysts, scientists, engineers	Data marketplace; data lake standard
Sensitive	Selective access	Sensitive data that requires elevated and selective access
Laboratory	Scientists, engineers	Exploratory analytics; sandbox area
Transient/Temp	Service accounts	Temp zone that supports ingestion of the data. A use case for this zone is that you may want to decompress data in this zone if you are moving large amounts of compressed data across networks. The data should be short-lived, hence the name Transient.
Master/Reference	Analysts, scientists, engineers	Reference data; archive data

Directories (Folders)

When designing a data lake folder structure, the hierarchy in Figure 3-6 is optimized for analytical querying. Each source system will be granted write permissions at the data source folder level with Default ACL specified. This will ensure permissions are inherited as new daily folders and files are created.

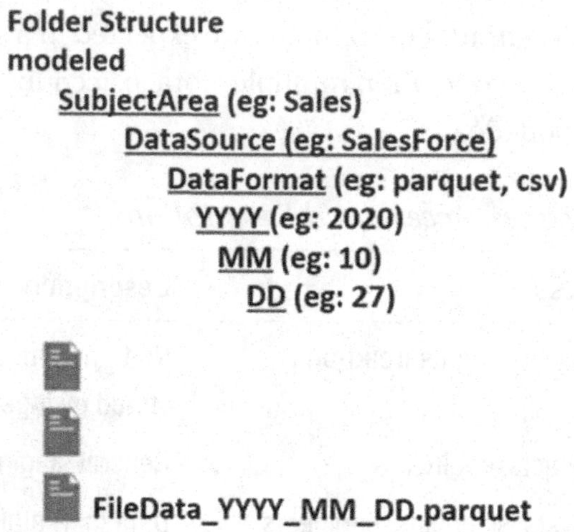

Folder Structure modeled

  SubjectArea (eg: Sales)

    DataSource (eg: SalesForce)

      DataFormat (eg: parquet, csv)

        YYYY (eg: 2020)

          MM (eg: 10)

            DD (eg: 27)

FileData_YYYY_MM_DD.parquet

Figure 3-6. *Sample data lake folder structure*

The hierarchical folder formats illustrated in Figure 3-6 can be dynamically parameterized and coded into an ETL solution within either Databricks or Data Factory to auto-create the folders and files based on the defined hierarchy. The following code snippet defines the same folder structure in Figure 3-6 and can be added to the configurations of an ADF pipeline. I will go over this in more detail in future chapters:

```
\Raw\DataSource\Entity\YYYY\MM\DD\File.extension
```

Sensitive sub-zones in the raw layer can be separated by top-level folders. This will allow you to define a separate lifecycle management policy. The following code shows a variation of the previous folder structure by introducing sub-zones within the Raw zone to further categorize the data sources:

```
\Raw\General\DataSource\Entity\YYYY\MM\DD\File.extension
\Raw\Sensitive\DataSource\Entity\YYYY\MM\DD\File.extension
```

Files

Azure Data Lake Storage Gen2 is optimized to perform better on larger files of around 65 MB–1 GB per file for Spark-based processing. Azure Data Factory compaction jobs can help achieve this. Additionally, the Optimize or Auto Optimize feature of Databricks Delta format can help with achieving this compaction. With Event Hubs, the Capture feature can be used to persist the data based on size or timing triggers.

For Curated or Modeled zones that require highly performant analytics that are read optimized, columnar formats such as Parquet and Databricks Delta would be ideal choices to take advantage of predicate pushdown and column pruning to save time and cost.

Table 3-2 captures some sample file types along with their capabilities and approximate compression ratio.

Table 3-2. *Data Lake File Types, Capabilities, and Compression Ratios*

Sample File Type	Capability	Compression Ratio
Avro	Compressible; splittable; stores schema within the file; good for unstructured and schema-differentiating data	~91.24%
Parquet	Columnar format; compressible	~97.5%
CSV/Text	Commonly used in nearly every organization; easily parsed; often a good use case for bulk processing; not always the best choice for Spark depending on use case	

Table 3-3 captures a few sample compression types along with their capabilities and sample uses.

Table 3-3. *Data Lake Compression Types, Capabilities, and Sample Uses*

Compression Type	Capability	Sample Use
Bzip2	High compression; low speed; works well for archival purposes, not HDFS queries	Archiving use cases; better compression than Gzip for some file types
Gzip	Mid-compression; mid-speed	Cold data in Avro/Parquet format that is infrequently accessed
LZO	High speed; lower compression; works well for text files	Hot data in Text format that is accessed frequently
Snappy	High speed; lower compression	Hot data in Avro/Parquet format that is accessed frequently

Security

Securing data is a critical component of designing a data lake. There are many permissions and controls that can be configured in your data lake. The following sections describe security features that must be considered when designing a data lake.

Control Plane Permissions

Control Plane Role-Based Access Control (RBAC) permissions are intended to give security principal rights only at the Azure resource level and do not include any data actions. Granting a user the "Reader" role will not grant access to the storage account data since additional ACLs or Data Plane RBAC permissions will be required. A good practice is to use Control Plane RBAC in combination with folder/file-level ACLs.

Data Plane Permissions

When Data Plane RBAC permissions are processed for a security principal, then all other ACLs will be ignored and will prevent assigning permissions on files and folders.

Data Plane RBAC permissions can be applied as low as the storage account level.

The list of built-in Data Plane RBAC roles that can be assigned include the following:

- **Storage Blob Data Owner:** Use to set ownership and manage POSIX access control for Azure Data Lake Storage Gen2.

- **Storage Blob Data Contributor:** Use to grant read/write/delete permissions to Blob Storage resources.

- **Storage Blob Data Reader:** Use to grant read-only permissions to Blob Storage resources.

- **Storage Queue Data Contributor:** Use to grant read/write/delete permissions to Azure queues.

- **Storage Queue Data Reader:** Use to grant read-only permissions to Azure queues.

- **Storage Queue Data Message Processor:** Use to grant peek, retrieve, and delete permissions to messages in Azure Storage queues.

- **Storage Queue Data Message Sender:** Use to grant add permissions to messages in Azure Storage queues.

POSIX-Like Access Control Lists

File- and folder-level access within ADLS Gen2 is granted by ACLs. Regardless of ACL permissions, the Control Plane RBAC permissions will be needed in combination with ACLs. As a best practice, it is advised to assign security principals an RBAC Reader role on the storage account/container level and to then proceed with the restrictive and selective ACLs on the file and folder levels.

The two types of ACLs include **Access ACLs**, which control access to a file or a folder, and **Default ACLs**, which are inherited by the assigned Access ACLs within the child file or folder.

Shared Access Signature

A shared access signature (SAS) supports limited access capabilities such as read, write, or update to containers for users. Additionally, timeboxes can be applied as to when the signature is valid for. This allows for temporary access to your storage account and easily manages different levels of access for users within or outside of your organization.

Data Encryption

Data is secured both in motion and at rest, and ADLS Gen2 manages data encryption, data decryption, and placement of the data automatically. ADLS Gen2 also offers functionality to allow a data lake administrator to manage encryption.

Azure Data Lake uses a Master Encryption Key, which is stored in an Azure Key Vault, to encrypt and decrypt data. User-managed keys provide additional control and flexibility, but unless there is a compelling reason, it is recommended to leave the encryption for the Data Lake service to manage.

Network Transport

When network rules are configured, only applications requesting data over the specified set of networks can access a storage account. Access to your storage account can be limited to requests originating from specified IP addresses or IP ranges or from a list of subnets in an Azure virtual network (VNet).

A private endpoint for your storage account can be created, which assigns a private IP address from your VNet to the storage account and secures all traffic between your VNet and the storage account over a private link.

Summary

In this chapter, I covered how to design, implement, and secure an Azure Data Lake Storage Gen2 account. As we begin to explore Azure Data Factory and Databricks ELT patterns in the subsequent chapters, having this understanding and baseline on how to implement and secure your data lake will provide useful considerations as we read from and/or write to the data lake.

PART II

Azure Data Factory for ELT

CHAPTER 4

Dynamically Load a SQL Database to Data Lake Storage Gen2

The process of moving data from either an on-premises SQL Server or an Azure SQL Server to Azure Data Lake Storage has become an ever-increasing and vital need for many businesses. Customers with numerous on-premises SQL Servers along with hundreds of databases within these servers are interested in leveraging Azure's data services to build ELT and/or ETL processes to fully load their Azure Data Lake Storage account with on-premises SQL Server databases and tables in a structured, partitioned, and repeatable process for one or many servers in their SQL Server ecosystem. Taking on-premises relational data, pushing them into files in the data lake, and then using distributed and multi-cloud computing technologies such as Databricks is a popular modern architectural pattern. This pattern empowers customers to increase their maturity level within their Azure cloud journey by introducing advanced analytics capabilities such as AI and ML that could not be achieved easily with their existing on-premises environments.

Azure Data Factory has been a critical ELT and ETL tool of choice for many data engineers that have been working with Azure's Data Platform. The ability to leverage dynamic SQL and parameters within ADF pipelines allows for seamless data engineering and scalability. In this chapter, you will gain an understanding of this process through a practical exercise by creating an end-to-end Data Factory pipeline to move all on-premises SQL Server objects including databases and tables to Azure Data Lake Storage Gen2 with a few pipelines that leverage dynamic parameters.

R. C. L'Esteve, *The Definitive Guide to Azure Data Engineering*, https://doi.org/10.1007/978-1-4842-7182-7_4

Azure Prerequisite Resources

In Chapter 3, I discussed how to design and implement an Azure Data Lake Storage Gen2 account. This chapter will build on those learnings by using the data lake as your target destination (sink) for the transit parquet files that will ultimately need to be loaded into an Azure SQL Database.

Parquet is a file format that is available within the Hadoop ecosystem and is designed for performant and efficient flat columnar storage format of data compared to row-based files like CSV or TSV files.

Rather than using simple flattening of nested namespaces, Parquet uses record shredding and assembly algorithms. Parquet features different ways for efficient data compression and encoding types and is optimized to work with complex data in bulk. This approach is optimal for queries that need to read certain columns from large tables. Parquet can only read the needed columns, which minimizes the I/O.

Figure 4-1 illustrates the data flow architecture from the source SQL database (Azure or on-premises) to the data lake. Additionally, the required components are also depicted in Figure 4-1.

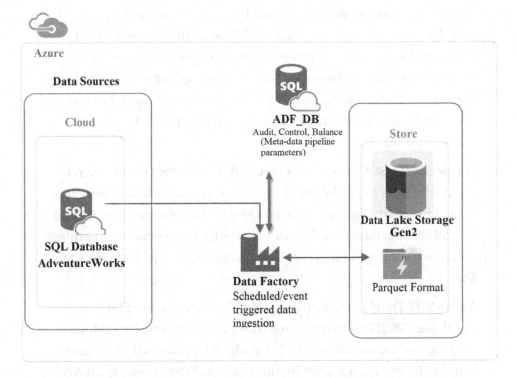

Figure 4-1. *SQL database to Data Lake Storage Gen2 architecture diagram*

The following list of Azure resources will be needed when designing and implementing the architecture in Figure 4-1. Microsoft offers a number of tutorials, demos, and resources describing how to create these resources either through the Azure Portal UI or through code. Read through the following list to understand the purpose of each of the components listed in Figure 4-1 and create them in your Azure subscription prior to building the ADF pipeline in this chapter:

- **Azure Data Lake Storage Gen2**: The data will be stored in Azure Data Lake Storage Gen2 (ADLS Gen2). For that, you will need to create an Azure Data Lake Storage Gen2 storage account. When creating this account, remember to apply the learnings from Chapter 3 and ensure that "Hierarchical Namespace" is enabled in the account creation settings so that a Gen2 account will be created.

- **Azure Data Factory V2**: Azure Data Factory (ADF) will be used as the ETL orchestration tool so you can create a data factory by using the Azure Data Factory UI.

- **Azure Data Factory self-hosted IR**: If you decide to go the on-premises SQL Server route, then a self-hosted IR will create a link between the on-premises resources and Azure services and allow you to connect to on-premises data in ADF. It is also important to note that the on-premises machine that will host the self-hosted IR must have the 64-bit Java Runtime Environment (JRE) installed on it to create parquet files.

- **On-premises SQL Server or Azure SQL Database**: For this exercise, the source data can either be stored in an on-premises SQL Server or in an Azure SQL Database for simplicity and to demonstrate the end-to-end architectural flow. You can always try SQL Server on-premises or in the cloud to obtain a Developer, Express, or Trial version of SQL Server.

- **Azure SQL Database (standard)**: Finally, use an Azure SQL Database (ASQLDB) to house the ADF pipeline parameter values, control, and logging tables. Additionally, this Azure SQL Database can also be used as a source for this exercise. When creating an Azure SQL database for this process, remember to grant your client IP address to the SQL Server's firewall rules. Additionally, remember to allow Azure services and resources to access this server.

Prepare and Verify SQL Server Database Objects

Prepare the process by importing a sample SQL database into an Azure SQL Database. There are many free sample versions of SQL Server databases available for download such as AdventureWorks or WideWorldImporters sample databases (`https://github.com/microsoft/sql-server-samples/tree/master/samples/databases`). After the sample database has been added, navigate to SQL Server Management Studio (SSMS) and connect to the on-premises SQL Server (or Azure SQL Database) containing two OLTP SQL databases as illustrated in Figure 4-2.

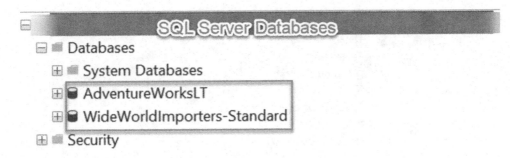

Figure 4-2. *SSMS view of OLTP source databases*

Figure 4-3 expands the details within the WideWorldImporters database to verify that there are tables in both databases.

Figure 4-3. *List of tables in source WideWorldImporters database*

Similarly, Figure 4-4 expands the details within the AdventureWorksLT database to verify that there are tables in both databases.

Figure 4-4. *List of tables in source AdventureWorksLT database*

Prepare and Verify Azure SQL Database Objects

Next, create an Azure SQL Database called ADF_DB along with a table to store the table names, catalog names, and process flags, which will drive the pipeline configurations at runtime. This table is also called a control table. The following code can be used to create the base table needed for this metadata-driven ELT process:

```
USE [ADF_DB]

go

SET ansi_nulls ON

go

SET quoted_identifier ON

go

CREATE TABLE [dbo].[pipeline_parameter1]
  (
     [parameter_id]   [INT] IDENTITY(1, 1) NOT NULL,
     [table_name]     [NVARCHAR](500) NULL,
```

```
  [table_catalog] [NVARCHAR](500) NULL,
  [process_type]  [NVARCHAR](500) NULL,
  PRIMARY KEY CLUSTERED ( [parameter_id] ASC )WITH (pad_index = OFF,
  statistics_norecompute = OFF, ignore_dup_key = OFF, allow_row_locks = on,
  allow_page_locks = on, optimize_for_sequential_key = OFF) ON [PRIMARY]
  )
ON [PRIMARY]

go
```

Figure 4-5 shows how these database objects will look in SSMS after the database and table are created.

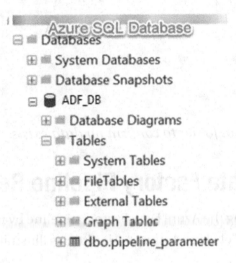

Figure 4-5. *View of ADF_DB database and table in SSMS*

Prepare an Azure Data Lake Storage Gen2 Container

An ADLS Gen2 container and folder for the root-level hierarchy will also be needed, which will be at the server level. Therefore, the root folder will be the name of the server, as illustrated in Figure 4-6.

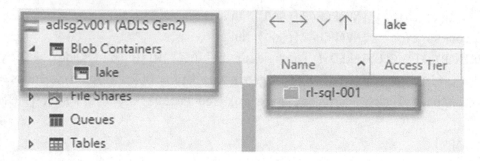

Figure 4-6. Azure Data Lake container and root folder

Figure 4-7 confirms that there is no existing data in the server-level folder.

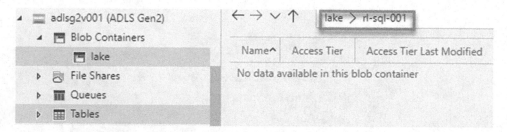

Figure 4-7. Drill into root folder to confirm no data exists.

Create Azure Data Factory Pipeline Resources

At this point, begin creating the Azure Data Factory pipeline by navigating to the Azure Data Factory resource and clicking Author & Monitor, as illustrated in Figure 4-8.

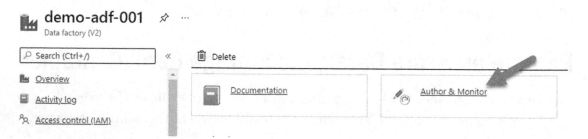

Figure 4-8. Author & Monitor for Azure Data Factory pipeline creation

Once the Azure Data Factory canvas loads, clicking "Create pipeline," as shown in Figure 4-9, will create a new pipeline and blank canvas.

Figure 4-9. *Create an Azure Data Factory pipeline.*

Create a Self-Hosted Integration Runtime

If you are linking an on-premises server to ADF, then you will need to create a self-hosted integration runtime (`https://docs.microsoft.com/en-us/azure/data-factory/create-self-hosted-integration-runtime`).

A self-hosted integration runtime can run Copy activities between a cloud data store and a data store in a private network. It also can dispatch transform activities against compute resources in an on-premises network or an Azure virtual network. The installation of a self-hosted integration runtime needs an on-premises machine or a virtual machine inside a private network.

Once the self-hosted IR is created, verify that it is in "Running" status, as depicted in Figure 4-10. A self-hosted integration runtime can run Copy activities between a cloud data store and a data store in a private network.

Figure 4-10. *Self-hosted integration runtime is in Running status*

Create Linked Services

After the self-hosted IR is created, all the required linked services can be created, which include SQL Server, Azure SQL Database, and Azure Data Lake Storage Gen2, as illustrated in Figure 4-11.

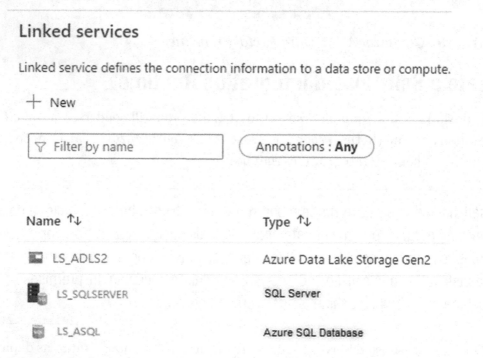

Figure 4-11. *Linked services that have been created*

Create Datasets

The ADF pipelines will require three datasets, which are listed in Figure 4-12. The first DS_ADLS2 dataset will be a connection to the sink Azure Data Lake Storage Gen2 account. The second DS_ASQLDB_PIPELINE_PARAMETER dataset will be a connection to the pipeline_parameter table in ADF_DB. And finally, the third DS_SQLSERVER dataset will be a connection to the source SQL Server database.

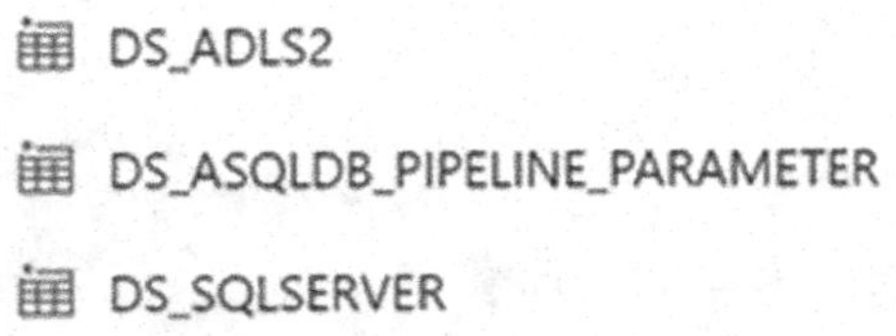

▦ DS_ADLS2

▦ DS_ASQLDB_PIPELINE_PARAMETER

▦ DS_SQLSERVER

Figure 4-12. *Datasets that have been created*

DS_ADLS2

Next, configure the ADLS Gen2 dataset as a Parquet format, which can be seen in Figure 4-13. Also, in the figure are the parameterized file path connections to allow us to partition the data by YY-MM-DD-HH listed in the "File path" section. This code is also provided on the following page. Also remember to set the compression type to "snappy" for improved performance. Ignore the yellow warning icon as this frequently appears for dynamic and parameterized content on occasion, especially when the string might be longer than expected.

Parquet
DS_ADLS2

General   **Connection**   Schema   Parameters

Linked service *   LS_ADLS2   ▼   *Test connection*   *Open*   + New

File path *   lake   /   @concat('rl-sql001/',dataset().table_catalog)   /   @{item().Table_Name}/@{formatDateTime(utcnow(),'yyyy')}/@{formatDateTime(utcnow(),'MM')}/@{formatDateTime(utcnow(),'dd')}/@{item().Table_Name}@{formatDateTime(utcnow(),'HH')}

⚠ Warning

Compression type   snappy   ▼

Figure 4-13. *ADLS2 dataset connection properties*

For reference, the following code has been used in the ADLS Gen2 dataset connection properties in Figure 4-13.

```
@concat('rl-sql001/',dataset().table_catalog)
```

```
@{item().Table_Name}/@{formatDateTime(utcnow(),'yyyy')}/@{formatDateTime
(utcnow(),'MM')}/@{formatDateTime(utcnow(),'dd')}/@{item().Table_Name}
@{formatDateTime(utcnow(),'HH')}
```

ADF allows you to interpret expressions within strings to easily enable calculations, parameters, and field values as part of your strings. Now, with string interpolation, you can produce super-easy string evaluations using expressions like these samples. Much of the code that is implemented in the ADF pipelines uses string interpolation.

Within this dataset, add the parameters shown in Figure 4-14, which will be used at a later stage.

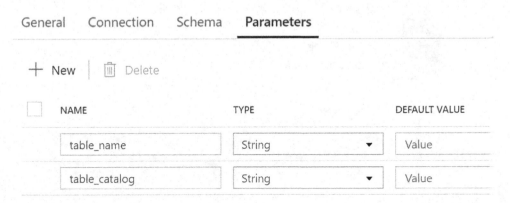

Figure 4-14. *ADLS2 parameters needed*

DS_SQLSERVER

Next, add a dataset connection to the on-premises SQL Server. Also leave the table as set to "None" to allow for traversing through all tables in the SQL Server, as shown in Figure 4-15.

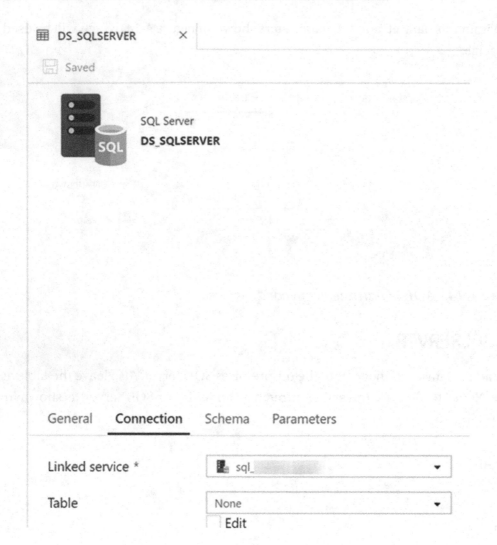

Figure 4-15. *SQL Server connection properties*

DS_ASQLDB_PIPELINE_PARAMETER

The final dataset will be for a connection to the pipeline_parameter table in an Azure SQL Database, shown in Figure 4-16. This completes all of the datasets that will be needed to continue creating the pipelines.

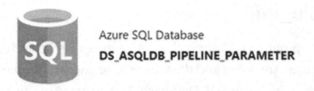

Azure SQL Database
DS_ASQLDB_PIPELINE_PARAMETER

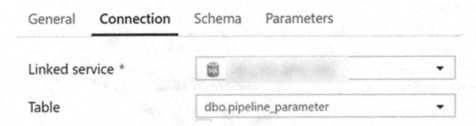

General   **Connection**   Schema   Parameters

Linked service *

Table   dbo.pipeline_parameter

Figure 4-16.   Azure SQL Database connection properties

Create Azure Data Factory Pipelines

Now that the datasets have been created, it's time to create the ADF pipeline to move data to the lake. This pipeline will leverage the self-hosted IR connection to the on-premises network to convert the SQL database tables and dynamically load them into the data lake as parquet files. Additionally, this pipeline will use the ADF_DB, which was created in the previous section of this chapter.

P_Insert_Base_Table_Info

The P_Insert_Base_Table_Info pipeline will query the on-premises `information_Schema.tables` as its source to get the table and database name and will then output the results to a basic parameter table in Azure SQL Database. The purpose of this process will be to populate the `pipeline_parameter` table to drive the rest of the pipelines via its metadata fields.

To create this pipeline, add a Copy activity to the pipeline canvas, shown in Figure 4-17, and set the source as the on-premises SQL Server dataset. Also add the query code contained in this section as the source. This query will query the specified database and list the tables that can be used in the rest of the process. For now, manually change the database name for every database that would need to be loaded and execute the pipeline. Figure 4-17 shows where this code block can be placed within the Copy activity.

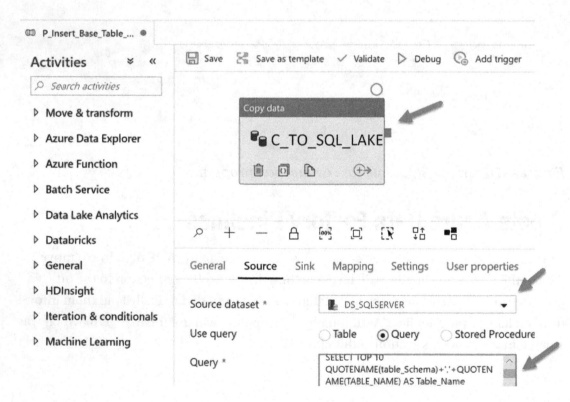

Figure 4-17. *Source query within the ADF Copy activity*

Here is the SQL query that has been added to the Query section in Figure 4-17:

```
USE adventureworkslt

SELECT Quotename(table_schema) + '.'
       + Quotename(table_name) AS Table_Name,
       table_catalog
FROM   information_schema.tables
WHERE  table_type = 'BASE TABLE'
```

Although we are manually pasting code into the source query section in ADF for the purposes of this exercise, as a best practice for production development and deployment, I would recommend creating stored procedures within your SQL database environment and then calling them in the ADF source as stored procedures rather than hard-coded queries. This will help with easier maintenance and management of the code. Also, when adding activities and designing ADF pipelines, be sure to name your ADF activities, pipelines, and datasets appropriately and sensibly.

As a best practice for naming ADF pipelines, activities, and datasets, I would recommend following the naming conventions listed in Table 4-1.

Table 4-1. *Azure Data Factory Naming Conventions*

Pipeline	PL_
Mapping Data Flows	MDF_
Copy activity	C_
Linked service	LS_
Dataset	DS_
ForEach loop	FE_
Lookup	L_
Stored procedure	SP_

Now that you have configured the source, it is time to set the sink by toggling to the Sink tab and choosing the `pipeline_parameter` dataset that you configured in the previous section, as shown in Figure 4-18.

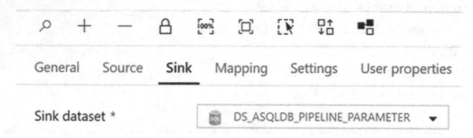

Figure 4-18. *Sink dataset properties*

Lastly, ensure that the source to destination mappings are accurate, as shown in Figure 4-19. Typically, a one-for-one naming convention between source and destination will auto-map, and you do also have the capability of manually altering this mapping.

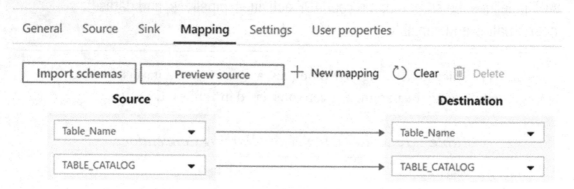

Figure 4-19. *ADF pipeline source to destination mapping*

The resulting `pipeline_parameter` Azure SQL table should look similar to the illustration shown in Figure 4-20.

	Table_Name	TABLE_CATALOG
1	[SalesLT].[Customer]	AdventureWorksL1
2	[SalesLT].[ProductModel]	AdventureWorksL1
3	[SalesLT].[ProductDescription]	AdventureWorksL1
4	[SalesLT].[Product]	AdventureWorksL1
5	[SalesLT].[ProductModelProductDescription]	AdventureWorksL1
6	[SalesLT].[ProductCategory]	AdventureWorksL1
7	[dbo].[BuildVersion]	AdventureWorksL1
8	[dbo].[ErrorLog]	AdventureWorksL1
9	[SalesLT].[Address]	AdventureWorksL1
10	[SalesLT].[CustomerAddress]	AdventureWorksL1
11	[SalesLT].[SalesOrderDetail]	AdventureWorksL1
12	[SalesLT].[SalesOrderHeader]	AdventureWorksL1

Figure 4-20. *ADF pipeline parameter table*

P_SQL_to_ADLS

In this next section, the goal is to create the SQL Server to ADLS Gen2 data orchestration pipeline. Add the Lookup and ForEach activities to the pipeline canvas, as shown in Figure 4-21.

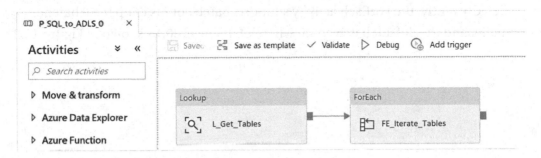

Figure 4-21. *ADF pipeline Lookup and ForEach activities*

The Lookup activity simply looks up the pipeline parameter table that was populated in the prior pipeline, as shown in the settings illustrated by Figure 4-22.

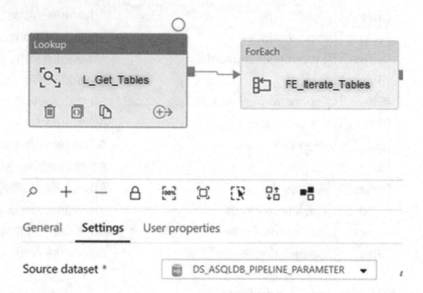

Figure 4-22. *ADF pipeline settings for the Lookup activity*

Next, within the settings of the ForEach activity, ensure that "Items" is set to the following:

```
@activity('Get-Tables').output.value
```

Also, ensure that "Sequential" remains unchecked so that the tables will execute in parallel. Currently, the ForEach activity supports batch counts of up to 50 batches sequentially, and the default batch count when left blank is 20, as shown in Figure 4-23.

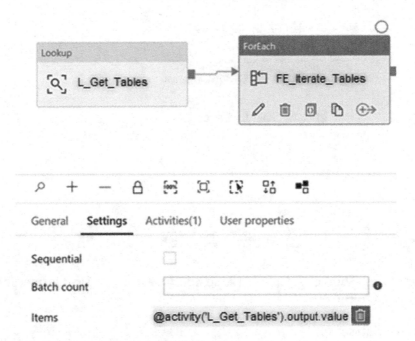

Figure 4-23. *ADF pipeline settings for the ForEach activity*

Also, within the Activities tab of the ForEach activity, add a Copy activity. Click Edit activities to view the details, as shown in Figure 4-24.

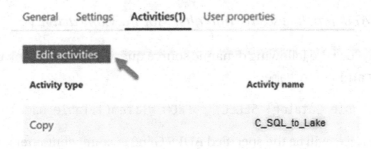

Figure 4-24. *ADF pipeline activities for the ForEach activity*

The source dataset is the on-premises SQL Server shown in Figure 4-25.

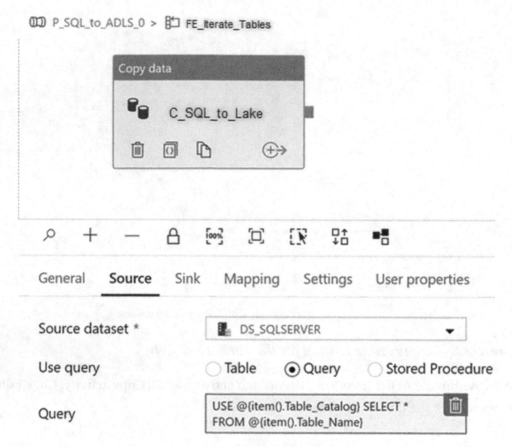

Figure 4-25. *ADF pipeline source and query for the Copy activity*

Additionally, use the following dynamic source query, which will look up the
Table_Catalog and Table_Name:

USE @{item().Table_Catalog} SELECT * FROM @{item().Table_Name}

Your sink dataset will be the specified ADLS Gen2 account container and folder
structure, which has been created as DS_ADLS2 in the previous section of this chapter.
Also add @{item().Table_Name} and @{item().Table_Catalog} to the dataset
properties that have been configured while creating the DS_ADLS2 dataset, as shown in
Figure 4-26.

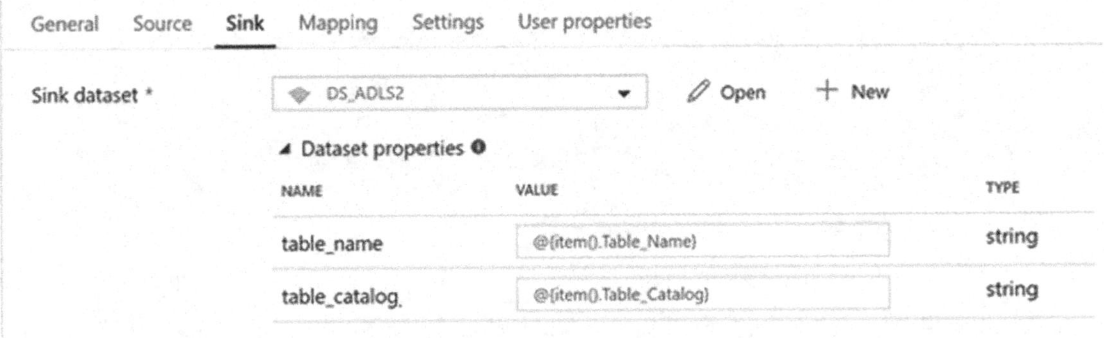

General	Source	**Sink**	Mapping	Settings	User properties

Sink dataset *   ◈ DS_ADLS2   ▼   🖉 Open   + New

▲ Dataset properties ❶

NAME	VALUE	TYPE
table_name	@{item().Table_Name}	string
table_catalog.	@{item().Table_Catalog}	string

Figure 4-26. *ADF pipeline sink dataset properties*

This pipeline is now complete and ready to be run.

Run the Data Factory Pipeline and Verify Azure Data Lake Storage Gen2 Objects

Run the ADF pipeline and notice from Figure 4-27 that there are two database-level folders that have been created for the two databases.

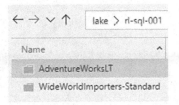

Figure 4-27. *ADLS2 database-level folders*

Notice also that the appropriate table-level folders have been created in your ADLS Gen2 account, as shown in Figure 4-28.

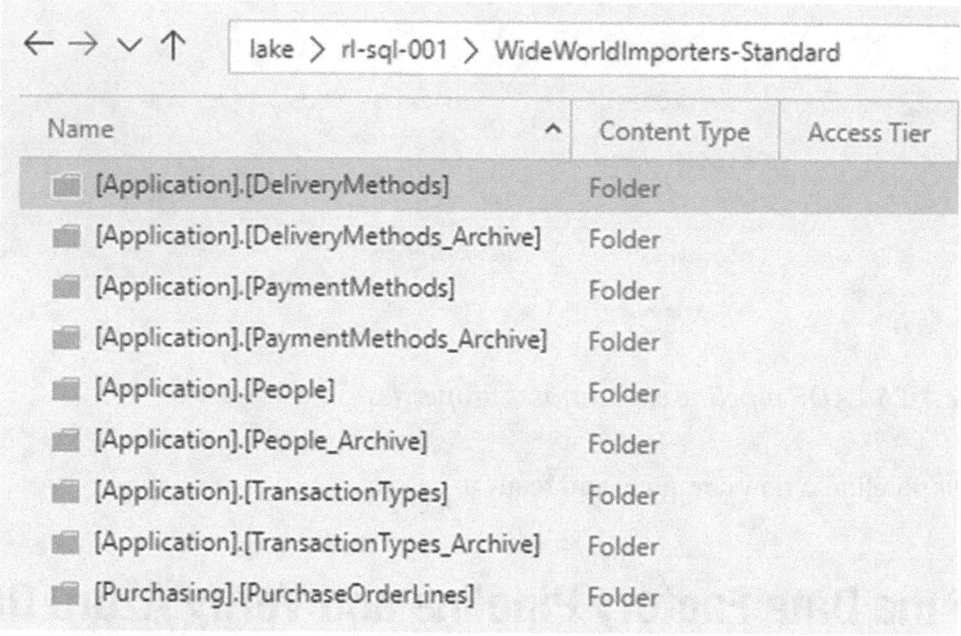

Figure 4-28. *ADLS2 table-level folders*

Additionally, upon drilling into the folders, notice from Figure 4-29 that the folders are appropriately time-partitioned and that the Parquet format table-level file has been created.

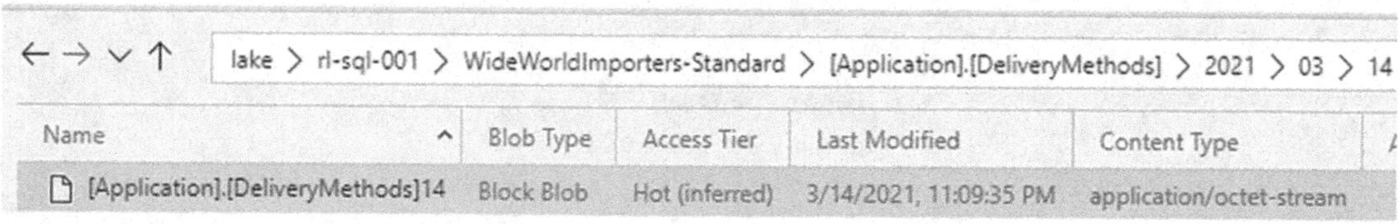

Figure 4-29. *ADLS Gen2 parquet files*

Summary

In this chapter, I introduced Azure Data Factory and its role in the modern cloud-based ETL landscape as a tool of choice for many data engineers that have been working with Azure's Data Platform. I also demonstrated a practical example on how to create an end-to-end Data Factory pipeline to move all on-premises SQL Server objects including databases and tables to Azure Data Lake Storage Gen2 with a few pipelines that leverage dynamic parameters.

This exercise demonstrated ADF's capabilities of building dynamic, metadata-driven ELT pipelines for high-volume, high-velocity, and wide-variety data ingestion. These pipelines can move data from a variety of source systems including on-premises databases to ADLS Gen2 and other low-cost storage accounts for further processing, cleansing, transformation, exploration, and advanced analytics use cases.

CHAPTER 5

Use COPY INTO to Load a Synapse Analytics Dedicated SQL Pool

In Chapter 4, I showed you how to load snappy compressed parquet files into Data Lake Storage Gen2 from an on-premises SQL Server. This chapter will introduce you to the fundamental mechanisms and technologies that will be used to fully load the parquet files from ADLS Gen2 into a Synapse Analytics dedicated SQL pool table. Also, this chapter will take a closer look at the COPY INTO command that offers numerous benefits including eliminating multiple steps in the data load process and reducing the number of database objects needed for the process. Additionally, COPY INTO does not require CONTROL access to the sink SQL pool table as with PolyBase and only requires INSERT and ADMINISTER DATABASE BULK OPERATIONS permissions. This chapter will demonstrate some common scenarios for using the COPY INTO command.

Azure Synapse Analytics is a limitless analytics service that brings together data integration, enterprise data warehousing, and big data analytics. It gives you the freedom to query data on your terms, using either serverless or dedicated resources at scale. Azure Synapse brings these worlds together with a unified experience to ingest, explore, prepare, manage, and serve data for immediate BI and machine learning needs. Dedicated SQL pool refers to the enterprise data warehousing features that are available in Azure Synapse Analytics. Dedicated SQL pool represents a collection of analytics resources that are provisioned when using Synapse SQL.

Figure 5-1 is similar to the architecture diagram presented in Figure 4-1 of Chapter 4, with the addition of a Synapse Analytics dedicated SQL pool as the final destination of the parquet files that we created and stored in ADLS Gen2.

83

© Ron C. L'Esteve 2021
R. C. L'Esteve, *The Definitive Guide to Azure Data Engineering*, https://doi.org/10.1007/978-1-4842-7182-7_5

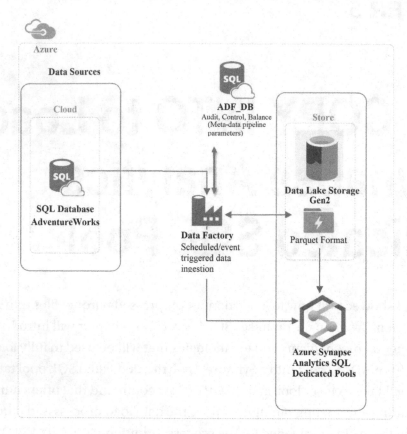

Figure 5-1. *Data Lake Storage Gen2 to Synapse dedicated SQL pool architecture diagram*

Prior to exploring the COPY INTO command in greater detail, you will need to create a Synapse Analytics dedicated SQL pool for the destination SQL data warehouse. Additionally, you will need to create a destination table that matches the column names, column order, and column data types. You can create a dedicated SQL pool in Synapse Analytics using Azure Portal.

Features of the COPY INTO Command

The COPY command feature in Synapse Analytics provides users a simple, flexible, and fast interface for high-throughput data ingestion for SQL workloads. The COPY INTO command supports the following arguments:

```
FILE_TYPE = {'CSV' | 'PARQUET' | 'ORC'}
FILE_FORMAT = EXTERNAL FILE FORMAT OBJECT
```

```
CREDENTIAL = (AZURE CREDENTIAL)
ERRORFILE = http(s)://storageaccount/container]/errorfile_directory[/]
ERRORFILE_CREDENTIAL = (AZURE CREDENTIAL)
MAXERRORS = max_errors
COMPRESSION = { 'Gzip' | 'DefaultCodec'|'Snappy'}
FIELDQUOTE = 'string_delimiter'
FIELDTERMINATOR =  'field_terminator'
ROWTERMINATOR = 'row_terminator'
FIRSTROW = first_row
DATEFORMAT = 'date_format'
ENCODING = {'UTF8'|'UTF16'}
IDENTITY_INSERT = {'ON' | 'OFF'}
```

Notice from this list that there are multiple options that can be selected and configured within each of these arguments to customize the configuration parameters.

Data Preparation Tips

Data preparation of the source data will be an important and necessary process prior to creating parquet files in ADLS Gen2. The following sections provide a few data preparation tips to ensure that the parquet files are ready to be run through the COPY INTO command.

Tip #1: Remove Spaces from the Column Names

Oftentimes, there are noticeable issues with column names containing spaces while loading parquet files to a Synapse Analytics dedicated SQL pool. These spaces in the column names can be handled by creating a view and assigning an alias to the columns containing spaces in their names if you are working with on-premises source systems.

Alternatively, as a more complex solution, column name spaces can be eliminated from multiple source tables by leveraging the sys columns and sys tables with the following script to remove spaces in column names in all tables:

```
SELECT 'EXEC SP_RENAME ''' + B.NAME + '.' + A.NAME
        + ''', ''' + Replace(A.NAME, ' ', '')
        + ''', ''COLUMN'''
```

```
FROM    sys.columns A
        INNER JOIN sys.tables B
                ON A.object_id = B.object_id
                    AND Objectproperty(b.object_id, N'IsUserTable') = 1
WHERE   system_type_id IN (SELECT system_type_id
                            FROM   sys.types)
        AND Charindex(' ', a.NAME) <> 0
```

If your source system is a cloud-based source, then you may also want to consider using Mapping Data Flows to remove spaces from your source columns dynamically and through patterns, rules, and derived columns. I will go over some of the additional capabilities of Mapping Data Flows in Chapters 11 and 12.

Tip #2: Convert VARCHAR(MAX) to VARCHAR(4000)

Oftentimes, there are capacity limits on data types and maximum values for various other components that need to be accounted for. Note that this may change and become a supported feature over time. The varchar(max) data type has historically been unsupported in Azure Synapse Analytics, and therefore, you can use the following CONVERT function on your source system to convert varchar(max) to varchar(4000) data types, as needed. This can be achieved by converting the source table's data type using the following CONVERT function into a select statement or a view:

```
CONVERT(VARCHAR(length) , varchar_column_name)
```

```
SELECT column1                            AS column1,
       column2                            AS column2,
       CONVERT(VARCHAR(4000), varchar_column) AS varchar_column
FROM   dbo.table_name
```

COPY INTO Using a Parquet File

The preferred method of using the COPY INTO command for big data workloads would be to read parquet (snappy compressed) files using snappy Parquet as the defined FILE_FORMAT. Additionally, for this scenario, use a Managed Identity credential.

The following is the COPY INTO SQL syntax for snappy parquet files that you must run in a Synapse Analytics dedicated SQL pool:

```
COPY INTO [Table1]
FROM 'https://lake.dfs.core.windows.net/lake/staging/Table1/parquet/*.parquet'
WITH (
    FILE_FORMAT = [snappyparquet],
    CREDENTIAL = (IDENTITY='Managed Identity')
)
```

Run the command and notice that the snappy parquet file was copied from ADLS Gen2 into a Synapse Analytics dedicated SQL pool table in around 30 seconds per 1 million rows.

Additionally, after performing the data preparation step, there were no errors with the following data types: DATETIME, INT, NVARCHAR(4000). Also, NULL ints, commas, and quotes in text fields were not an issue with this snappy Parquet format.

COPY INTO Using a CSV File

Certain scenarios may require the source files to be in CSV format. For this scenario, there is a bit more setup required on the source dataset.

Begin by configuring a CSV dataset in Data Factory and select the following connection properties, as illustrated in Figure 5-2:

- **Column delimiter**: Comma (,).

- **Row delimiter**: Auto detect.

- **Encoding**: Default(UTF-8). This will need to be set for CSV files. Alternatively, "Encoding" can be specified in the COPY INTO command syntax.

- **Escape character**: '' (note that this setting will allow double quotes and commas in text fields).

- **Quote character**: Double quote (") (note that this setting will allow double quotes and commas in text fields).

- **Null value**: @concat('') (note that this setting will allow NULL int data types).

General   **Connection**   Schema   Parameters

Linked service *	[_____ ▼] ✐ Test connection ✐ Open ＋ New

File path *	[_____] / [@{item().dst_folder}_____] / [@{item().dst_name}/csv/@{formatDateTime
	                                 ⚠ Warning (utcnow(),'yyyy-MM-
	                                                          dd')}/@{item().dst_name}.csv
	                                                             ⚠ Warning]

Compression type	[none ▼]
Column delimiter	[Comma (,) ▼] ☐ Edit
Row delimiter	[Auto detect (\r,\n, or \r\n) ▼] ☐ Edit
Encoding	[Default(UTF-8) ▼]
Escape character	[·_____] ☑ Edit
Quote character	[Double quote (') ▼] ☐ Edit
First row as header	☐
Null value	[@concat('')_____]

Figure 5-2. *CSV connection configuration properties in ADF*

Ensure that the ADF source Connection tab contains all of the configured properties. The code used for the dynamic file paths in Figure 5-2 is listed in the following:

```
@{item().dst_folder}
```

```
@{item().dst_name}/csv/@{formatDateTime(utcnow(),'yyyy-MM-dd')}/ @{item().
dst_name}.csv
```

The following sample code shows the COPY INTO SQL syntax for CSV files that can be run in a Synapse Analytics dedicated SQL pool. Note that ENCODING is being specified as UTF8 in the syntax along with the comma as our field terminator:

```
COPY INTO [Table1]
FROM 'https://sdslake.dfs.core.windows.net/lake/staging/Table1/csv/*.csv'
WITH (
    FILE_TYPE = 'CSV',
    CREDENTIAL = (IDENTITY='Managed Identity'),
    ENCODING = 'UTF8',
    FIELDTERMINATOR = ','
)
```

Similar to COPY INTO using snappy Parquet syntax, run the command and notice that the CSV file was copied from ADLS Gen2 into an Azure Synapse Analytics dedicated SQL pool table in around 12 seconds for 300K rows. Additionally, after performing the data preparation step, no errors were encountered with the following data types: DATETIME, INT, NVARCHAR(4000). Also, after configuring the CSV dataset properties, NULL ints, commas, and quotes in text fields were not an issue with this CSV file type.

Using COPY INTO from Data Factory

To use the COPY INTO command from Data Factory, create a Synapse Analytics dedicated SQL pool dataset, along with a new pipeline containing a Copy activity, as illustrated in Figure 5-3. Set the source as the dataset containing the ADLS Gen2 storage account and set the sink as the Synapse Analytics dedicated SQL pool dataset.

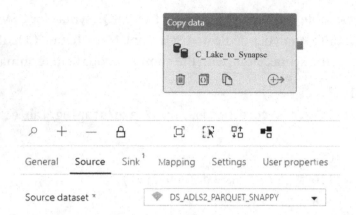

Figure 5-3. *ADF source settings for COPY INTO configuration*

Figure 5-4 shows the various copy methods that will be displayed once the sink dataset is configured to a Synapse Analytics dedicated SQL pool dataset.

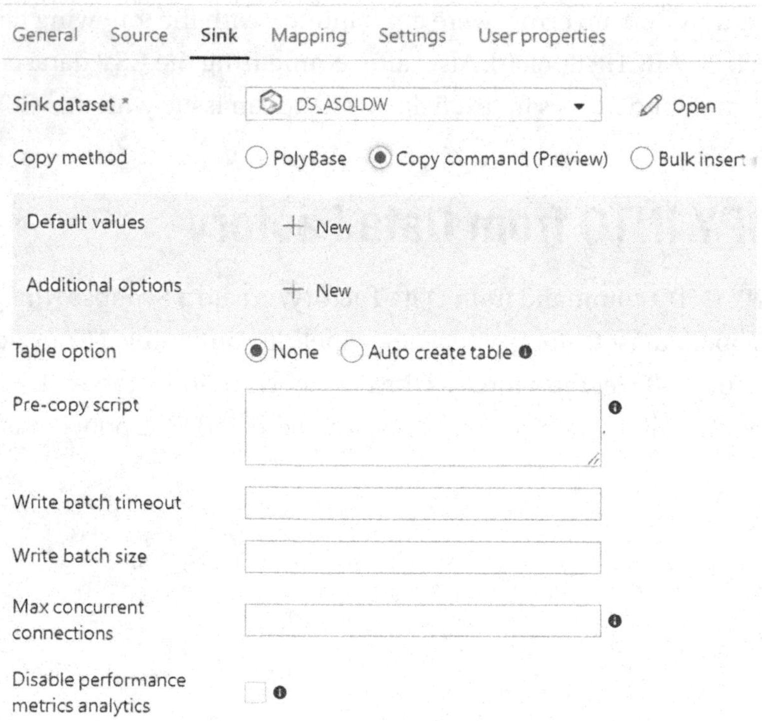

General   Source   **Sink**   Mapping   Settings   User properties

Sink dataset *   DS_ASQLDW   ▾   ✎ Open

Copy method   ○ PolyBase   ◉ Copy command (Preview)   ○ Bulk insert

   Default values   + New

   Additional options   + New

Table option   ◉ None   ○ Auto create table ❶

Pre-copy script

Write batch timeout

Write batch size

Max concurrent connections

Disable performance metrics analytics   ☐ ❶

Figure 5-4. *ADF sink settings for COPY INTO configuration*

Notice the options include PolyBase, Copy command, and Bulk insert. By selecting "Copy command," notice that there is an option to add a pre-copy script in the event that you might need to truncate a staging table prior to a full reload. Additionally, there is an "Auto create table" option. When the ADF pipeline is run, the snappy parquet file from ADLS Gen2 will be loaded to the Synapse Analytics dedicated SQL pool from the Data Factory pipeline.

Summary

In this chapter, I have discussed the fundamental mechanisms and technologies that are used to fully load the parquet files from ADLS Gen2 into a Synapse Analytics dedicated SQL pool table. Additionally, I demonstrated usage of the COPY INTO command, which eliminates multiple steps in the data load process and reduces the number of database objects needed for the process.

## CHAPTER 6

Load Data Lake Storage Gen2 Files into a Synapse Analytics Dedicated SQL Pool

Chapter 4 showed you how to create a dynamic, parameterized, and metadata-driven process to fully load data from an on-premises SQL Server to Azure Data Lake Storage Gen2. This chapter will demonstrate how to fully load all of the snappy compressed parquet data files from ADLS Gen2 into an Azure dedicated SQL pool.

Azure Data Factory's sink Copy activity allows three different copy methods for loading data into an Azure dedicated SQL pool, which is part of the Azure Synapse Analytics ecosystem. This chapter will explore these three methods using a dynamic and parameterized ADF pipeline:

1) PolyBase

2) Copy command

3) Bulk insert

First, there is some prep work to do by creating the datasets and the pipelines demonstrated in the first few sections of the chapter. Toward the end of the chapter, the three methods for loading data that have been listed previously will be discussed and demonstrated.

© Ron C. L'Esteve 2021
R. C. L'Esteve, *The Definitive Guide to Azure Data Engineering*, https://doi.org/10.1007/978-1-4842-7182-7_6

Recreate the Pipeline Parameter Table

Begin by recreating the pipeline_parameter table in the ADF_DB that you had created in Chapter 4 to make it much more robust in preparation for the ADF pipeline that will be built in this chapter.

The following is the code to recreate the table. It first drops the table if the table exists. Then it creates the table anew:

```
USE [ADF_DB]

go

/****** Object:  Table [dbo].[pipeline_parameter]    ******/
IF EXISTS (SELECT *
           FROM    sys.objects
           WHERE   object_id = Object_id(N'[dbo].[pipeline_parameter]')
                   AND type IN ( N'U' ))
  DROP TABLE [dbo].[pipeline_parameter]

go

/****** Object:  Table [dbo].[pipeline_parameter]  ******/
SET ansi_nulls ON

go

SET quoted_identifier ON

go

CREATE TABLE [dbo].[pipeline_parameter]
  (
     [parameter_id]                 [INT] IDENTITY(1, 1) NOT NULL,
     [server_name]                  [NVARCHAR](500) NULL,
     [src_type]                     [NVARCHAR](500) NULL,
     [src_schema]                   [NVARCHAR](500) NULL,
     [src_db]                       [NVARCHAR](500) NULL,
     [src_name]                     [NVARCHAR](500) NULL,
     [dst_type]                     [NVARCHAR](500) NULL,
```

```
    [dst_schema]                            [NVARCHAR](500) NULL,
    [dst_name]                              [NVARCHAR](500) NULL,
    [include_pipeline_flag]                 [NVARCHAR](500) NULL,
    [partition_field]                       [NVARCHAR](500) NULL,
    [process_type]                          [NVARCHAR](500) NULL,
    [priority_lane]                         [NVARCHAR](500) NULL,
    [pipeline_date]                         [NVARCHAR](500) NULL,
    [pipeline_status]                       [NVARCHAR](500) NULL,
    [load_synapse]                          [NVARCHAR](500) NULL,
    [load_frequency]                        [NVARCHAR](500) NULL,
    [dst_folder]                            [NVARCHAR](500) NULL,
    [file_type]                             [NVARCHAR](500) NULL,
    [lake_dst_folder]                       [NVARCHAR](500) NULL,
    [spark_flag]                            [NVARCHAR](500) NULL,
    [dst_schema]                            [NVARCHAR](500) NULL,
    [distribution_type]                     [NVARCHAR](500) NULL,
    [load_sqldw_etl_pipeline_date]          [DATETIME] NULL,
    [load_sqldw_etl_pipeline_status]        [NVARCHAR](500) NULL,
    [load_sqldw_curated_pipeline_date]      [DATETIME] NULL,
    [load_sqldw_curated_pipeline_status]    [NVARCHAR](500) NULL,
    [load_delta_pipeline_date]              [DATETIME] NULL,
    [load_delta_pipeline_status]            [NVARCHAR](500) NULL,
    PRIMARY KEY CLUSTERED ( [parameter_id] ASC )WITH (statistics_
    norecompute =
    OFF, ignore_dup_key = OFF) ON [PRIMARY]
)
ON [PRIMARY]

go
```

Notice from the columns listed in the code that there are quite a few new metadata fields that have been added and can be captured within the ADF pipelines by creating dynamic datasets and pipelines.

Create the Datasets

In this next section, create a source dataset for the ADLS Gen2 snappy compressed parquet files and a sink dataset for the Azure dedicated SQL pool.

Begin by creating three datasets and name the datasets as follows:

1) DS_ADLS_TO_SYNAPSE

2) DS_ADLS_TO_SYNAPSE_MI

3) DS_SYNAPSE_ANALYTICS_DW

Then the following subsections will show how to create each one.

DS_ADLS_TO_SYNAPSE

Start by creating a source ADLS Gen2 dataset with parameterized paths. Remember that the pipeline_date has been added to the pipeline_parameter table that you had created in Chapter 4 since the pipeline_date captures the date when the data was loaded to ADLS Gen2. In this step, you are loading data from ADLS Gen2 into a Synapse Analytics dedicated SQL pool. You could either rerun the pipeline from Chapter 4 or manually enter a date in this pipeline_date column, which would ideally contain the latest folder date. Chapter 8 will discuss how to automate the insertion of the max folder date into this pipeline_date column to ensure this column always has the latest and max folder date that can be passed into the parameterized ADF pipeline. This will be achieved by using a Stored Procedure activity that runs immediately following the success of a Copy activity.

Figure 6-1 illustrates how to set up the parameterized connection properties for reading the source ADLS Gen2 parquet directories and folders that are driven by the pipeline_parameter control table.

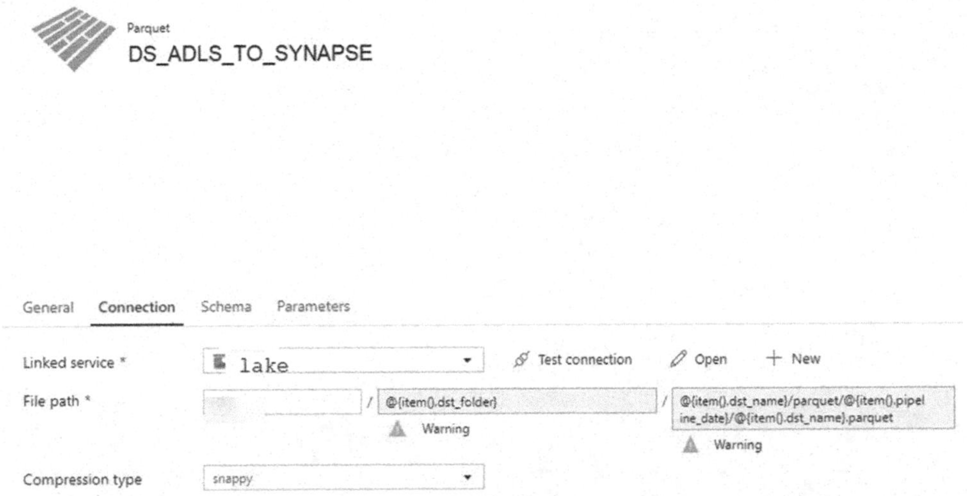

Parquet
DS_ADLS_TO_SYNAPSE

General   **Connection**   Schema   Parameters

Linked service *   ▪ lake   ▾   ✗ Test connection   ✐ Open   + New

File path *   [   ] / @{item().dst_folder}   / @{item().dst_name}/parquet/@{item().pipel
                                        ⚠ Warning   ine_date}/@{item().dst_name}.parquet
                                                                           ⚠ Warning

Compression type   snappy   ▾

Figure 6-1. *ADF parameterized connections for ADLS Gen2 parquet folders and files*

Here is the code that has been added to the File path section in Figure 6-1:

```
@{item().dst_folder}
```

```
@{item().dst_name}/parquet/ @{item().pipeline_date}/ @{item().dst_name}.parquet
```

Figure 6-2 shows how to add the parameters that will be needed.

Figure 6-2. *ADF parameters for ADLS Gen2 parquet folders and files*

The linked service details are illustrated in Figure 6-3. An Azure Key Vault is being used to store the credential secrets. This will be relevant in the later sections when the pipelines begin to run and if any authentication errors are noticed.

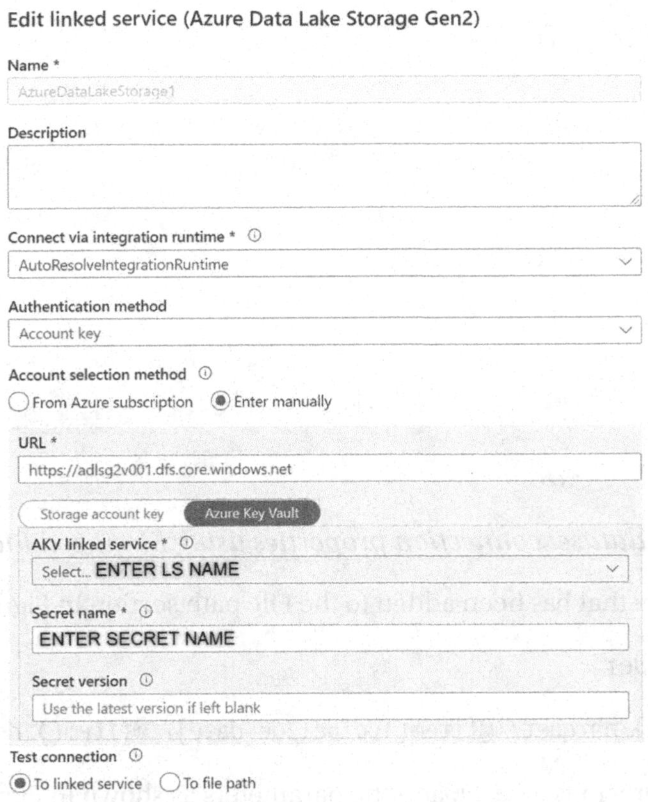

Edit linked service (Azure Data Lake Storage Gen2)

Name *

AzureDataLakeStorage1

Description

Connect via integration runtime * ⓘ

AutoResolveIntegrationRuntime   ⌄

Authentication method

Account key   ⌄

Account selection method ⓘ

◯ From Azure subscription   ● Enter manually

URL *

https://adlsg2v001.dfs.core.windows.net

| Storage account key | Azure Key Vault |

AKV linked service * ⓘ

Select... ENTER LS NAME   ⌄

Secret name * ⓘ

ENTER SECRET NAME

Secret version ⓘ

Use the latest version if left blank

Test connection ⓘ

● To linked service   ◯ To file path

Figure 6-3. *ADF linked service connection properties using Azure Key Vault*

DS_ADLS_TO_SYNAPSE_MI

The ADF dataset connection in Figure 6-4 uses Managed Identity connection credentials. The difference between this dataset shown in Figure 6-4 and the last one is that this linked service connection does not use an Azure Key Vault. Use this to test and switch between the Key Vault connection and non-Key Vault connection when errors are noticed later.

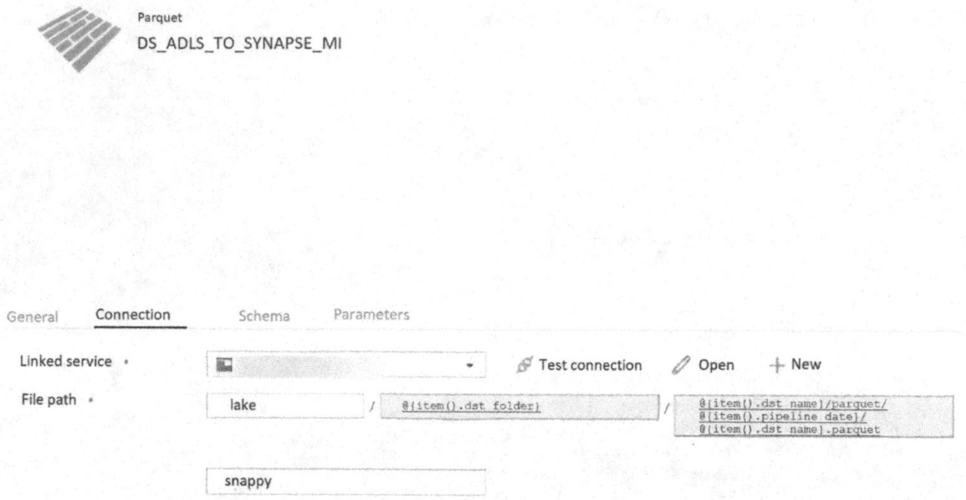

Parquet
DS_ADLS_TO_SYNAPSE_MI

| General | Connection | Schema | Parameters |

Linked service •   [▼]   ⌀ Test connection   ⌀ Open   + New

File path • | lake | / | @{item().dst_folder} | / | @{item().dst_name}/parquet/
@{item().pipeline_date}/
@{item().dst_name}.parquet |

| snappy |

Figure 6-4. *ADF dataset connection properties using Managed Identity*

Here is the code that has been added to the File path section in Figure 6-4:

`@{item().dst_folder}`

`@{item().dst_name}/parquet/ @{item().pipeline_date}/ @{item().dst_name}.parquet`

Similar to the previous dataset, add the parameters as shown in Figure 6-5.

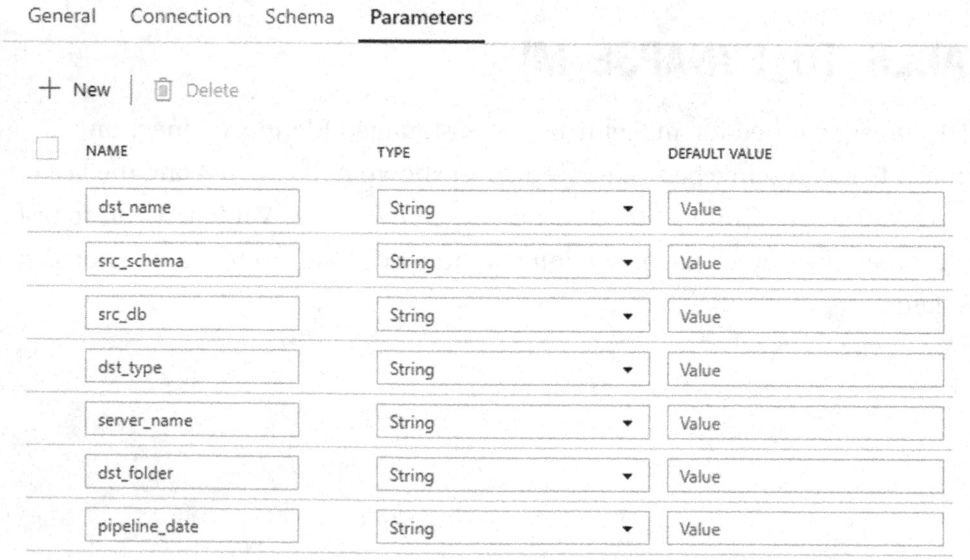

| General | Connection | Schema | **Parameters** |

+ New | 🗑 Delete

NAME	TYPE	DEFAULT VALUE
dst_name	String ▼	Value
src_schema	String ▼	Value
src_db	String ▼	Value
dst_type	String ▼	Value
server_name	String ▼	Value
dst_folder	String ▼	Value
pipeline_date	String ▼	Value

Figure 6-5. *ADF parameters for ADLS Gen2 parquet folders and files – Managed Identity*

The linked service details are in Figure 6-6. An Azure Key Vault is not being used here. Again, this will be relevant in the later sections when the pipelines are executed and if any authentication errors are noticed.

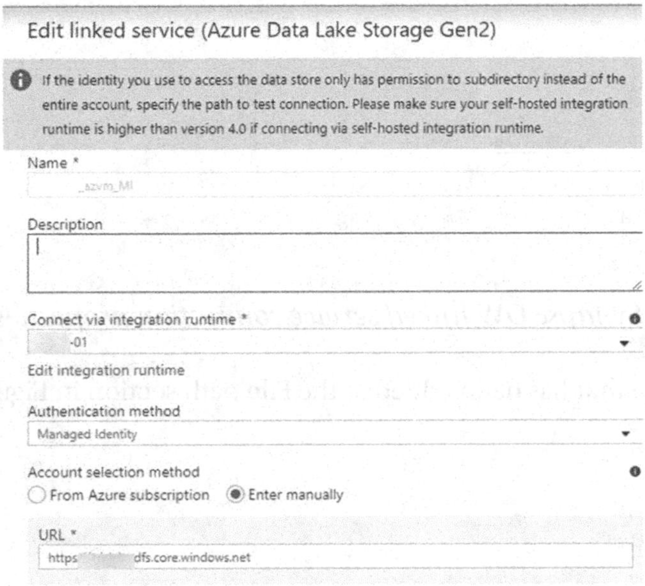

Figure 6-6. *ADF linked service connections*

In this section, a new ADF linked service connection has been created using Managed Identity.

DS_SYNAPSE_ANALYTICS_DW

The sink connection will be to an Azure Synapse Analytics dedicated SQL pool, shown in Figure 6-7. Also, parameters are being used to specify the schema and table name from the `pipeline_parameter` table. This will be a good feature when the ForEach loop activity will be used to create multiple tables using the same sink dataset.

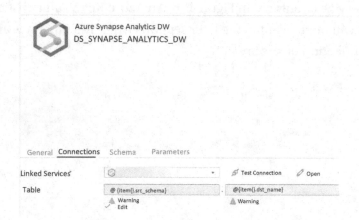

Figure 6-7. *ADF Synapse DW linked service connection properties*

Here is the code that has been added to the File path section in Figure 6-7:

```
@{item().src_schema}
```

```
@{item().dst_name}
```

Create the Pipeline

Now that the datasets have been created, also create a new pipeline. When doing so, add a Lookup activity connected to a ForEach loop activity as shown in Figure 6-8.

Figure 6-8. *ADF pipeline canvas containing Lookup and ForEach loop activities*

The lookup query shown in Figure 6-9 will get a list of tables that will need to be loaded to the Azure Synapse Analytics dedicated SQL pool. Note that currently there is a filter applied to the query, which would only include records WHERE load_synapse = 1.

Figure 6-9. *ADF Lookup activity query setting*

The code snippet that is included in Figure 6-9 is the following:

```
SELECT [server_name],
       [src_type],
       [src_schema],
       [src_db],
       [src_name],
       [dst_type],
       [dst_name],
```

```
        [include_pipeline_flag],
        [partition_field],
        [process_type],
        [priority_lane],
        [pipeline_date],
        [pipeline_status],
        [dst_folder],
        [file_type]
FROM    [dbo].[pipeline_parameter]
WHERE   load_synapse = 1
```

Within the settings of the ForEach loop activity, add the output value of the Lookup activity, shown in Figure 6-10. Remember to leave the "Sequential" box unchecked to ensure multiple tables will process in parallel. The default "Batch count" if left blank is 20, and the max is 50.

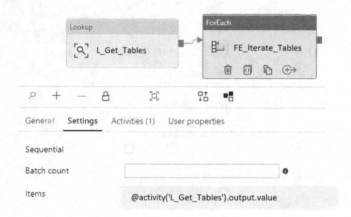

Figure 6-10. *ADF ForEach activity settings*

Also add one Copy activity to the ForEach loop activity, as shown in Figure 6-11. Click the pencil icon to view the Copy activity.

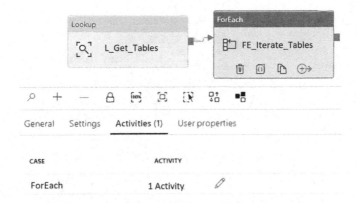

Figure 6-11. *ADF ForEach activities*

The source is set to DS_ADLS_TO_SYNAPSE, which uses an Azure Key Vault in the linked service connection. Add the dynamic parameters that will be needed. Note that the parameters were defined in the dataset. Figure 6-12 shows how and where to add these values.

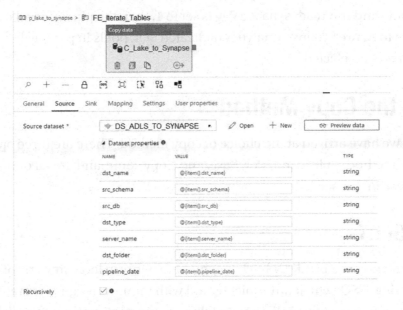

Figure 6-12. *ADF Copy activity source dataset properties*

Finally, choose the DS_SYNAPSE_ANALYTICS_DW dataset as the sink and select "Bulk insert" with the "Auto create table" option enabled, as shown in Figure 6-13.

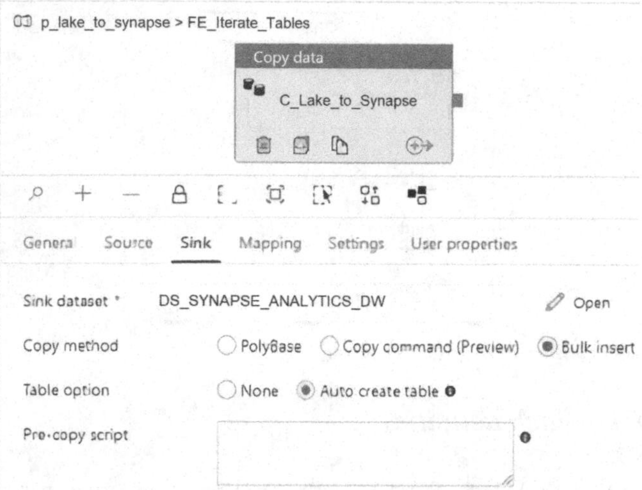

Figure 6-13. *ADF Copy activity sink dataset properties*

Based on the current configurations of the pipeline, since it is driven by the `pipeline_parameter` table, when (n) number of tables/records are added to the pipeline parameter table and the `load_synapse` flag is set to 1, then the pipeline will execute and load all tables to Azure Synapse Analytics dedicated SQL pools in parallel based on the copy method that is selected.

Choose the Copy Method

Now, finally, we have arrived at the choice of copy method. There are three options for the sink copy method. Bulk insert, PolyBase, and Copy command are all options that you will learn to use in this section.

BULK INSERT

SQL Server provides the BULK INSERT statement to perform large imports of data into SQL Server using T-SQL efficiently, quickly, and with minimal logging operations.

Within the Sink tab of the ADF Copy activity, set the copy method to Bulk insert. "Auto create table" automatically creates the table if it does not exist using the schema from the source file. This isn't supported when the sink specifies a stored procedure or the Copy activity is equipped with the staging settings. For this scenario, the source file is a parquet snappy compressed file that does not contain incompatible data types such as VARCHAR(MAX), so there should be no issues with the "Auto create table" option.

Note that the pre-copy script will run before the table is created, so in a scenario using "Auto create table" when the table does not exist, run it without the pre-copy script first to prevent errors and then add the pre-copy script back once the table has been created for ongoing full loads. Figure 6-14 shows the sink settings along with where to add any pre-copy script such as a TRUNCATE script.

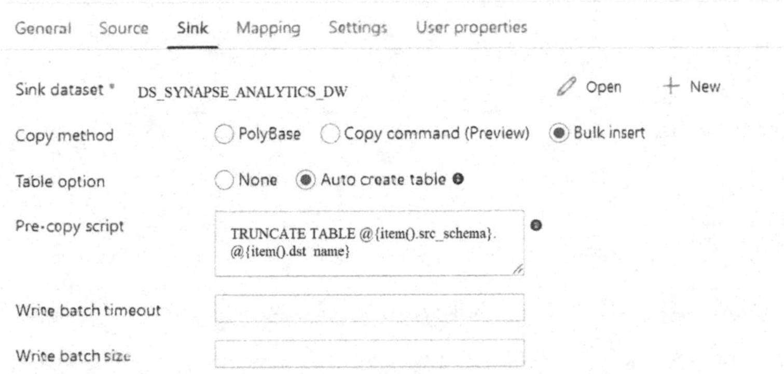

Figure 6-14. *ADF Copy activity sink pre-copy script*

If the default Auto create table option does not meet the distribution needs for custom distributions based on tables, then there is "Add dynamic content" that can be leveraged to use a distribution method specified in the pipeline parameter table per table.

Here is the code that has been added to the Pre-copy script section in Figure 6-14:

```
TRUNCATE TABLE @{item().src_schema}.@{item().dst_name}
```

After running the pipeline, it succeeded using the Bulk insert copy method, as shown in the activity run monitor illustrated in Figure 6-15.

Figure 6-15. *ADF pipeline success for Bulk insert*

Figure 6-16 shows the details of the Bulk insert Copy pipeline status.

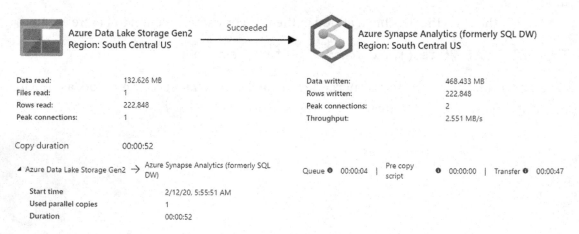

Figure 6-16. *ADF pipeline run details for Bulk insert*

After querying the Synapse table, notice that there is the same number of rows in the table, as shown in Figure 6-17.

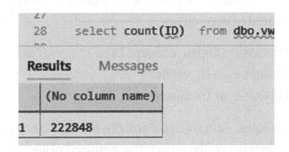

Figure 6-17. *Query Synapse Analytics dedicated SQL pool tables to verify Bulk insert ADF pipeline results*

The Bulk insert method also works for an on-premises SQL Server as the source with a Synapse Analytics dedicated SQL pool being the sink.

PolyBase

Using PolyBase is an efficient way to load a large amount of data into Azure Synapse Analytics with high throughput. You'll see a large gain in throughput by using PolyBase instead of the default Bulk insert mechanism.

For this next exercise, select PolyBase, shown in Figure 6-18, to test this copy method.

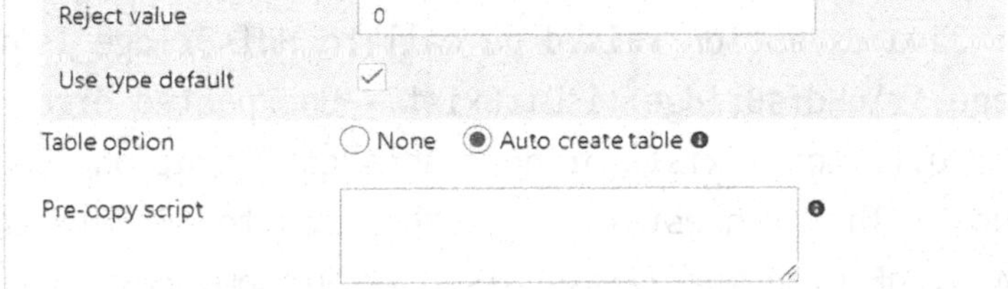

General	Source	**Sink**	Mapping	Settings	User properties

Sink dataset *   DS_SYNAPSE_ANALYTICS_DW   ✎ Open

Copy method   ⦿ PolyBase ◯ Copy command (Preview) ◯ Bulk insert

Reject type   Value   ▼

Reject value   0

Use type default   ✓

Table option   ◯ None ⦿ Auto create table ❶

Pre-copy script   [   ] ❶

Figure 6-18. *ADF pipeline sink dataset properties to select PolyBase*

109

PolyBase will need Managed Identity credentials to provision Azure AD and grant Data Factory full access to the database.

For more details on verifying the access, review and run the following queries on a Synapse Analytics dedicated SQL pool:

```
select * from sys.database_scoped_credentials
select * from sys.database_role_members
select * from sys.database_principals
```

Also, when external tables, data sources, and file formats need to be created, the following queries can help with verifying that the required objects have been created:

```
select * from sys.external_tables
select * from sys.external_data_sources
select * from sys.external_file_formats
```

After configuring the pipeline and running it, you might notice the pipeline fail with the following error:

```
"ErrorCode=FailedDbOperation,'Type=Microsoft.DataTransfer.Common.Shared.
HybridDeliveryException,Message=Error happened when loading data into SQL
Data Warehouse.,Source=Microsoft.DataTransfer.ClientLibrary,''Type=System.
Data.SqlClient.SqlException,Message=External file access failed due to
internal error: 'Error occurred while accessing HDFS: Java exception
raised on call to HdfsBridge_IsDirExist. Java exception message:\r\
nHdfsBridge::isDirExist - Unexpected error encountered checking whether
directory exists or not: AbfsRestOperationException: Operation failed: \"This
request is not authorized to perform this operation.\", 403, HEAD, https://
lake.dfs.core.windows.net/lake     //?upn=false&action=getAccessControl&time
out=90',Source=.Net SqlClient Data Provider,SqlErrorNumber=105019,Class=16,
ErrorCode=-2146232060,State=1,Errors=[{Class=16,Number=105019,State=1,Message
=External file access failed due to internal error: 'Error occurred while
accessing HDFS: Java exception raised on call to HdfsBridge_IsDirExist. Java
exception message:\r\nHdfsBridge::isDirExist - Unexpected error encountered
checking whether directory exists or not: AbfsRestOperationException:
Operation failed: \"This request is not authorized to perform this
operation.\", 403, HEAD, https://lake.dfs.core.windows.net/lake
//?upn=false&action=getAccessControl&timeout=90',},],'",
```

After researching the error, the reason is because the original Azure Data Lake Storage linked service from source dataset DS_ADLS_TO_SYNAPSE is using an Azure Key Vault to store authentication credentials, which is an unsupported Managed Identity authentication method at this time for using PolyBase and Copy command.

Change the source dataset to DS_ADLS_TO_SYNAPSE_MI, which no longer uses an Azure Key Vault, and notice in Figure 6-19 that the pipeline succeeds using the PolyBase copy method.

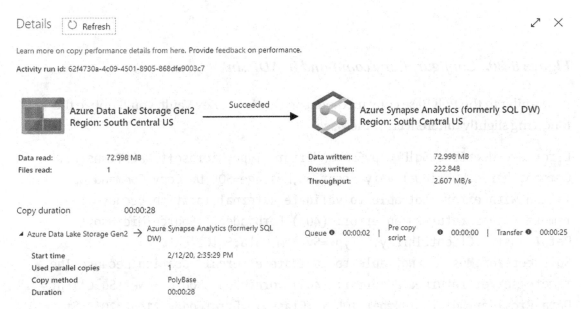

Figure 6-19. *ADF pipeline execution showing success after changing to Managed Identity*

Copy Command

Copy command will function similar to PolyBase, so the permissions needed for PolyBase will be more than sufficient for Copy command as well. For more information on COPY INTO, revisit Chapter 5, which covers details on permissions, use cases, and the SQL syntax for COPY INTO. Figure 6-20 shows how to configure Copy command in the ADF sink activity.

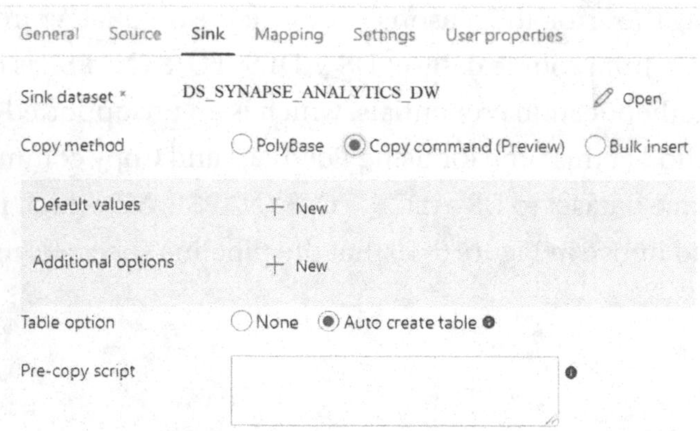

General   Source   **Sink**   Mapping   Settings   User properties

Sink dataset *   DS_SYNAPSE_ANALYTICS_DW   ✎ Open

Copy method   ○ PolyBase ● Copy command (Preview) ○ Bulk insert

Default values   + New

Additional options   + New

Table option   ○ None ● Auto create table ❶

Pre-copy script

Figure 6-20. *Configure Copy command in ADF sink*

Similar to the PolyBase copy method using an Azure Key Vault, you'll notice the following slightly different error message:

```
ErrorCode=UserErrorSqlDWCopyCommandError,'Type=Microsoft.DataTransfer.
Common.Shared.HybridDeliveryException,Message=SQL DW Copy Command operation
failed with error 'Not able to validate external location because The
remote server returned an error: (403) Forbidden.',Source=Microsoft.
DataTransfer.ClientLibrary,''Type=System.Data.SqlClient.
SqlException,Message=Not able to validate external location because The
remote server returned an error: (403) Forbidden.,Source=.Net SqlClient
Data Provider,SqlErrorNumber=105215,Class=16,ErrorCode=-2146232060,State
=1,Errors=[{Class=16,Number=105215,State=1,Message=Not able to validate
external location because The remote server returned an error: (403)
Forbidden.,},],'", "failureType": "UserError", "target": "Copy data1",
"details": []
```

Switch to the linked service that does not use an Azure Key Vault, and notice that the pipeline succeeds, as shown in Figure 6-21.

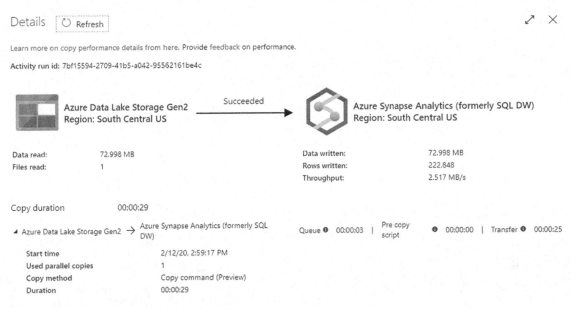

Figure 6-21. *ADF pipeline execution results*

Note that it is important to consider scheduling and triggering your ADF pipelines once they have been created and tested. Triggers determine when the pipeline execution will be fired, based on the trigger type and criteria defined in the trigger. There are three main types of Azure Data Factory triggers: the **Schedule** trigger that executes the pipeline on a wall-clock schedule, the **Tumbling window** trigger that executes the pipeline on a periodic interval and retains the pipeline state, and the **event-based** trigger that responds to a blob-related event. Additionally, ADF features alerts to monitor pipeline and trigger failures and send notifications via email, text and more.

Summary

In this chapter, I showed you how to create a source Azure Data Lake Storage Gen2 dataset and a sink Synapse Analytics dedicated SQL pool dataset along with an Azure Data Factory pipeline driven by a parameter table. You also learned how to load snappy compressed parquet files into a Synapse Analytics dedicated SQL pool by using three copy methods: Bulk insert, PolyBase, and Copy command. This chapter taught you about the various ingestion options that are available in Azure Data Factory.

## CHAPTER 7

Create and Load Synapse Analytics Dedicated SQL Pool Tables Dynamically

In Chapter 6, you learned how to load data lake files into a Synapse Analytics dedicated SQL pool using Data Factory by using the COPY INTO command as one such load option. Now that you have designed and developed a dynamic process to auto-create and load the ETL schema tables into a Synapse Analytics dedicated SQL pool with snappy compressed parquet files, let's explore options for creating and loading tables into a curated schema, where you can dynamically define schemas and distribution types at runtime to create curated schema tables. Note that in many modern cloud-based data architectural patterns, the staging and curation is happening more within the data lake. However, this chapter will demonstrate the vast capabilities of persisting large volumes of tables into a data warehouse by using a few simple ADF pipelines. Despite all of the benefits of Delta Lake, which we will cover in greater detail in Chapter 15, customers are still interested in persisting their final production-ready, trusted, and curated data into a SQL data warehouse for a number of reasons, including ease of analytical querying, ease of connectivity to Power BI and other reporting tools, and more.

In Chapter 4, I introduced the concept of a pipeline parameter table to track and control all SQL Server tables, servers, schemas, and more. Essentially, this pipeline parameter table is set up to drive the Data Factory orchestration process. To solve for dynamically being able to define distribution types along with curated schemas, I will introduce a few new columns to this pipeline parameter table: [`distribution_type`], [`dst_schema`], and [`dst_name`]. These new columns can be used within the Data Factory pipeline to dynamically create and load curated tables from the ETL schema.

© Ron C. L'Esteve 2021
R. C. L'Esteve, *The Definitive Guide to Azure Data Engineering*, https://doi.org/10.1007/978-1-4842-7182-7_7

Dynamically Create and Load New Tables Using an ADF Pre-copy Script

The ADF pipeline process to load tables from the source Data Lake Storage account into Synapse Analytics DW tables will begin with a Lookup activity to the pipeline parameter table shown in Figure 7-1 using a query in which you can specify your flags and filters appropriately.

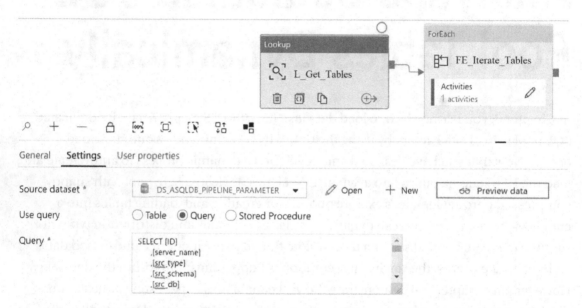

Figure 7-1. *ADF Lookup settings showing query to select from pipeline_parameter table*

Note that Figure 7-1 adds the source SQL select statement as a query for the purpose of this exercise. As a best practice, I would recommend considering converting this SQL statement to a stored procedure instead and then calling the code through the ADF pipeline by setting the source query to Stored Procedure. This will allow for easier maintenance of the code outside of the ADF environment.

Once you add the source lookup query, the filter `pipeline_status = 'success'` allows for tracking if the files successfully made it to the lake, and this is done via a SQL stored procedure. Also, it is important to note that there are quite a few columns in this pipeline parameter that help track steps throughout the end-to-end process. For the purpose of this exercise, there is an interest in columns [dst_schema], [dst_schema], and [distribution_type].

Add the following code to the source query section of the ADF Lookup activity shown in Figure 7-1:

```
SELECT  [id],
        [server_name],
        [src_type],
        [src_schema],
        [src_db],
        [src_name],
        [dst_type],
        [dst_name],
        [include_pipeline_flag],
        [partition_field],
        [process_type],
        [priority_lane],
        [pipeline_date],
        [pipeline_status],
        [load_synapse],
        [load_frequency],
        [dst_folder],
        [file_type],
        [lake_dst_folder],
        [spark_flag],
        [data_sources_id],
        [dst_schema],
        [distribution_type],
        [load_sqldw_etl_pipeline_date],
        [load_sqldw_etl_pipeline_status],
        [load_sqldw_curated_pipeline_date],
        [load_sqldw_curated_pipeline_status],
        [load_delta_pipeline_date],
        [load_delta_pipeline_status]
FROM    [dbo].[pipeline_parameter]
```

```
WHERE  load_synapse = 1
       AND pipeline_status = 'success'
       AND include_pipeline_flag = 1
       AND process_type = 'full'
       AND load_frequency = 'daily'
```

As an example, the dst_schema and distribution_type in the pipeline_parameter table may look like the following illustration in Figure 7-2.

dst_schema	distribution_type
ref	replicate
ref	replicate
ref	replicate
ref	replicate
ref	replicate
svc	round robin
svc	round robin
svc	round robin
ref	replicate
svc	round robin

Figure 7-2. *dst_schema and distribution_type in the pipeline_parameter table*

As you move on to the ForEach loop activity shown in Figure 7-3, ensure that the Items field within the Settings tab is filled in correctly to get the output of the Lookup activity.

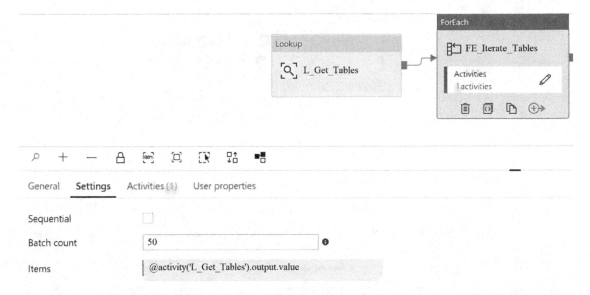

Figure 7-3. *ADF ForEach settings for the pipeline*

Here is the code that you will need to add to the Items field within the ForEach loop activity in Figure 7-3:

```
@activity('L_Get_Tables').output.value
```

Drill into the ForEach activities. There is the Copy data activity shown in Figure 7-4 along with the required dataset properties.

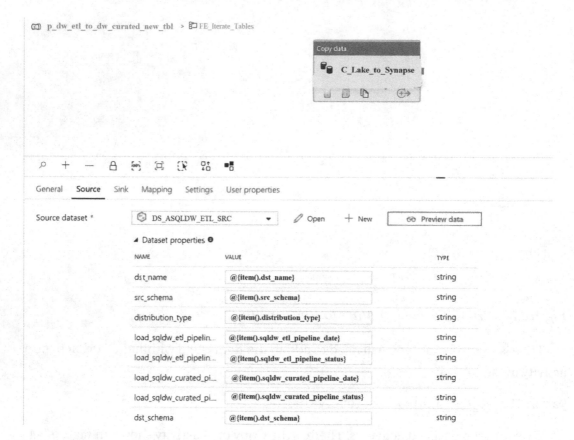

Figure 7-4. *ADF Copy activity source settings and dataset properties*

The names and values of the source dataset properties as shown in Figure 7-4 can be found in the following:

Name	Value
dst_name	@{item().dst_name}
src_schema	@{item().src_schema}
distribution_type	@{item().distribution_type}
load_sqldw_etl_pipeline_date	@{item().load_sqldw_etl_pipeline_date}
load_sqldw_etl_pipeline_status	@{item().load_sqldw_etl_pipeline_status}
load_sqldw_curated_pipeline_date	@{item().load_sqldw_curated_pipeline_date}
load_sqldw_curated_pipeline_status	@{item().load_sqldw_curated_pipeline_status}
dst_schema	@{item().dst_schema}

Configure the source dataset connection for the Synapse Analytics dedicated SQL pool ETL schema as shown in Figure 7-5. Notice that the etl schema is hard-coded. However, the table name is coming from the `pipeline_parameter` table.

Azure Synapse Analytics (formerly SQL DW)
DS_ASQLDW_ETL_SRC

General	**Connection**	Schema	Parameters

Linked service *   [🔷 _____ ▼]   ✐ Test connection   ✐ Open   +

Table   [etl _____] . [@{item().dst_name}]

Figure 7-5. *ADF source dataset connection properties*

Here is the code that you will need to enter in the Table connection setting in Figure 7-5:

```
etl.@{item().dst_name}
```

The sink dataset shown in Figure 7-6 is defined as the curated schema where you will need to parametrize the destination schema and name. Note that the source dataset contains parameters that will be needed from the source schema. However, the sink dataset does not contain any parameters.

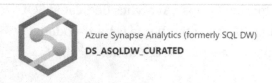

Azure Synapse Analytics (formerly SQL DW)
DS_ASQLDW_CURATED

General   **Connection**   Schema   Parameters

Linked service *   [datadw_azvm ▼]   _⌀_ Test connection   _⌀_ Open   +

Table   [@{item().dst_schema}] . [@{item().dst_name}]

Figure 7-6. *ADF sink dataset connection properties*

Here is the code that you will need to enter into the Table connection setting in
Figure 7-6:

```
@{item().dst_schema}.@{item().dst_name}
```

After creating the datasets, take a closer look at the pre-copy script. Note that Bulk
insert is being used as the copy method since the data currently exists in the ETL schema
in a Synapse Analytics dedicated SQL pool and must be loaded to a curated schema.

Also set the table option to "None" as shown in Figure 7-7, since the tables will be
created using the following pre-copy script, which is basically a dynamic Create Table
as Select (CTAS) syntax that references the destination schema and name along with the
distribution type from the `pipeline_parameter` table, specifically the sections shown
in Figure 7-2. Additionally, SELECT TOP (0) is being used in the script because we only
want to create the tables using this step and load them using the ADF Copy activity.

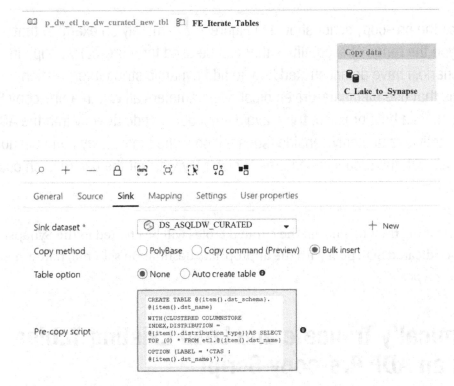

Figure 7-7. *ADF Copy data sink settings*

Here is the code that is used in the sink pre-copy script in Figure 7-7:

```
CREATE TABLE @{item().dst_schema}.@{item().dst_name}
WITH
    (
    CLUSTERED COLUMNSTORE INDEX,
    DISTRIBUTION = @{item().distribution_type}
    )
AS SELECT TOP (0) * FROM etl.@{item().dst_name}
OPTION (LABEL = 'CTAS : @{item().dst_name}');
```

Note that the pre-copy script shown in Figure 7-7 is merely an example that shows you the range of capabilities that can be used for a pre-copy script. In this scenario, I have demonstrated how to add dynamic string interpolation functions that use metadata-driven pipeline parameters all within a pre-copy SQL statement. As a best practice, try to avoid embedding code directly into the ADF pipeline activities and only consider such options when there may be noticeable limitations with the product's features, which may warrant the use of such custom configurations.

After running the pipeline, all the curated tables will be created in the Synapse Analytics dedicated SQL pool with the appropriate destination schema, name, and distribution type.

Dynamically Truncate and Load Existing Tables Using an ADF Pre-copy Script

In a scenario where you may need to dynamically truncate and load existing tables rather than recreate the tables, approach that task by simply truncating the destination table as shown in Figure 7-8. This approach would be the only notable change from the previous pipeline.

Figure 7-8. *ADF Copy data sink settings with pre-copy script changed to* *TRUNCATE*

Here is the code that is used in the sink pre-copy script in Figure 7-8:

```
TRUNCATE TABLE @{item().dst_schema}.@{item().dst_name}
```

Dynamically Drop, Create, and Load Tables Using a Stored Procedure

Finally, let's explore an option to use stored procedures from the Synapse Analytics dedicated SQL pool to drop and create the curated tables.

The pipeline design will be very similar to the previous pipelines by starting with a lookup and then flowing into a ForEach loop activity, shown in Figure 7-9.

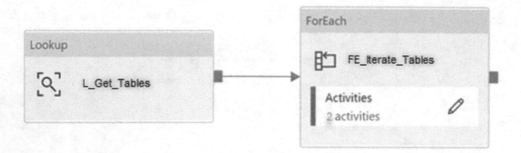

Figure 7-9. *ADF pipeline flow containing Lookup and ForEach activities*

Within the ForEach loop activity, there is one Stored Procedure activity called CTAS from the pipeline parameter shown in Figure 7-10. This stored procedure has been created within the Synapse Analytics dedicated SQL pool and is based on a dynamic Create Table as Select (CTAS) statement for which I will provide the code further in this section. Additionally, the destination name and schema have been defined as stored procedure parameters whose values are coming from the pipeline parameter table and are being passed to the stored procedure, shown in Figure 7-10.

Figure 7-10. *ADF stored procedure details and parameters*

It uses dynamic parameters that can be passed from the `pipeline_parameter` table to the stored procedure, which will be called by the ADF pipeline. Here is the source code that has been used in the `[etl].[Ctas_from_pipeline_parameter]` ADF stored procedure that is being called in the ForEach loop activity as shown in Figure 7-10.

```
SET ansi_nulls ON

go

SET quoted_identifier ON

go
```

```
CREATE PROC [etl].[Ctas_from_pipeline_parameter] @schema
            [VARCHAR](
255),
                                                @name               [VARCHAR](
255),
                                                @distribution_type [VARCHAR](
255)
AS
  BEGIN
      DECLARE @table VARCHAR(255)
      DECLARE @table_stage VARCHAR(255)
      DECLARE @table_etl VARCHAR(255)
      DECLARE @sql VARCHAR(max)

      SET @table = @schema + '.' + @name
      SET @table_stage = @table + '_stage'
      SET @table_etl = 'etl.' + @name
      SET @sql = 'if object_id (''' + @table_stage
                + ''',''U'') is not null drop table '
                + @table_stage + '; CREATE TABLE ' + @table_stage
                + ' WITH ( DISTRIBUTION = ' + @distribution_type
                + ' ,CLUSTERED COLUMNSTORE INDEX ) AS SELECT  * FROM      '
                + @table_etl + '; if object_id ('''
                + @table
                + ''',''U'') is not null drop table '
                + @table + '; RENAME OBJECT ' + @table_stage + ' TO '
                + @name + ';'

      EXEC(@sql)
  END

go
```

After heading over to SSMS and then scripting the stored procedure, notice that the preceding script is doing the following

- Declaring and setting the distribution_type along with ETL, curated, and schema/table names dynamically

- Dropping the curated_stage table if it exists

- Setting the SQL syntax to create the stage table as selecting all data from the etl table with the distribution type dynamically set

- Dropping the actual/original curated table

- Renaming the curated_stage to the actual/original curated table

In a scenario where you might be interested in renaming the original curated table, rather than dropping the original curated table, use this script in the Stored Procedure activity within the ADF pipeline:

```
SET ansi_nulls ON

go

SET quoted_identifier ON

go

CREATE PROC [etl].[Ctas_from_pipeline_parameter] @schema
          [VARCHAR](
255),
                                          @name            [VARCHAR](
255),
                                          @distribution_type [VARCHAR](
255)
AS
  BEGIN
      DECLARE @table VARCHAR(255)
      DECLARE @table_stage VARCHAR(255)
      DECLARE @table_drop VARCHAR(255)
      DECLARE @table_etl VARCHAR(255)
      DECLARE @schematable_drop VARCHAR(255)
      DECLARE @sql VARCHAR(max)
```

129

```
    SET @table = @schema + '.' + @name
    SET @table_stage = @table + '_stage'
    SET @table_drop = @name + '_drop'
    SET @table_etl = 'etl.' + @name
    SET @schematable_drop = @table + '_drop'
    SET @sql = 'if object_id (''' + @table_stage
             + ''','''U''') is not null drop table '
             + @table_stage + '; CREATE TABLE ' + @table_stage
             + ' WITH ( DISTRIBUTION = ' + @distribution_type
             + ' ,CLUSTERED COLUMNSTORE INDEX ) AS SELECT  * FROM   '
             + @table_etl + '; if object_id ('''
             + @table
             + ''','''U''') is not null rename object '
             + @table + ' TO ' + @table_drop + '; RENAME OBJECT '
             + @table_stage + ' TO ' + @name + '; if object_id ('''
             + @schematable_drop
             + ''','''U''') is not null drop table '
             + @schematable_drop + ';'

        EXEC(@sql)
    END

go
```

Lastly, it is important to note that within the Synapse Analytics dedicated SQL pool, if you are attempting to drop or rename a table that has dependencies linked to a materialized view that has been created, then the drop and rename script might fail.

Summary

In this chapter, I have outlined steps on how to dynamically create and load new tables into a Synapse Analytics dedicated SQL pool by using ADF's pre-copy script within the Copy activity. Additionally, I covered how to dynamically truncate and load existing tables by using ADF's pre-copy script, and finally I demonstrated how to dynamically drop, create, and load Synapse Analytics dedicated SQL pool tables by using a SQL stored procedure stored in a Synapse Analytics dedicated SQL pool.

Some of the examples that I have demonstrated in this chapter might not be fully applicable to your specific scenario or use case, but may help with further deepening your understanding of the capabilities of ADF and how to build and leverage customized and dynamic SQL scripts and stored procedures to fit specific use cases that might not be available through the out-of-the-box features within ADF. I hope you will find some of these examples helpful. In the next two chapters, you will learn more about how to make these ADF pipelines that you have learned about and built more robust by building custom audit and error logging processes to capture and persist the pipeline-related metrics within your SQL database tables after the ADF pipeline completes running.

CHAPTER 8

Build Custom Logs in SQL Database for Pipeline Activity Metrics

In the previous chapters, I demonstrated how to load data from ADLS Gen2 and then into a Synapse Analytics dedicated SQL pool using Data Factory. In this chapter, I will demonstrate how to leverage the pipelines that have been built to implement a process for tracking the log activity for pipelines that run and persist the data.

Azure Data Factory is a robust cloud-based ELT tool that is capable of accommodating multiple scenarios for logging pipeline audit data including out-of-the-box services such as Log Analytics, Azure Monitor, and more, along with more custom methods of extracting the pipeline metrics and passing them to another custom process. In this chapter, I will show you how to implement three of these possible custom logging options, which are the following:

1. **Option 1 – Create a Stored Procedure activity**: Updating pipeline status and datetime columns in a static pipeline parameter table using ADF can be achieved by using a Stored Procedure activity.

2. **Option 2 – Create a CSV log file in Data Lake Storage Gen2**: Generating a metadata CSV file for every parquet file that is created and storing the logs as CSV files in hierarchical folders in ADLS Gen2 can be achieved using a Copy data activity.

3. **Option 3 – Create a log table in Azure SQL Database**: Creating a pipeline log table in Azure SQL Database and storing the pipeline activity as records in the table can be achieved by using a Copy data activity.

133

© Ron C. L'Esteve 2021
R. C. L'Esteve, *The Definitive Guide to Azure Data Engineering*, https://doi.org/10.1007/978-1-4842-7182-7_8

Figure 8-1 illustrates these three options and also shows a visual representation of the data flow from the Copy-Table activity to creating the various logging methods.

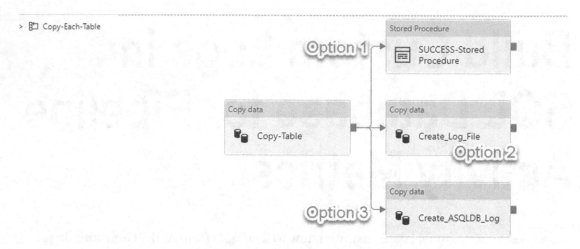

Figure 8-1. *ADF options for logging custom pipeline data*

Option 1: Create a Stored Procedure Activity

The Stored Procedure activity will be used to invoke a stored procedure in a Synapse Analytics dedicated SQL pool.

For this scenario, you might want to maintain your `pipeline_status` and `pipeline_date` details as columns in your `adf_db.dbo.pipeline_parameter` table rather than having a separate log table. The downside to this method is that it will not retain historical log data but will simply update the values in the original `pipeline_parameter` table based on a lookup of the incoming files to records in the `pipeline_parameter` table. This gives a quick, yet not necessarily robust, method of viewing the status and load date across all items in the pipeline parameter table. Specific use cases for this approach might be to log the max folder date based on the incoming folder and file names and timestamps.

Begin by adding a Stored Procedure activity to your preliminary Copy data activity, which is within the ForEach loop activity, shown in Figure 8-2, to ensure that the process iterates and logs each table using the stored procedure. Notice the green arrow and line connector that indicates the stored procedure must run on the success of the Copy data activity.

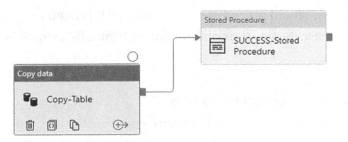

Figure 8-2. *ADF option 1 for logging data using a stored procedure*

Next, add the following stored procedure to the database where your pipeline parameter table resides. This procedure simply looks up the destination table name in the pipeline parameter table and updates the status and datetime for each table once the Copy data activity is successful. Similarly, you could just as easily add a new stored procedure to handle errors by linking the stored procedure to the failure constraint after the Copy activity completes:

```
SET quoted_identifier ON

go

CREATE PROCEDURE [dbo].[Sql2adls_data_files_loaded] @dst_name NVARCHAR(500)
AS
    SET nocount ON
    -- turns off messages sent back to client after DML is run, keep this
       here

    DECLARE @Currentday DATETIME = Getdate();

    UPDATE [dbo].[pipeline_parameter]
    SET    pipeline_status = 'success',
           pipeline_date = @Currentday
    WHERE  dst_name = @dst_name;

go
```

After creating your stored procedure, confirm that it has been created in your corresponding database. Notice the confirmation in Figure 8-3, which is a screen capture within SSMS.

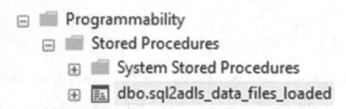

Figure 8-3. *SSMS view showing stored procedure used by ADF option 1*

Next, return to the Data Factory pipeline and configure the Stored Procedure activity. In the Stored Procedure tab, select the stored procedure that was just created. Also add a new stored procedure parameter that references the destination name, which was configured in the Copy activity, as illustrated in Figure 8-4.

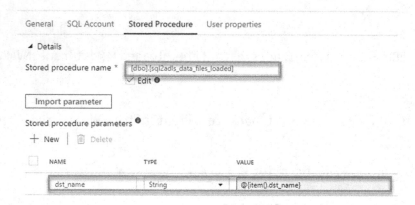

Figure 8-4. *ADF stored procedure details and parameters*

After saving, publishing, and running the pipeline, notice that the `pipeline_date` and `pipeline_status` columns have been updated as a result of the ADF Stored Procedure activity, shown in the `pipeline_parameter` table view in Figure 8-5. This is meant to be a lightweight testing pattern, which provides you with details on the status and load date for each table within a centralized location. I've noticed that ADF does not always provide robust details related to the problematic tables or columns.

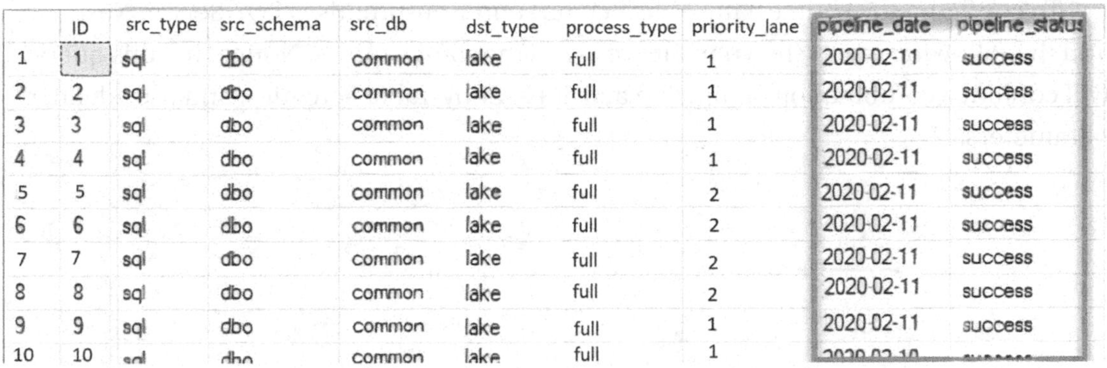

	ID	src_type	src_schema	src_db	dst_type	process_type	priority_lane	pipeline_date	pipeline_status
1	1	sql	dbo	common	lake	full	1	2020-02-11	success
2	2	sql	dbo	common	lake	full	1	2020-02-11	success
3	3	sql	dbo	common	lake	full	1	2020-02-11	success
4	4	sql	dbo	common	lake	full	1	2020-02-11	success
5	5	sql	dbo	common	lake	full	2	2020-02-11	success
6	6	sql	dbo	common	lake	full	2	2020-02-11	success
7	7	sql	dbo	common	lake	full	2	2020-02-11	success
8	8	sql	dbo	common	lake	full	2	2020-02-11	success
9	9	sql	dbo	common	lake	full	1	2020-02-11	success
10	10	sql	dbo	common	lake	full	1	2020-02-10	success

Figure 8-5. *SSMS view of updated pipeline_date and pipeline_status columns from ADF stored procedure*

Option 2: Create a CSV Log File in Data Lake Storage Gen2

Since your Copy-Table activity is generating snappy parquet files into hierarchical ADLS Gen2 folders, you may also want to create a CSV file containing pipeline run details for every parquet file in your ADLS Gen2 account. For this scenario, you could set up a Data Factory Event Grid trigger to listen for metadata files and then trigger a process to transform the source table and load it into a curated zone to somewhat replicate a near-real-time ELT process.

Begin designing the ADF pipeline to create a CSV log file in your ADLS Gen2 account by adding a Copy activity for creating the log files and connecting it to the Copy-Table activity, shown in Figure 8-6. Similar to the previous process, which looped through each table, this process will generate a CSV extension metadata file in a metadata folder per table.

Figure 8-6. *ADF option 2 for logging data using a Copy data activity to create a CSV file*

To configure the source dataset, select the source on-premises SQL Server. Next, add the following query shown in Figure 8-7 as the source query. Notice that this query will contain a combination of pipeline activities, Copy-Table activities, and user-defined parameters.

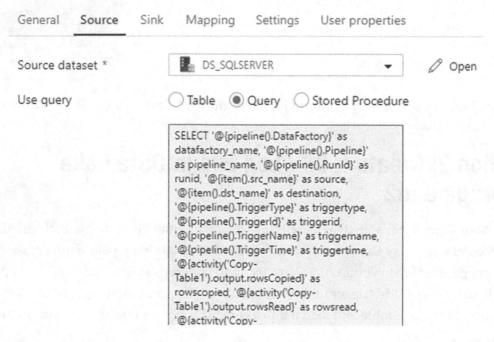

Figure 8-7. *ADF option 2 source query to extract pipeline activity metrics*

Recall the notes from previous chapters that call out the embedding of source queries as code into the ADF pipeline merely for visual demonstration purposes. It is always a better and more efficient process to add the source query as a stored procedure and then call the stored procedure. This will help with ease of maintenance of the code in a centralized location.

Here is the corresponding source code used in Figure 8-7:

```
SELECT '@{pipeline().DataFactory}'              AS datafactory_name,
       '@{pipeline().Pipeline}'                 AS pipeline_name,
       '@{pipeline().RunId}'                    AS runid,
       '@{item().src_name}'                     AS source,
```

```
'@{item().dst_name}'                          AS destination,
'@{pipeline().TriggerType}'                   AS triggertype,
'@{pipeline().TriggerId}'                      AS triggerid,
'@{pipeline().TriggerName}'                    AS triggername,
'@{pipeline().TriggerTime}'                    AS triggertime,
'@{activity('copy-TABLE').output.rowsCopied}'  AS rowscopied,
'@{activity('copy-TABLE').output.rowsRead}'    AS rowsread,
'@{activity('copy-TABLE').output.usedParallelCopies}'
                              AS no_parallelcopies,
'@{activity('copy-TABLE').output.copyDuration}'
                              AS copyduration_in_secs,
'@{activity('copy-TABLE').output.effective
IntegrationRuntime}'          AS effectiveintegrationruntime,
'@{activity('copy-TABLE').output.executionDetails[0].
source.type}'                 AS source_type,
'@{activity('copy-TABLE').output.executionDetails[0].
sink.type}'                   AS sink_type,
'@{activity('copy-TABLE').output.executionDetails[0].
status}'                      AS execution_status,
'@{activity('copy-TABLE').output.executionDetails[0].
start}'                       AS copyactivity_start_time,
'@{utcnow()}'                 AS copyactivity_end_time,
'@{activity('copy-TABLE').output.executionDetails[0].detailed
Durations.queuingDuration}'  AS copyactivity_queuingduration_in_
secs,
'@{activity('copy-TABLE').output.executionDetails[0].detailed
Durations.timeToFirstByte}'  AS copyactivity_timetofirstbyte_in_
secs,
'@{activity('copy-TABLE').output.executionDetails[0].detailed
Durations.transferDuration}' AS copyactivity_transferduration_in_
secs
```

The sink will be a CSV dataset with a CSV extension, as shown in Figure 8-8.

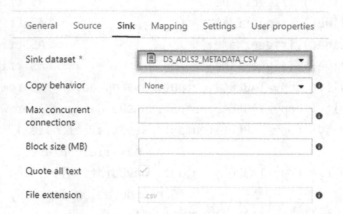

Figure 8-8. *ADF sink dataset for CSV*

Figure 8-9 shows the connection configuration to use for the CSV dataset.

DelimitedText
DS_ADLS2_METADATA_CSV

General   **Connection**   Schema

Linked service *	[ ] lake	🖋 Test connection ✏ open
File path *	[ ] / @{item().dst_folder}	/ @{item().server_name}/@{item().src_db}/ @{item().src_schema}/@{item().dst_name}/metadata/ @{formatDateTime(utcnow(),'yyyy-MM-dd')}/ @{item().dst_name}.csv

Compression type	none
Column delimiter	Comma (,)
	☐ Edit
Row delimiter	Auto detect (\r,\n, or \r\n)
	☐ Edit
Encoding	Default (UTF-8)
Escape character	Backslash (\)
	☐ Edit
Quote character	Double quote (")

Figure 8-9. *ADF sink dataset connection properties for CSV*

The following parameterized path will ensure that the file is generated in the correct folder structure. Here is the code shown in Figure 8-9:

```
@{item().server_name}/@{item().src_db}/@{item().src_schema}/@{item().dst_
name}/metadata/@{formatDateTime(utcnow(),'yyyy-MM-dd')}/@{item().dst_name}.
csv
```

After saving, publishing, and running the pipeline, notice how the metadata folder has been created in the following folder structure:

```
Server>database>schema>date>Destination_table location
```

Figure 8-10 shows this folder as you will see it in ADSL Gen2.

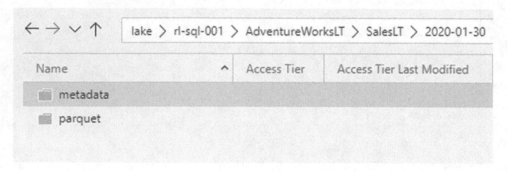

Figure 8-10. *ADLS Gen2 folders generated by ADF pipeline option 2*

Open the metadata folder and notice that there are CSV folders that have been created per day that the pipeline runs, as shown in Figure 8-11.

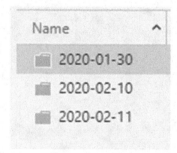

Figure 8-11. *ADLS Gen2 folders by timestamp created by ADF pipeline option 2*

Finally, notice in Figure 8-12 that a metadata .csv file with the name of the table has been created.

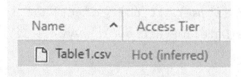

Figure 8-12. *ADLS Gen2 metadata file containing ADF pipeline metrics from option 2*

Download and open the file and notice that all of the query results have been populated in the .csv file, as shown in Figure 8-13.

File   Edit   Format   View   Help
>lidated","PipelineActivity","202c35c0-601f-425f-93f7-0cca33b326c7","202c35c0-601f-425f-93f7-0cca33b326c7","2020-01-21T20:40:45.9674820Z","216629","216629",

Figure 8-13. *ADLS Gen2 metadata file containing details for ADF pipeline metrics from option 2*

Option 3: Create a Log Table in Azure SQL Database

The final scenario in this chapter involves creating a log table in your database where the parameter table resides and then writing the data as records into the table. For this option, start by adding a Copy data activity connected to the Copy-Table activity, shown in Figure 8-14.

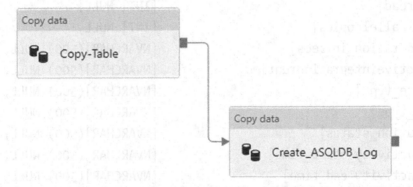

Figure 8-14. *ADF option 3 for logging data using a Copy data activity written to a log table*

Next, create the following table in your ADF_DB database. This table will store and capture the pipeline and Copy activity details:

```
SET ansi_nulls ON

go

SET quoted_identifier ON

go
```

```
CREATE TABLE [dbo].[pipeline_log]
  (
      [log_id]                                   [INT] IDENTITY(1, 1) NOT NULL,
      [parameter_id]                             [INT] NULL,
      [datafactory_name]                         [NVARCHAR](500) NULL,
      [pipeline_name]                            [NVARCHAR](500) NULL,
      [runid]                                    [NVARCHAR](500) NULL,
      [source]                                   [NVARCHAR](500) NULL,
      [destination]                              [NVARCHAR](500) NULL,
      [triggertype]                              [NVARCHAR](500) NULL,
      [triggerid]                                [NVARCHAR](500) NULL,
      [triggername]                              [NVARCHAR](500) NULL,
      [triggertime]                              [NVARCHAR](500) NULL,
      [rowscopied]                               [NVARCHAR](500) NULL,
      [dataread]                                 [INT] NULL,
      [no_parallelcopies]                        [INT] NULL,
      [copyduration_in_secs]                     [NVARCHAR](500) NULL,
      [effectiveintegrationruntime]              [NVARCHAR](500) NULL,
      [source_type]                              [NVARCHAR](500) NULL,
      [sink_type]                                [NVARCHAR](500) NULL,
      [execution_status]                         [NVARCHAR](500) NULL,
      [copyactivity_start_time]                  [NVARCHAR](500) NULL,
      [copyactivity_end_time]                    [NVARCHAR](500) NULL,
      [copyactivity_queuingduration_in_secs]     [NVARCHAR](500) NULL,
      [copyactivity_transferduration_in_secs]    [NVARCHAR](500) NULL,
      CONSTRAINT [PK_pipeline_log] PRIMARY KEY CLUSTERED ( [log_id] ASC )
      WITH (
      statistics_norecompute = OFF, ignore_dup_key = OFF) ON [PRIMARY]
  )
ON [PRIMARY]

go
```

Similar to the last pipeline option, configure the on-premises SQL Server as the source and use the query code provided in option 2 as the source query, shown in Figure 8-15.

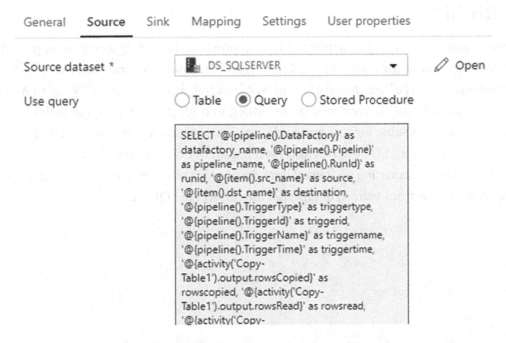

Figure 8-15. *ADF option 3 source query to extract pipeline activity metrics*

The sink will be a connection to the SQL database pipeline log table that was created earlier.

After saving, publishing, and running the pipeline, notice that the pipeline Copy activity records have been captured in the `dbo.pipeline_log` table, as shown in Figure 8-16.

triggername	triggertime	rowscopied	rowsread
Manual	2020-02-01T06:00:43.1588038Z	9	9
ea7a222e-5bc1-4fa0-87a1-20f826ef96b8	2020-02-03T20:00:40.6652657Z	717539307	717539307
721b51b0-cdc0-42c5-85c7-731aae142ae3	2020-02-04T15:22:20.0586528Z	53775693	53775693
427ca039-329b-40cf-a5cf-4acd95fa096e	2020-02-04T20:18:23.5079153Z	1933547826	1933547826
Manual	2020-02-10T21:19:01.5711991Z	222848	222848
Manual	2020-02-10T21:39:52.3051061Z	1001	1001
Manual	2020 02 10T21:39:52.3051061Z	227	227
acdebc48-b17b-4c77-8651-0cbe07ff7e39	2020-02-10T23:08:49.3034994Z	15	15

Figure 8-16. *SSMS view of pipeline_log table results, which logged ADF option 3 pipeline data to SQL DW*

Summary

In this chapter, I demonstrated how to log ADF pipeline data using a custom approach of extracting metrics for each table or file from the ADF pipeline and then either writing the results to the `pipeline_parameter` table, `pipeline_log` table, or a CSV file in ADLS Gen2. This custom logging process can be added to multiple ADF pipeline activity steps for a robust and reusable logging and auditing process that can be easily queried in SSMS through custom SQL queries and even linked to Power BI dashboards and reports to promote robust reporting on logged data. In the next chapter, I will demonstrate how to log error details from your ADF pipeline into an Azure SQL table.

## CHAPTER 9

Capture Pipeline Error Logs in SQL Database

In Chapter 8, I discussed a variety of methods for capturing Azure Data Factory pipeline logs and persisting the data either to a SQL table or within Azure Data Lake Storage Gen2. While this process of capturing pipeline log data is valuable when the pipeline activities succeed, this chapter will cover how to capture and persist Azure Data Factory pipeline errors to a SQL table within the ADF_DB that was created in previous chapters. Additionally, we will recap the pipeline parameter process that I had discussed in the previous chapters to demonstrate how the `pipeline_errors`, `pipeline_log`, and `pipeline_parameter` tables relate to each other.

To recap, the tables needed for this process include

- pipeline_parameter
- pipeline_log
- pipeline_errors

Figure 9-1 illustrates how these tables are interconnected to each other.

© Ron C. L'Esteve 2021
R. C. L'Esteve, *The Definitive Guide to Azure Data Engineering*, https://doi.org/10.1007/978-1-4842-7182-7_9

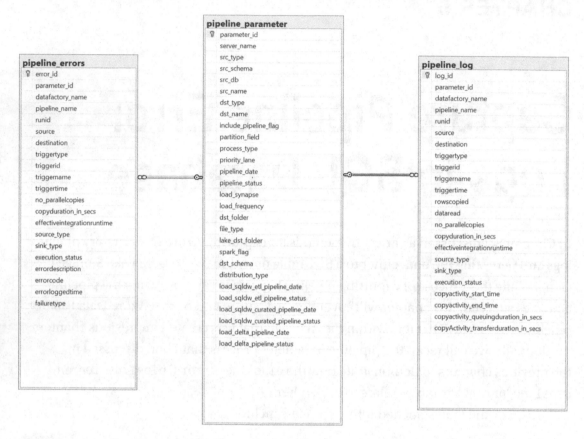

Figure 9-1. *Diagram depicting relationship between pipeline errors, log, and parameter tables*

Create a Parameter Table

To recap, we created a few variations of the `pipeline_parameter` table in previous chapters. The following script will create an updated version of the `pipeline_parameter` table containing a few additional columns, along with the column `parameter_id` as the primary key. Recall from the exercises in prior chapters that this table drives the metadata ETL approach:

```
SET ANSI_NULLS ON
GO

SET QUOTED_IDENTIFIER ON
go
```

```
CREATE TABLE [dbo].[pipeline_parameter]
   (
       [parameter_id]                          [INT] IDENTITY(1, 1) NOT NULL,
       [server_name]                           [NVARCHAR](500) NULL,
       [src_type]                              [NVARCHAR](500) NULL,
       [src_schema]                            [NVARCHAR](500) NULL,
       [src_db]                                [NVARCHAR](500) NULL,
       [src_name]                              [NVARCHAR](500) NULL,
       [dst_type]                              [NVARCHAR](500) NULL,
       [dst_name]                              [NVARCHAR](500) NULL,
       [include_pipeline_flag]                 [NVARCHAR](500) NULL,
       [partition_field]                       [NVARCHAR](500) NULL,
       [process_type]                          [NVARCHAR](500) NULL,
       [priority_lane]                         [NVARCHAR](500) NULL,
       [pipeline_date]                         [NVARCHAR](500) NULL,
       [pipeline_status]                       [NVARCHAR](500) NULL,
       [load_synapse]                          [NVARCHAR](500) NULL,
       [load_frequency]                        [NVARCHAR](500) NULL,
       [dst_folder]                            [NVARCHAR](500) NULL,
       [file_type]                             [NVARCHAR](500) NULL,
       [lake_dst_folder]                       [NVARCHAR](500) NULL,
       [spark_flag]                            [NVARCHAR](500) NULL,
       [dst_schema]                            [NVARCHAR](500) NULL,
       [distribution_type]                     [NVARCHAR](500) NULL,
       [load_sqldw_etl_pipeline_date]          [DATETIME] NULL,
       [load_sqldw_etl_pipeline_status]        [NVARCHAR](500) NULL,
       [load_sqldw_curated_pipeline_date]      [DATETIME] NULL,
       [load_sqldw_curated_pipeline_status]    [NVARCHAR](500) NULL,
       [load_delta_pipeline_date]              [DATETIME] NULL,
       [load_delta_pipeline_status]            [NVARCHAR](500) NULL,
       PRIMARY KEY CLUSTERED ( [parameter_id] ASC )WITH (statistics_norecompute =
       OFF, ignore_dup_key = OFF) ON [PRIMARY]
   )
ON [PRIMARY]

go
```

Create a Log Table

This next script will create the pipeline_log table for capturing the Data Factory success logs. In this table, column log_id is the primary key, and column parameter_id is a foreign key with a reference to column parameter_id from the parameter table:

```
SET ansi_nulls ON

go

SET quoted_identifier ON

go

CREATE TABLE [dbo].[pipeline_log]
  (
    [log_id]                                [INT] IDENTITY(1, 1) NOT NULL,
    [parameter_id]                          [INT] NULL,
    [datafactory_name]                      [NVARCHAR](500) NULL,
    [pipeline_name]                         [NVARCHAR](500) NULL,
    [runid]                                 [NVARCHAR](500) NULL,
    [source]                                [NVARCHAR](500) NULL,
    [destination]                           [NVARCHAR](500) NULL,
    [triggertype]                           [NVARCHAR](500) NULL,
    [triggerid]                             [NVARCHAR](500) NULL,
    [triggername]                           [NVARCHAR](500) NULL,
    [triggertime]                           [NVARCHAR](500) NULL,
    [rowscopied]                            [NVARCHAR](500) NULL,
    [dataread]                              [INT] NULL,
    [no_parallelcopies]                     [INT] NULL,
    [copyduration_in_secs]                  [NVARCHAR](500) NULL,
    [effectiveintegrationruntime]           [NVARCHAR](500) NULL,
    [source_type]                           [NVARCHAR](500) NULL,
    [sink_type]                             [NVARCHAR](500) NULL,
    [execution_status]                      [NVARCHAR](500) NULL,
    [copyactivity_start_time]               [NVARCHAR](500) NULL,
    [copyactivity_end_time]                 [NVARCHAR](500) NULL,
    [copyactivity_queuingduration_in_secs]  [NVARCHAR](500) NULL,
```

```
        [copyactivity_transferduration_in_secs] [NVARCHAR](500) NULL,
        CONSTRAINT [PK_pipeline_log] PRIMARY KEY CLUSTERED ( [log_id] ASC )WITH (
        statistics_norecompute = OFF, ignore_dup_key = OFF) ON [PRIMARY]
    )
ON [PRIMARY]

go

ALTER TABLE [dbo].[pipeline_log]
    WITH CHECK ADD FOREIGN KEY([parameter_id]) REFERENCES
    [dbo].[pipeline_parameter] ([parameter_id]) ON UPDATE CASCADE

go
```

Create an Errors Table

Creating an errors table will require executing this next script, which will create a
pipeline_errors table that will be used to capture the Data Factory error details from
failed pipeline activities. In this table, column error_id is the primary key, and column
parameter_id is a foreign key with a reference to column parameter_id from the
pipeline_parameter table:

```
SET ansi_nulls ON

go

SET quoted_identifier ON

go

CREATE TABLE [dbo].[pipeline_errors]
    (
        [error_id]                  [INT] IDENTITY(1, 1) NOT NULL,
        [parameter_id]              [INT] NULL,
        [datafactory_name]          [NVARCHAR](500) NULL,
        [pipeline_name]             [NVARCHAR](500) NULL,
        [runid]                     [NVARCHAR](500) NULL,
        [source]                    [NVARCHAR](500) NULL,
        [destination]               [NVARCHAR](500) NULL,
```

```
    [triggertype]                   [NVARCHAR](500) NULL,
    [triggerid]                     [NVARCHAR](500) NULL,
    [triggername]                   [NVARCHAR](500) NULL,
    [triggertime]                   [NVARCHAR](500) NULL,
    [no_parallelcopies]             [INT] NULL,
    [copyduration_in_secs]          [NVARCHAR](500) NULL,
    [effectiveintegrationruntime]   [NVARCHAR](500) NULL,
    [source_type]                   [NVARCHAR](500) NULL,
    [sink_type]                     [NVARCHAR](500) NULL,
    [execution_status]              [NVARCHAR](500) NULL,
    [errordescription]              [NVARCHAR](max) NULL,
    [errorcode]                     [NVARCHAR](500) NULL,
    [errorloggedtime]               [NVARCHAR](500) NULL,
    [failuretype]                   [NVARCHAR](500) NULL,
    CONSTRAINT [PK_pipeline_error] PRIMARY KEY CLUSTERED ( [error_id]
    ASC )WITH
    (statistics_norecompute = OFF, ignore_dup_key = OFF) ON [PRIMARY]
  )
ON [PRIMARY]
textimage_on [PRIMARY]

go

ALTER TABLE [dbo].[pipeline_errors]
  WITH CHECK ADD FOREIGN KEY([parameter_id]) REFERENCES
  [dbo].[pipeline_parameter] ([parameter_id]) ON UPDATE CASCADE

go
```

Create a Stored Procedure to Update the Log Table

Now that you have all the necessary SQL tables in place, begin creating a few necessary stored procedures by using the following script, which will create a stored procedure to update the pipeline_log table with data from the successful pipeline run. Note that this stored procedure will be called from the Data Factory pipeline at runtime:

```
SET ansi_nulls ON

go

SET quoted_identifier ON

go

CREATE PROCEDURE [dbo].[usp_updatelogtable] @datafactory_name
VARCHAR(250),
                                    @pipeline_name
VARCHAR(250),
                                    @runid
VARCHAR(250),
                                    @source
VARCHAR(300),
                                    @destination
VARCHAR(300),
                                    @triggertype
VARCHAR(300),
                                    @triggerid
VARCHAR(300),
                                    @triggername
VARCHAR(300),
                                    @triggertime
VARCHAR(300),
                                    @rowscopied
VARCHAR(300),
                                    @dataread
INT,
                                    @no_parallelcopies
INT,
                                    @copyduration_in_secs
VARCHAR(300),
                                    @effectiveintegrationruntime
VARCHAR(300),
                                    @source_type
```

```
VARCHAR(300),
                                        @sink_type

VARCHAR(300),
                                        @execution_status

VARCHAR(300),
                                        @copyactivity_start_time

VARCHAR(500),
                                        @copyactivity_end_time

VARCHAR(500),
                                        @copyactivity_queuingduration_
                                        in_secs

VARCHAR(500),
@copyactivity_transferduration_in_secs VARCHAR(500)
AS
INSERT INTO [pipeline_log]
([datafactory_name],
[pipeline_name],
[runid],
[source],
[destination],
[triggertype],
[triggerid],
[triggername],
[triggertime],
[rowscopied],
[dataread],
[no_parallelcopies],
[copyduration_in_secs],
[effectiveintegrationruntime],
[source_type],
[sink_type],
[execution_status],
[copyactivity_start_time],
[copyactivity_end_time],
[copyactivity_queuingduration_in_secs],
```

```
[copyactivity_transferduration_in_secs])
VALUES      ( @datafactory_name,
@pipeline_name,
@runid,
@source,
@destination,
@triggertype,
@triggerid,
@triggername,
@triggertime,
@rowscopied,
@dataread,
@no_parallelcopies,
@copyduration_in_secs,
@effectiveintegrationruntime,
@source_type,
@sink_type,
@execution_status,
@copyactivity_start_time,
@copyactivity_end_time,
@copyactivity_queuingduration_in_secs,
@copyactivity_transferduration_in_secs )

go
```

Create a Stored Procedure to Update the Errors Table

Next, run the following script, which will create a stored procedure to update the pipeline_errors table with detailed error data from the failed pipeline run. Note that this stored procedure will be called from the Data Factory pipeline at runtime:

```
SET ANSI_NULLS ON
GO

SET QUOTED_IDENTIFIER ON
GO
```

```sql
CREATE PROCEDURE [dbo].[usp_updateerrortable]
      @datafactory_name [nvarchar](500) NULL,
      @pipeline_name [nvarchar](500) NULL,
      @runid [nvarchar](500) NULL,
      @source [nvarchar](500) NULL,
      @destination [nvarchar](500) NULL,
      @triggertype [nvarchar](500) NULL,
      @triggerid [nvarchar](500) NULL,
      @triggername [nvarchar](500) NULL,
      @triggertime [nvarchar](500) NULL,
      @no_parallelcopies [int] NULL,
      @copyduration_in_secs [nvarchar](500) NULL,
      @effectiveintegrationruntime [nvarchar](500) NULL,
      @source_type [nvarchar](500) NULL,
      @sink_type [nvarchar](500) NULL,
      @execution_status [nvarchar](500) NULL,
      @errordescription [nvarchar](max) NULL,
      @errorcode [nvarchar](500) NULL,
      @errorloggedtime [nvarchar](500) NULL,
      @failuretype [nvarchar](500) NULL
AS
INSERT INTO [pipeline_errors]

(

    [datafactory_name],
      [pipeline_name],
      [runid],
      [source],
      [destination],
      [triggertype],
      [triggerid],
      [triggername],
      [triggertime],
      [no_parallelcopies],
```

```
        [copyduration_in_secs],
        [effectiveintegrationruntime],
        [source_type],
        [sink_type],
        [execution_status],
        [errordescription],
        [errorcode],
        [errorloggedtime],
        [failuretype]
)
VALUES
(
        @datafactory_name,
        @pipeline_name,
        @runid,
        @source,
        @destination,
        @triggertype,
        @triggerid,
        @triggername,
        @triggertime,
        @no_parallelcopies,
        @copyduration_in_secs,
        @effectiveintegrationruntime,
        @source_type,
        @sink_type,
        @execution_status,
        @errordescription,
        @errorcode,
        @errorloggedtime,
        @failuretype
)

GO
```

Create a Source Error

In Chapter 4, you initiated the process of building an ADF pipeline to load a source SQL Server table to Data Lake Storage Gen2. Based on this process, let's test a known error within the Data Factory pipeline and process by recreating an error for the purpose of this exercise. Typically, a varchar(max) data type containing at least 8,000+ characters will fail when being loaded into a Synapse Analytics dedicated SQL pool since varchar(max) is an unsupported data type. This seems like a good use case for an error test.

Begin by creating a table (e.g., SalesLT.Address) that will store a large block of text that will ultimately cause an error when being loaded into a Synapse Analytics dedicated SQL pool. The SalesLT.Address illustrated in Figure 9-2 contains the Description column that has the varchar(max) data type.

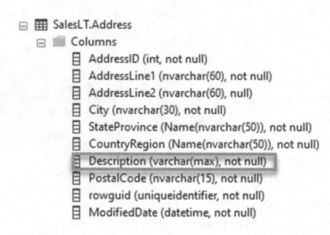

Figure 9-2. *SSMS view of SalesLT.Address*

Within SalesLT.Address, add a large block of text. For example purposes, you could enter some random text that is greater than 8,001 words, similar to what is shown in the sample query output shown in Figure 9-3. Confirm that the Description column contains 8,001 words, which is sure to fail the Azure Data Factory pipeline due to the 8,000-character length limit from the Synapse Analytics dedicated SQL pool destination side, and it will trigger a record to be created in the pipeline_errors table.

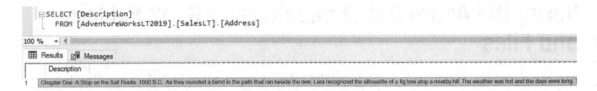

Figure 9-3. *SSMS view of SalesLT.Address Description column containing large block of sample text*

Add Records to a Parameter Table

After identifying the source SQL tables to run through the process, add them to the pipeline_parameter table, shown in Figure 9-4. For this exercise, add the SalesLT. Address table that was created in the previous step that contains the large block of 8,001 characters, along with a regular table (e.g., SalesLT.Customer) that we would expect to succeed, to demonstrate both a successful and failed end-to-end logging process.

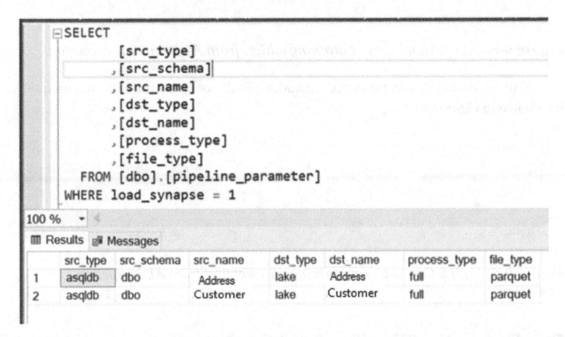

Figure 9-4. *SSMS view of records added to pipeline_parameter table to test errors in ADF pipeline*

Verify the Azure Data Lake Storage Gen2 Folders and Files

After running the pipeline to load the SQL tables to Azure Data Lake Storage Gen2, Figure 9-5 shows that the destination ADLS Gen2 container now has both of the tables in snappy compressed Parquet format.

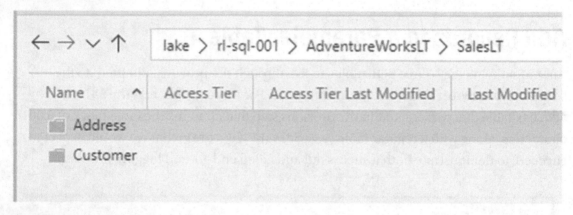

Figure 9-5. *ADLS Gen2 view containing folders from ADF pipeline execution*

As an additional verification step, the Address folder contains the expected parquet file shown in Figure 9-6.

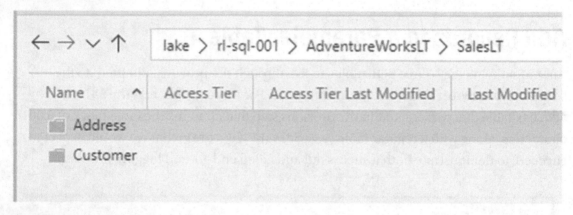

Figure 9-6. *ADLS Gen2 view containing parquet files from ADF pipeline execution*

Configure the Pipeline Lookup Activity

It's now time to build and configure the ADF pipeline. In Chapter 7, I covered the details on how to build an ADF pipeline to load a Synapse Analytics dedicated SQL pool with parquet files stored in ADLS Gen2. As a recap of the process, the select query within the lookup gets the list of parquet files that need to be loaded to a Synapse Analytics dedicated SQL pool and then passes them to the ForEach loop, which will load the parquet files to a Synapse Analytics dedicated SQL pool, as shown in Figure 9-7.

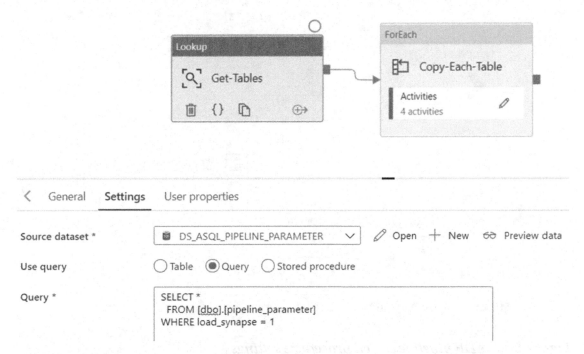

Figure 9-7. *ADF pipeline Lookup activity settings and query to get pipeline_ parameter listed tables*

Configure the Pipeline ForEach Loop Activity

The ForEach loop activity contains the Copy-Table activity that takes the parquet files and loads them to a Synapse Analytics dedicated SQL pool while auto-creating the tables. If the Copy-Table activity succeeds, it will log the pipeline run data to the `pipeline_log` table. However, if the Copy-Table activity fails, it will log the pipeline error details to the `pipeline_errors` table.

Configure a Stored Procedure to Update the Log Table

Notice also that the `UpdateLogTable` stored procedure that was created earlier will be called by the success Stored Procedure activity shown in Figure 9-8.

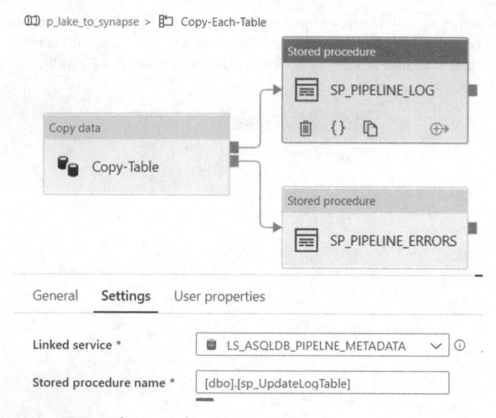

Figure 9-8. *ADF pipeline stored procedure settings*

Figure 9-9 shows the stored procedure parameters that will update the `pipeline_log` table and can be imported directly from the stored procedure.

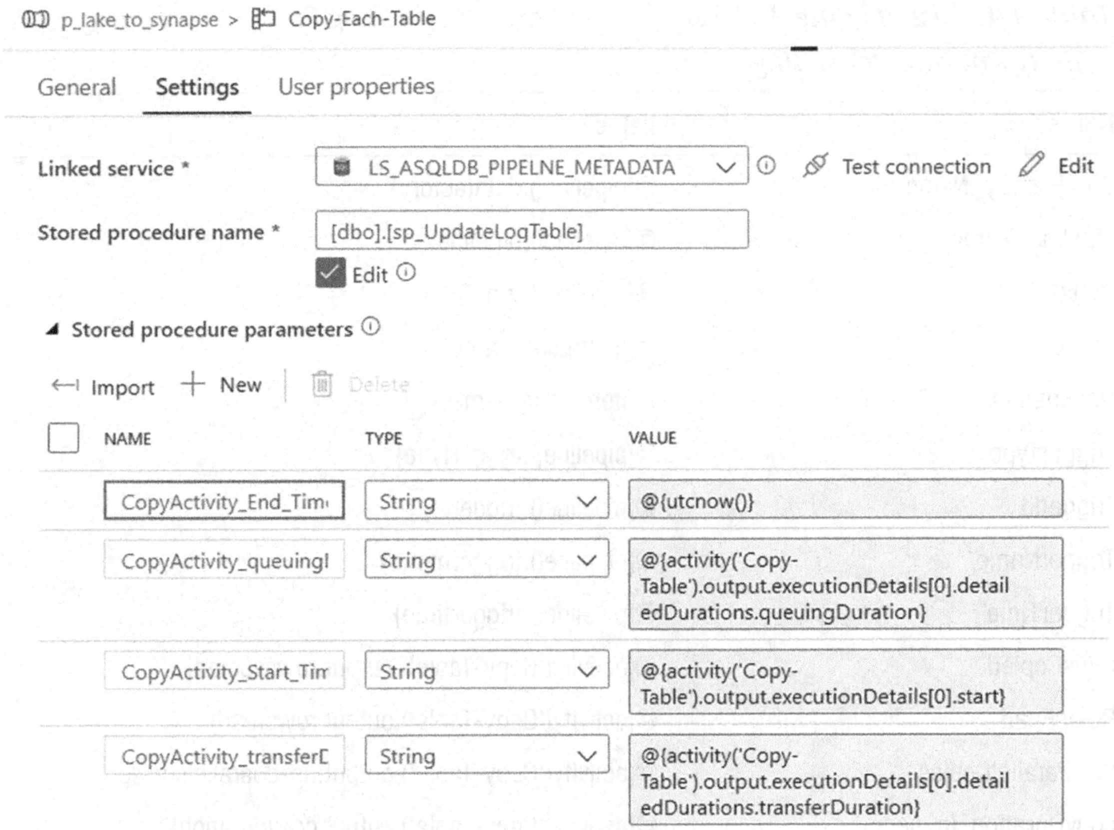

⫘ p_lake_to_synapse > 🔲 Copy-Each-Table

General   **Settings**   User properties

Linked service *   🗄 LS_ASQLDB_PIPELNE_METADATA   ⌄ ⓘ   ✎ Test connection   ✏ Edit

Stored procedure name *   [dbo].[sp_UpdateLogTable]

☑ Edit ⓘ

◢ Stored procedure parameters ⓘ

←ᴵ Import   ✛ New | 🗑 Delete

	NAME	TYPE		VALUE
☐	CopyActivity_End_Tim⟩	String	⌄	@{utcnow()}
	CopyActivity_queuingⅠ	String	⌄	@{activity('Copy-Table').output.executionDetails[0].detailedDurations.queuingDuration}
	CopyActivity_Start_Tim⟩	String	⌄	@{activity('Copy-Table').output.executionDetails[0].start}
	CopyActivity_transferⅮ	String	⌄	@{activity('Copy-Table').output.executionDetails[0].detailedDurations.transferDuration}

Figure 9-9. *ADF pipeline stored procedure parameters*

The following values listed in Table 9-1 will need to be entered as the Stored Procedure activity parameter values as shown partially in Figure 9-9.

Table 9-1. *List of values to be entered into the Pipeline Log Stored Procedure activity's parameter settings*

Name	Value
DataFactory_Name	@{pipeline().datafactory}
Pipeline_Name	@{pipeline().pipeline}
RunId	@{pipeline().runid}
Source	@{item().src_name}
Destination	@{item().dst_name}
TriggerType	@{pipeline().triggertype}
TriggerId	@{pipeline().triggerid}
TriggerName	@{pipeline().triggername}
TriggerTime	@{pipeline().triggertime}
rowsCopied	@{activity('Copy-Table').output.rowscopied}
RowsRead	@{activity('Copy-Table').output.rowsread}
No_ParallelCopies	@{activity('Copy-Table').output.usedparallelcopies}
copyDuration_in_secs	@{activity('Copy-Table').output.copyduration}
effectiveIntegrationRuntime	@{activity('Copy-Table').output.effectiveintegrationruntime}
Source_Type	@{activity('Copy-Table').output.executiondetails[0].source.type}
Sink_Type	@{activity('Copy-Table').output.executiondetails[0].sink.type}
Execution_Status	@{activity('Copy-Table').output.executiondetails[0].status}
CopyActivity_Start_Time	@{activity('Copy-Table').output.executiondetails[0].start}
CopyActivity_End_Time	@{utcnow()}
CopyActivity_queuingDuration_in_secs	@{activity('Copy-Table').output.executionDetails[0].detaileddurations.queuingDuration}
CopyActivity_transferDuration_in_secs	@{activity('Copy-Table').output.executionDetails[0].detaileddurations.transferDuration}

Configure a Stored Procedure to Update the Errors Table

The last stored procedure within the ForEach loop activity is the UpdateErrorTable stored procedure that was created earlier and will be called by the failure Stored Procedure activity, shown in Figure 9-10.

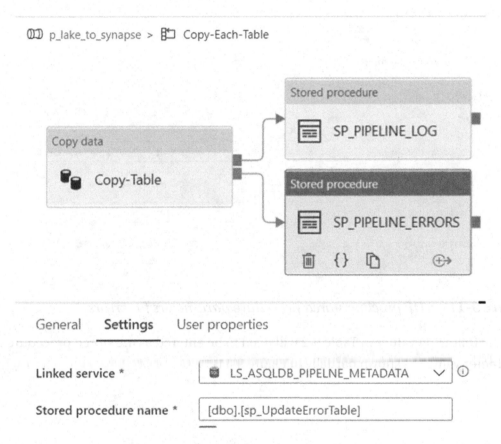

Figure 9-10. *ADF pipeline stored procedure for errors*

Figure 9-11 shows the stored procedure parameters that will update the pipeline_ errors table and can be imported directly from the stored procedure.

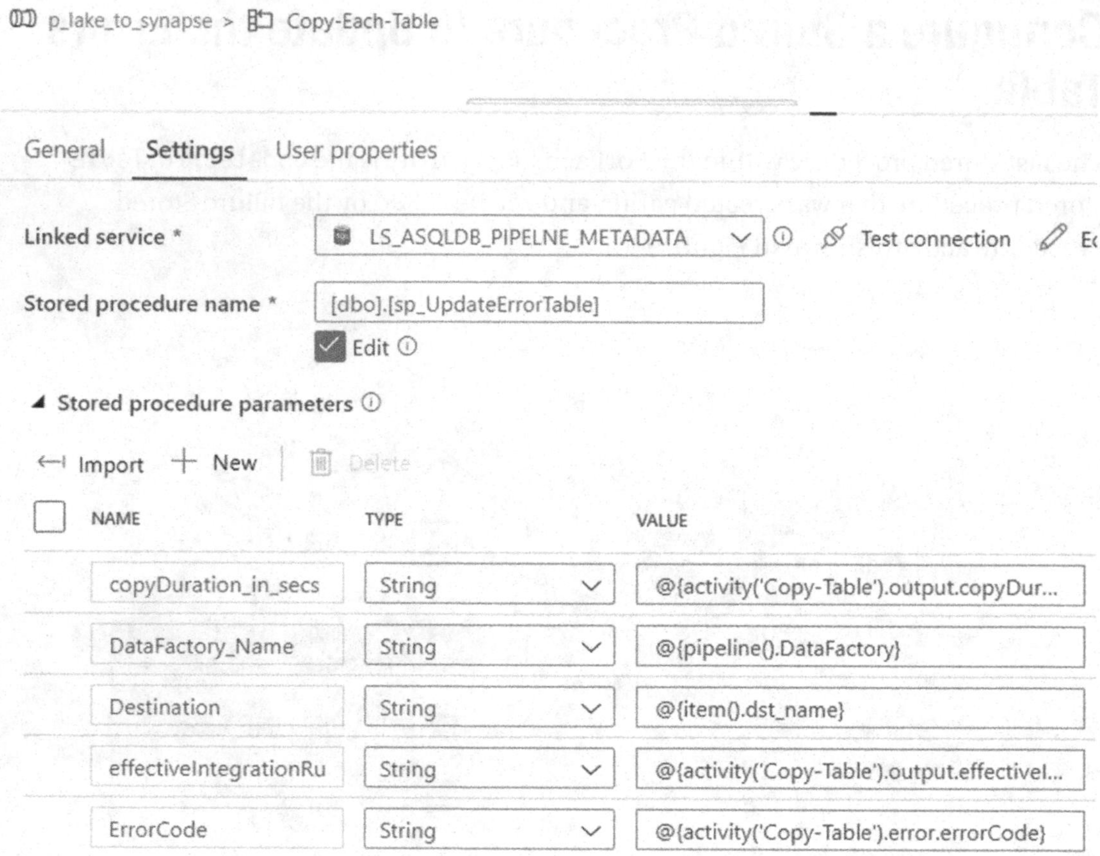

Figure 9-11. *ADF pipeline stored procedure parameters for errors*

The following values in Table 9-2 will need to be entered as the stored procedure parameter values as shown partially in Figure 9-11.

Table 9-2. *List of values to be entered into the Pipeline Errors Stored Procedure activity's parameter settings*

Name	Value
DataFactory_Name	@{pipeline().datafactory}
Pipeline_Name	@{pipeline().pipeline}
RunId	@{pipeline().runid}
Source	@{item().src_name}
Destination	@{item().dst_name}
TriggerType	@{pipeline().triggertype}
TriggerId	@{pipeline().triggerid}
TriggerName	@{pipeline().triggername}
TriggerTime	@{pipeline().triggertime}
No_ParallelCopies	@{activity('Copy-Table').output.usedparallelcopies}
copyDuration_in_secs	@{activity('Copy-Table').output.copyduration}
effectiveIntegrationRuntime	@{activity('Copy-Table').output.effectiveintegrationruntime}
Source_Type	@{activity('Copy-Table').output.executiondetails[0].source.type}
Sink_Type	@{activity('Copy-Table').output.executiondetails[0].sink.type}
Execution_Status	@{activity('Copy-Table').output.executiondetails[0].status}
ErrorCode	@{activity('Copy-Table').error.errorcode}
ErrorDescription	@{activity('Copy-Table').error.message}
ErrorLoggedTIme	@utcnow()
FailureType	@concat(activity('Copy-Table').error.message,'failuretype:', activity('Copy-Table').error.failuretype)

Run the Pipeline

Now that you have configured the pipeline, go ahead and run the pipeline. Notice from the Debug mode output log in Figure 9-12 one table succeeded and the other failed, as expected.

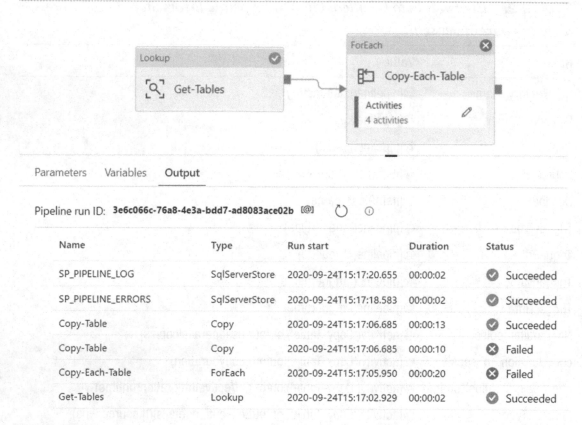

Figure 9-12. *ADF pipeline run output containing activity status*

Verify the Results

Finally, verify the results in the pipeline_log table. Notice in Figure 9-13 the pipeline_log table has captured one log containing the source SalesLT.Customer.

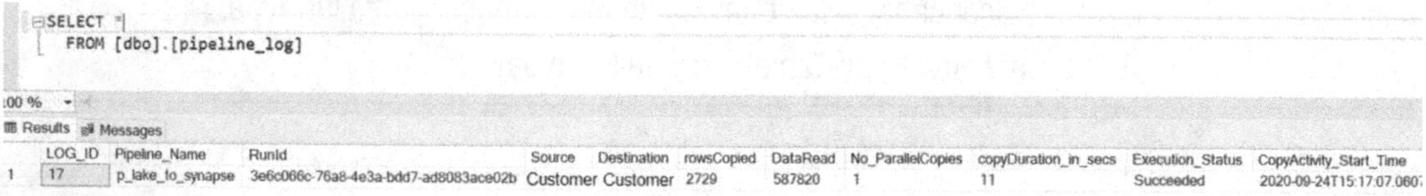

Figure 9-13. *SSMS view of pipeline_log table showing successful ADF pipeline*

The pipeline_errors table has one record for SalesLT.Address, along with detailed error codes, descriptions, messages, and more, shown in Figure 9-14.

Figure 9-14. *SSMS view of pipeline_errors table showing failed ADF pipeline*

As a final check, upon navigating to the Synapse Analytics dedicated SQL pool, notice from Figure 9-15 that both tables have been auto-created, despite the fact that one failed and one succeeded.

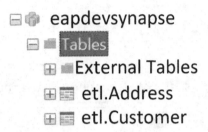

Figure 9-15. *View of auto-created Synapse Analytics dedicated SQL pool tables*

It is important to note that data was only loaded in `etl.Customer` since `etl.Address` failed and subsequently had no data loaded into it. Figure 9-16 shows a `select` * of data from `etl.Address` to confirm that the table does not contain any data.

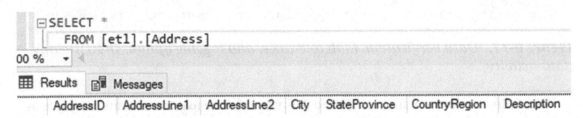

Figure 9-16. *Select * of auto-created etl.Address table, which contains no data as result of failed ADF pipeline*

Other ADF Logging Options

The preceding sections described a more custom method of logging and handling errors within ADF pipelines. Within the Copy data activity shown in Figure 9-17, there is a built-in method of verifying data consistency, managing fault tolerance, and enabling logging.

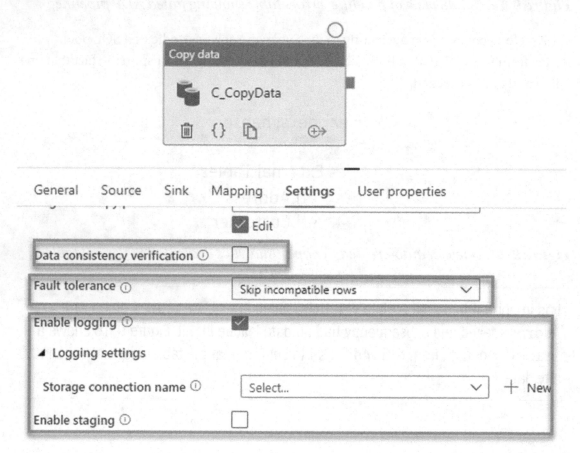

Figure 9-17. *Data verification, fault tolerance, and logging options in ADF Copy data activity*

The following list contains details and definitions of the features that are available as optional configuration properties within the Settings tab of the Copy data activity within ADF. These features are meant to provide additional out-of-the-box data verification, validation, and logging methods, and they can selectively be enabled and/or disabled within the ADF pipeline:

- **Data consistency verification**: When selecting this option, the Copy activity will do additional data consistency verification between source and destination store after data movement. The verification includes file size check and checksum verification for binary files and row count verification for tabular data.

- **Fault tolerance**: When selecting this option, you can ignore some errors that occurred in the middle of the copy process, for example, incompatible rows between source and destination store, file being deleted during data movement, etc.

- **Enable logging**: When selecting this option, you can log copied files, skipped files, and rows into a blob storage.

Similarly, within Mapping Data Flows, there is also an option within the sink settings, which in the case of Figure 9-18 is using an Azure SQL DB. Notice the various options including the error row handling, transaction commit, and report success on error options.

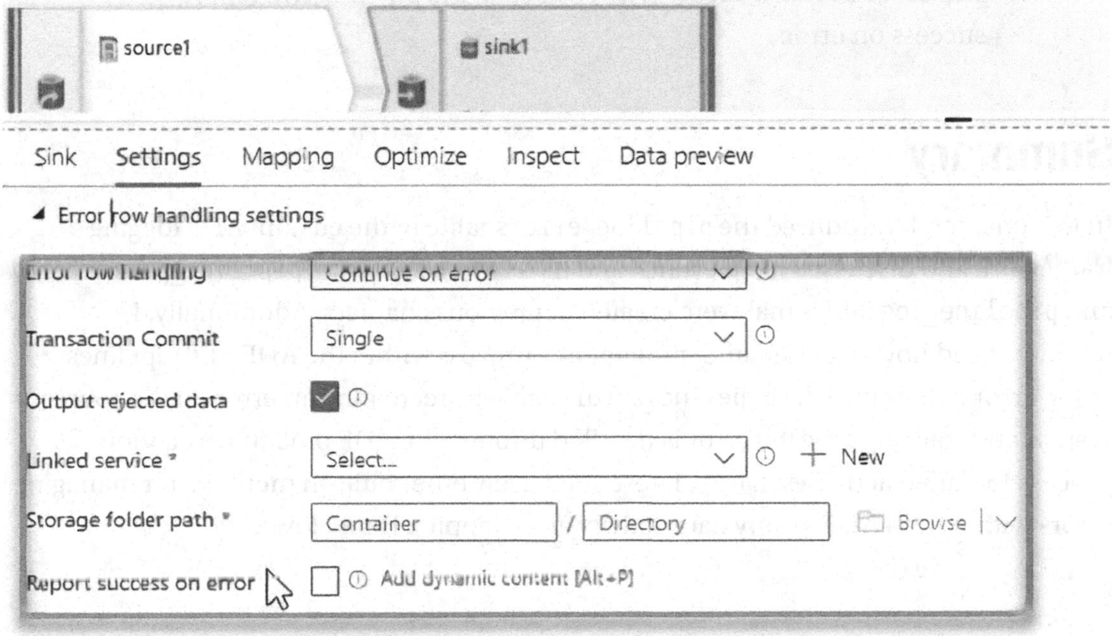

Figure 9-18. *Error row handling in ADF Mapping Data Flows*

The following list contains details and definitions of the features that are available as optional configuration properties within the Settings tab of the sink Copy data activity within ADF. These features are meant to provide additional out-of-the-box error handling, transaction commits, and more, and they can selectively be enabled and/or disabled within the ADF pipeline:

- **Error row handling**: "Continue on error" will prevent the pipeline from failing – for example, in the event that there are millions of rows being loaded with a few rows that fail within the process.

- **Transaction Commit**: Allows both "Single" and "Batch" commits in the context of SQL transaction commits. For example, Batch will have better performance since data will be loaded in batches as opposed to a single commit.

- **Output rejected data**: This option, similar to the regular ADF Copy data activity, when selected will prompt you to log the rejected data to a storage account.

- **Report success on error**: When selected, the pipeline will report success on error.

Summary

In this chapter, I introduced the `pipeline_errors` table in the custom ADF logging framework and discussed its relevance and its relationship to the `pipeline_parameter` and `pipeline_log` tables that were created in previous chapters. Additionally, I demonstrated how to design and implement a process within the ADF ELT pipelines to log error details into the `pipeline_errors` table by recreating an error on the source system and tracking how the error is handled through the ADF pipeline execution process for failed activities. Lastly, I described a few other built-in methods for managing errors within both ADF's Copy data activity and Mapping Data Flows.

CHAPTER 10

Dynamically Load a Snowflake Data Warehouse

Many organizations and customers are considering a Snowflake data warehouse as an alternative to Azure's Synapse Analytics dedicated SQL pool. While applying the same patterns that were introduced in Chapter 7, this chapter will focus on showing you how to load data into a Snowflake DW by using both Databricks and Data Factory.

There is more than one option for dynamically loading ADLS Gen2 data into a Snowflake DW within the modern Azure Data Platform. Some of these options, which you will learn about in this chapter, include

- Parameterized Databricks notebooks within an ADF pipeline

- Data Factory's regular Copy activity

- Data Factory's Mapping Data Flows

After going through detailed end-to-end exercises for these various options, I will discuss recommendations based on capabilities and limitations of the options from the various ADF pipeline execution runs.

The architectural diagram in Figure 10-1 illustrates the Azure resources being evaluated for dynamically loading data from a SQL database (AdventureWorks) into an ADLS Gen2 account using Data Factory for the ELT and a Snowflake database (ADF_DB) for the control and logging tables. Note that ADF_DB has been created in previous chapters containing the metadata pipeline parameter and errors tables. As you progress through this chapter, these various options to load data into a Snowflake DW will be explored for dynamically loading the ADLS Gen2 files into Snowflake.

© Ron C. L'Esteve 2021
R. C. L'Esteve, *The Definitive Guide to Azure Data Engineering*, https://doi.org/10.1007/978-1-4842-7182-7_10

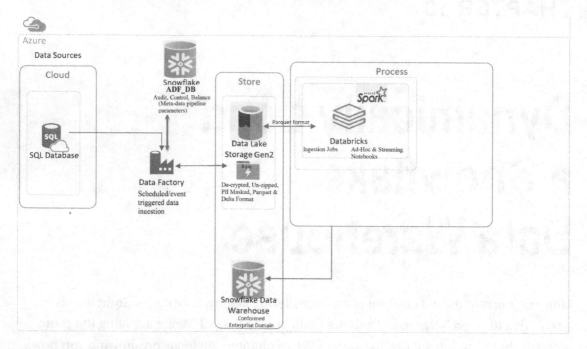

Figure 10-1. *Azure data flow architecture for dynamically loading data in Snowflake DW*

Linked Services and Datasets

Initially, there will be a few linked services and datasets that you will need to create in ADF to build the pipeline. The linked services will include the source, sink, control database, and Databricks notebook, which will be part of this ADF pipeline process.

Base Linked Services

To begin the process, please create the following linked services shown in Figure 10-2 in Data Factory.

▓ LS_ADLS2_ADVENTUREWORKS	Azure Data Lake Storage Gen2
▓ LS_ASQL_ADVENTUREWORKS	Azure SQL Database
❄ LS_SNOW_ADFDB	Snowflake
❄ LS_SNOW_ADVENTUREWORKS	Snowflake
◈ AzureDatabricks1	Azure Databricks

Figure 10-2. *Base linked services needed for loading data into Snowflake*

An ADLS Gen2 linked service, shown in Figure 10-3, will serve as the container and landing zone for the snappy compressed files.

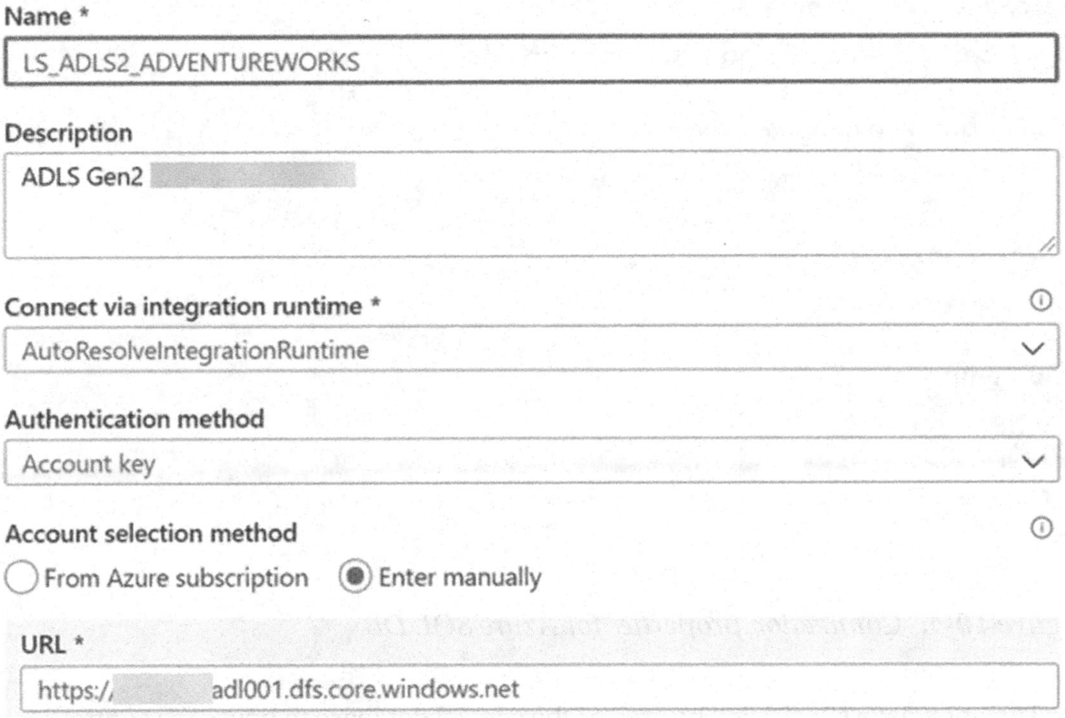

Name *

LS_ADLS2_ADVENTUREWORKS

Description

ADLS Gen2 ▓▓▓▓▓▓

Connect via integration runtime *   ⓘ

AutoResolveIntegrationRuntime   ⌄

Authentication method

Account key   ⌄

Account selection method   ⓘ

◯ From Azure subscription   ⦿ Enter manually

URL *

https://▓▓▓▓adl001.dfs.core.windows.net

Figure 10-3. *Connection properties for ADLS Gen2*

A sample AdventureWorks Azure SQL Database will serve as the source SQL database, and Figure 10-4 shows the connection properties for the database used to generate this chapter's examples.

Name *

LS_ASQL_ADVENTUREWORKS

Description

Sample Adventureworks database for testing only

Connect via integration runtime *

AutoResolveIntegrationRuntime

(Connection string)   (Azure Key Vault)

Account selection method

◯ From Azure subscription   ◉ Enter manually

Fully qualified domain name *

sql001.database.windows.net

Database name *

AdventureWorksLT2019

Authentication type *

SQL authentication

User name *

admin

Figure 10-4. *Connection properties for Azure SQL DB*

The Snowflake ADF_DB will be used as the control database to house the control, logging, and audit tables within the ADF pipelines. Figure 10-5 shows some example connection properties.

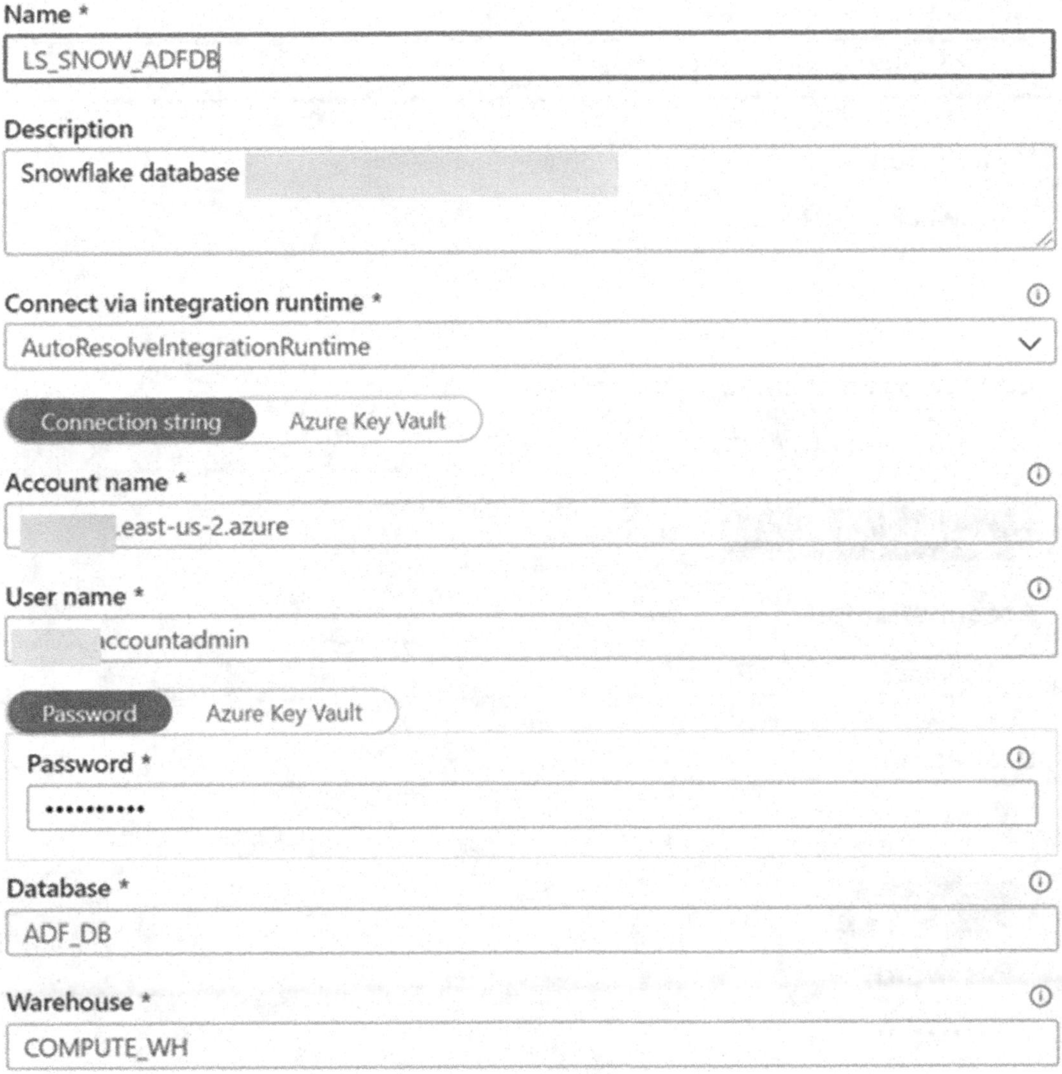

Name *

LS_SNOW_ADFDB

Description

Snowflake database

Connect via integration runtime *   ⓘ

AutoResolveIntegrationRuntime   ⌄

(Connection string)   Azure Key Vault

Account name *   ⓘ

.east-us-2.azure

User name *   ⓘ

.ccountadmin

(Password)   Azure Key Vault

Password *   ⓘ

••••••••••

Database *   ⓘ

ADF_DB

Warehouse *   ⓘ

COMPUTE_WH

Figure 10-5. *Connection properties for Snowflake control tables*

Figure 10-6 shows the Snowflake AdventureWorks DB that will serve as the destination/target Snowflake DB where all AdventureWorks tables will land.

Name *

LS_SNOW_ADVENTUREWORKS

Description

Snowflake destination db

Connect via integration runtime *

AutoResolveIntegrationRuntime

Connection string   Azure Key Vault

Account name *

.east-us-2.azure

User name *

accountadmin

Password   Azure Key Vault

Password *

••••••••••

Database *

ADVENTUREWORKS

Warehouse *

COMPUTE_WH

Figure 10-6. *Connection properties for Snowflake destination DW*

The Databricks linked service shown in Figure 10-7 is created to process the Databricks notebook containing Scala code that pushes ADLS Gen2 files to Snowflake target tables.

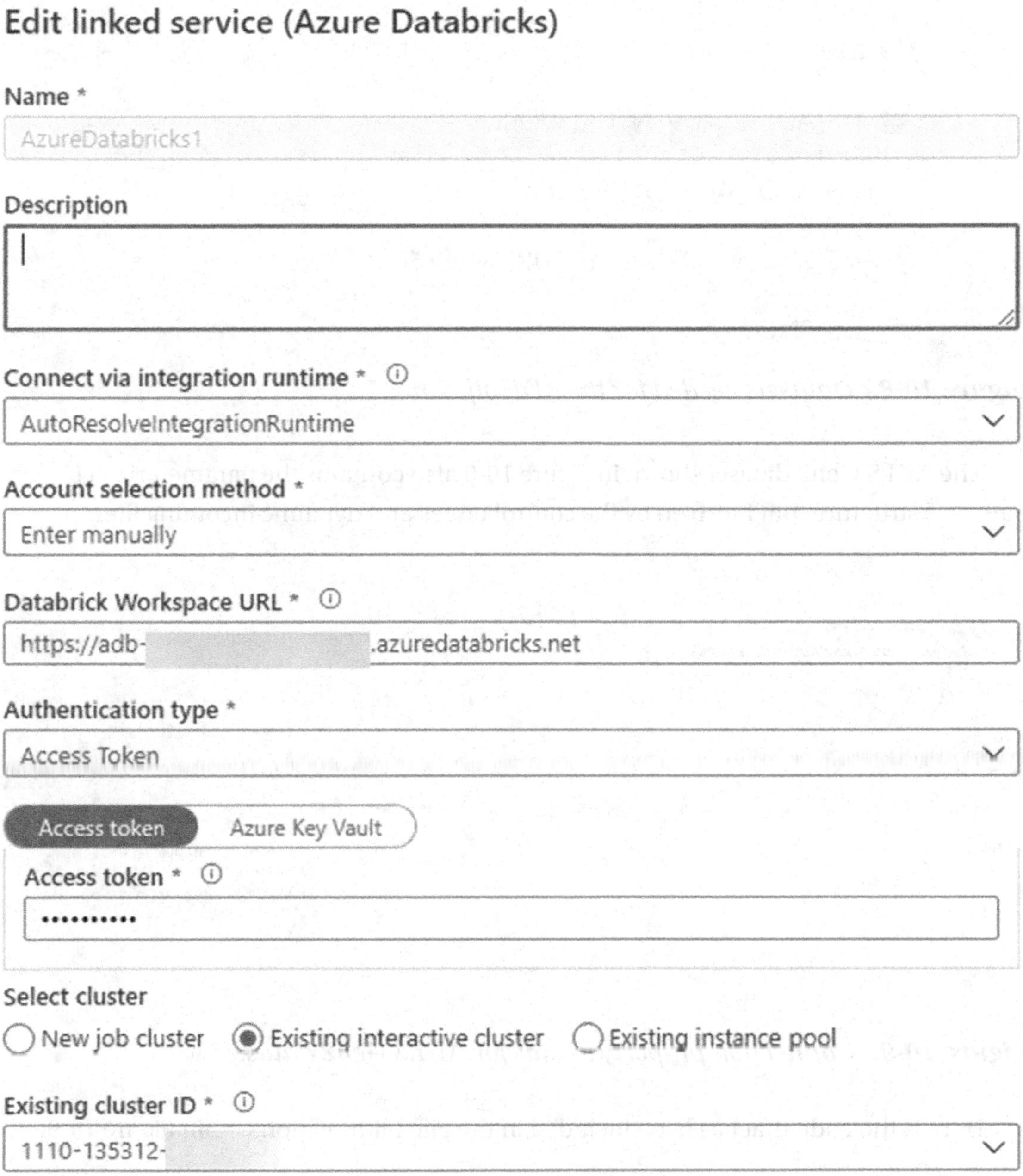

Edit linked service (Azure Databricks)

Name *

AzureDatabricks1

Description

|

Connect via integration runtime * ⓘ

AutoResolveIntegrationRuntime   ⌄

Account selection method *

Enter manually   ⌄

Databrick Workspace URL * ⓘ

https://adb-   .azuredatabricks.net

Authentication type *

Access Token   ⌄

(Access token   Azure Key Vault)

Access token * ⓘ

••••••••••

Select cluster

◯ New job cluster   ◉ Existing interactive cluster   ◯ Existing instance pool

Existing cluster ID * ⓘ

1110-135312-   ⌄

Figure 10-7. *Connection properties for Azure Databricks*

179

Datasets

Once the linked services are created, the datasets in Figure 10-8 will need to also be created. These are the datasets that will be used in the pipelines.

Figure 10-8. Datasets needed for the ADF pipeline

The ADLS Gen2 dataset shown in Figure 10-9 also contains the parameterized folder path and structure that is driven by the control tables and dynamic incoming files.

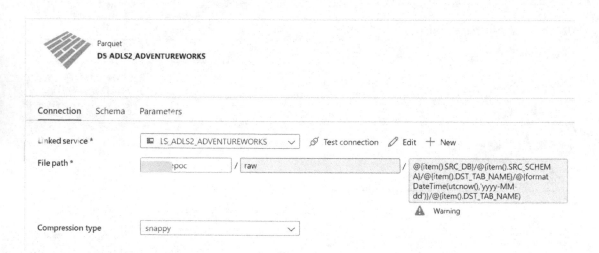

Figure 10-9. Connection property details for ADLS Gen2 dataset

Here is the code that has been included in the File path section within Figure 10-9:

```
poc/raw/@{item().SRC_DB}/@{item().SRC_SCHEMA}/@{item().DST_TAB_NAME}
/@{formatDateTime(utcnow(),'yyyy-MM-dd')}/ @{item().SAT_TAB_NAME}
```

The parameters in Figure 10-10 will need to be set. These parameters are populated by the Snowflake control tables that contain the detailed metadata information, which are called as dynamic parameters within the ADF pipelines and datasets.

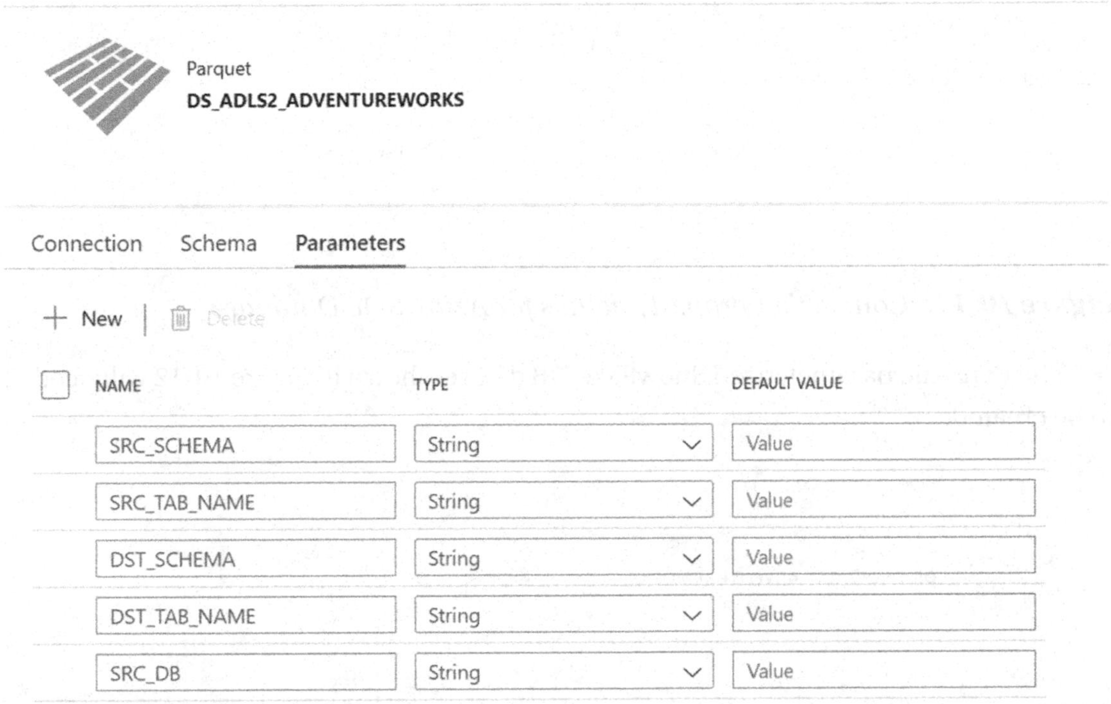

Parquet
DS_ADLS2_ADVENTUREWORKS

Connection   Schema   **Parameters**

+ New   🗑 Delete

NAME	TYPE	DEFAULT VALUE
SRC_SCHEMA	String ⌄	Value
SRC_TAB_NAME	String ⌄	Value
DST_SCHEMA	String ⌄	Value
DST_TAB_NAME	String ⌄	Value
SRC_DB	String ⌄	Value

Figure 10-10. *Parameter property details for ADLS Gen2 dataset*

Next, please create the AdventureWorks dataset shown in Figure 10-11. This is our source database connection to the AdventureWorks2019LT database, which will reside on an Azure SQL Database. We will use this database to extract the tables and load them to Data Lake Storage Gen2, and we'll then move them to Snowflake DW using the Databricks notebook, which will be orchestrated and called by ADF.

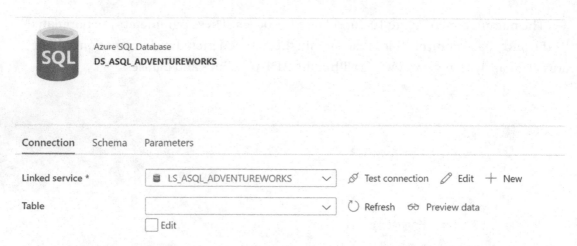

Figure 10-11. *Connection property details for Azure SQL Database*

The dynamic parameterized Snowflake DB dataset shown in Figure 10-12 will need to be created.

Figure 10-12. *Connection property details for dynamic Snowflake DW destination tables*

Also, be sure to create an ADF_DB in the target Snowflake database containing the control tables that can then be used as datasets shown in Figure 10-13.

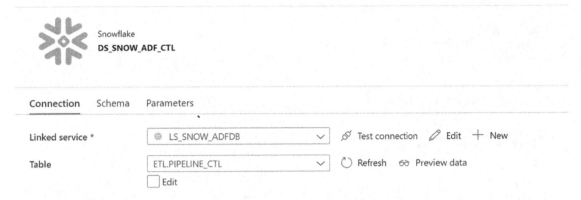

Figure 10-13. *Connection property details for Snowflake DW control database and tables*

The ADF_DB control table in Snowflake contains the following schema shown in Figure 10-14 and can be further updated and edited as desired.

Snowflake
DS_SNOW_ADF_CTL

Connection   **Schema**   Parameters

Import schema   Clear

Column name	Type
PARAMETER_ID	NUMBER
SERVER_NAME	VARCHAR
SRC_TYPE	VARCHAR
SRC_SCHEMA	VARCHAR
SRC_DB	VARCHAR
DEST_DB	VARCHAR
SRC_TAB_NAME	VARCHAR
DST_TYPE	VARCHAR
DST_TAB_NAME	VARCHAR
INCLUDE_PIPELINE_FLAG	VARCHAR
PROCESS_TYPE	VARCHAR
LOAD_SNOWFLAKE	VARCHAR
LOAD_FREQUENCY	VARCHAR
DST_FOLDER	VARCHAR
FILE_TYPE	VARCHAR
LAKE_DST_FOLDER	VARCHAR
DST_SCHEMA	VARCHAR
DISTRIBUTION_TYPE	VARCHAR
ASQL_TO_LAKE_PIPELINE_DATE	TIMESTAMP_NTZ
ASQL_TO_LAKE_PIPELINE_STATUS	VARCHAR
LOAD_SNOW_ETL_PIPELINE_DATE	TIMESTAMP_NTZ
LOAD_SNOW_ETL_PIPELINE_STATUS	VARCHAR
LOAD_SNOW_CURATED_PIPELINE_DATE	TIMESTAMP_NTZ
LOAD_SNOW_CURATED_PIPELINE_STATUS	VARCHAR
LOAD_DELTA_PIPELINE_DATE	TIMESTAMP_NTZ
LOAD_DELTA_PIPELINE_STATUS	VARCHAR
PARTITION_FIELD	VARCHAR
PRIORITY_LANE	VARCHAR
SPARK_FLAG	VARCHAR
SWIM_LANE	NUMBER

Figure 10-14. *Snowflake ADF_DB control table Schema*

Snowflake Control Database and Tables

In Chapter 4, I showed you how to implement control databases and tables that are used to control the ADF pipelines through a metadata-driven and parameterized process. In this chapter, follow the same concept of control databases and tables in the Snowflake DW. The SnowSQL syntax will have some variations when compared to T-SQL, and I will provide the SnowSQL code that you will need to execute on the Snowflake DW.

Figure 10-15 shows the ADF_DB pipeline control table called ETL_PIPELINE_CTL.

Table: ADF_DB.ETL_PIPELINE_CTL   Data   Details

Filter result...

Row	PARAMETER_ID	SERVER_NAME	SRC_TYPE	SRC_SCHEMA	SRC_DB	SRC_TAB_NAME ↓	DST_TYPE	DST_TAB_NAME	
9	1	gze2np1asq'001	ASQL	SALESLT	AdventureWorksLT2019	SalesOrderHeader	ADLS2	SalesOrderHeader	\
10	1	gze2np1asq'001	ASQL	SALESLT	AdventureWorksLT2019	SalesOrderDetail	ADLS2	SalesOrderData	\
8	1	gze2np1asq'001	ASQL	SALESLT	AdventureWorksLT2019	ProductModelProductDescription	ADLS2	ProductModelProductDescription	\
7	1	gze2np1asq'001	ASQL	SALESLT	AdventureWorksLT2019	ProductModel	ADLS2	ProductModel	\
6	1	gze2np1asq'001	ASQL	SALESLT	AdventureWorksLT2019	ProductDescription	ADLS2	ProductDescription	\
	1	gze2np1asq'001	ASQL	SALESLT	AdventureWorksLT2019	ProductCategory	ADLS2	ProductCategory	\
3	1	gze2np1asq'001	ASQL	SALESLT	AdventureWorksLT2019	Product	ADLS2	Product	\
4	1	gze2np1asq'001	ASQL	SALESLT	AdventureWorksLT2019	CustomerAddress	ADLS2	CustomerAddress	\
2	1	gze2np1asq'001	ASQL	SALESLT	AdventureWorksLT2019	Customer	ADLS2	Customer	\
1	1	gze2np1asq'001	ASQL	SALESLT	AdventureWorksLT2019	Address	ADLS2	Address	\

Figure 10-15. *Snowflake ADF_DB.ETL_PIPELINE_CTL control table*

Additionally, the following is the script used to create the control and audit tables in Snowflake. Be sure to run the script in the ETL schema of your ADF_DB database. The script will create two tables – PIPELINE_CTL and AUDIT_TAB:

```
CREATE OR replace TABLE pipeline_ctl
  (
  parameter_id                  number(38,0) NOT NULL autoincrement,
  server_name                   varchar(500),
  src_type                      varchar(500),
  src_schema                    varchar(500),
  src_db                        varchar(500),
```

```
  src_tab_name                         varchar(500),
  dst_type                             varchar(500),
  dst_tab_name                         varchar(500),
  include_pipeline_flag                varchar(500),
  process_type                         varchar(500),
  load_snowflake                       varchar(500),
  load_frequency                       varchar(500),
  dst_folder                           varchar(500),
  file_type                            varchar(500),
  lake_dst_folder                      varchar(500),
  dst_schema                           varchar(500),
  distribution_type                    varchar(500),
  asql_to_lake_pipeline_date           timestamp_ntz(9),
  asql_to_lake_pipeline_status         varchar(500),
  load_snow_etl_pipeline_date          timestamp_ntz(9),
  load_snow_etl_pipeline_status        varchar(500),
  load_snow_curated_pipeline_date      timestamp_ntz(9),
  load_snow_curated_pipeline_status    varchar(500),
  load_delta_pipeline_date             timestamp_ntz(9),
  load_delta_pipeline_status           varchar(500),
  partition_field                      varchar(500),
  priority_lane                        varchar(500),
  spark_flag                           varchar(500),
  swim_lane                            int,
  PRIMARY KEY (parameter_id)
   );

CREATE OR replace TABLE audit_tab
   (
  pipeline_name                        varchar(100),
  db_name                              varchar(20),
  sch_name                             varchar(20),
  table_name                           varchar(50),
  source_count                         number(38,0),
  adls_count                           number(38,0),
```

```
snowflake_count                        number(38,0),
load_time timestamp_ntz(9)             DEFAULT CURRENT_TIMESTAMP()
  );
```

Pipelines

Now that you have created and populated the base metadata-driven ETL control tables, begin creating the ADF pipelines by following a few key steps. First, you'll need to design and execute the ADF pipeline to load Azure SQL Database to ADLS Gen2. The next step will be to load ADLS Gen2 to Snowflake. Within the second step of loading ADLS Gen2 to Snowflake, you will learn about a few different options to achieve this ingestion task.

Step 1: Design and Execute an ADF Pipeline to Load Azure SQL Database to Data Lake Storage Gen2

The movement of data from Azure SQL DB to ADLS Gen2 is documented in this section. As a reference, this process has been discussed and demonstrated in detail in Chapter 4 for loading data into a Synapse Analytics dedicated SQL pool, while this chapter focuses on Snowflake as the target data warehouse.

Once the dataset and linked services are created, the following Lookup activity shown in Figure 10-16 will look up the list of tables defined in the control table and pass them to a ForEach loop activity, which will loop over the list of source tables (up to a 50 batch count in parallel at a time). Ensure that the Lookup activity is configured as follows within the General tab.

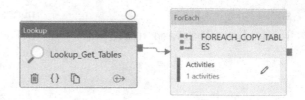

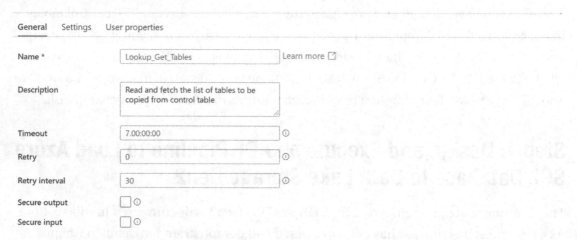

Figure 10-16. *ADF pipeline Lookup and outer ForEach loop activity general properties*

Also ensure that the Lookup activity is configured as shown in Figure 10-17 within the Settings tab.

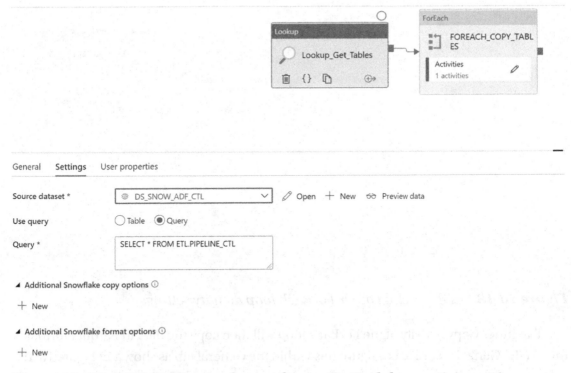

Figure 10-17. *ADF pipeline Lookup and outer ForEach loop activity settings*

Finally, ensure that the outer ForEach loop activity is configured within the Settings tab as shown in Figure 10-18.

Figure 10-18. *ADF pipeline outer ForEach loop activity settings*

The inner Copy activity of the ForEach loop will then copy the data in Parquet format into ADLS Gen2. Be sure to configure this within the General tab as shown in Figure 10-19.

PL_____SQL_TO_LAKE_TO_SNOW > ⋮⊐ FOREACH_COPY_TABLES

General   Source   Sink   Mapping   Settings   User properties

Name *	CP_ASQL_TO_LAKE	Learn more ⬀
Description	Copy data from Azure sql to Gen2	
Timeout	7.00:00:00	ⓘ
Retry	0	ⓘ
Retry interval	30	ⓘ
Secure output	☐ ⓘ	
Secure input	☐ ⓘ	

Figure 10-19. *ADF pipeline Copy activity general connection properties*

Also ensure that the Copy activity's source dataset properties are configured as in Figure 10-20.

Figure 10-20. *ADF pipeline Copy activity source dataset properties*

Here is the code that has been added to the source query section in Figure 10-20:

```
SELECT * FROM @item().SRC_SCHEMA}.@item().DST_TAB_NAME}
```

The last configuration step would be to ensure that the Copy activity's sink dataset properties are configured as in Figure 10-21.

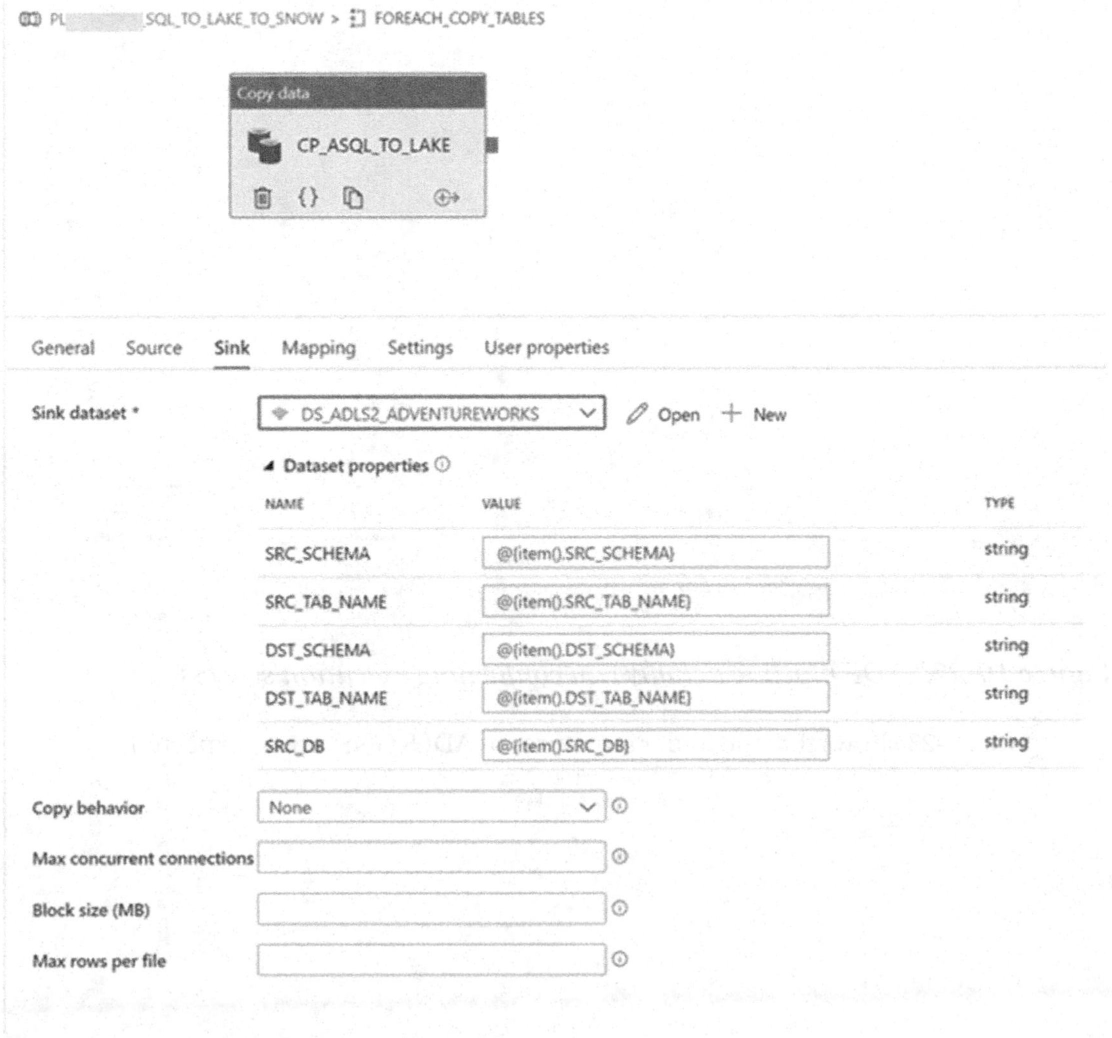

PL░░░░░░SQL_TO_LAKE_TO_SNOW > FOREACH_COPY_TABLES

Copy data

CP_ASQL_TO_LAKE

General   Source   **Sink**   Mapping   Settings   User properties

Sink dataset *   DS_ADLS2_ADVENTUREWORKS   ⌄   ✎ Open   + New

▲ Dataset properties ⓘ

NAME	VALUE	TYPE
SRC_SCHEMA	@{item().SRC_SCHEMA}	string
SRC_TAB_NAME	@{item().SRC_TAB_NAME}	string
DST_SCHEMA	@{item().DST_SCHEMA}	string
DST_TAB_NAME	@{item().DST_TAB_NAME}	string
SRC_DB	@{item().SRC_DB}	string

Copy behavior   None   ⌄ ⓘ

Max concurrent connections [   ] ⓘ

Block size (MB)   [   ] ⓘ

Max rows per file   [   ] ⓘ

Figure 10-21. *ADF pipeline Copy activity sink dataset properties*

Once the pipeline executes successfully, we can see that all tables were successfully loaded to ADLS Gen2 in Parquet format. Figure 10-22 shows the success confirmations from a successful pipeline activity run.

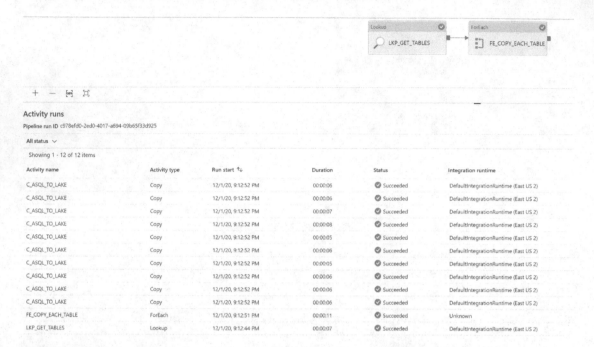

Figure 10-22. *ADF pipeline activity run confirming execution success*

Figure 10-23 shows that the folder structures in ADLS Gen2 are as expected.

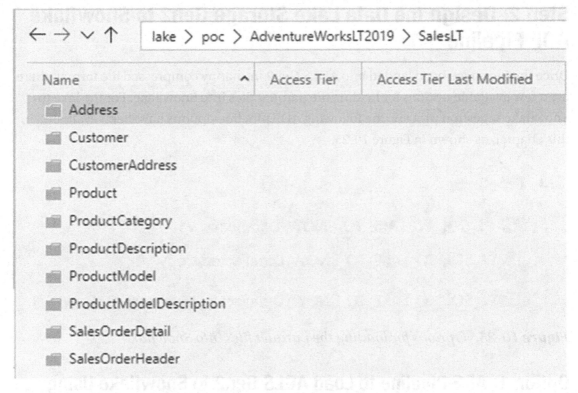

Figure 10-23. *ADLS Gen2 folders created from ADF pipeline*

The files are as expected in ADLS Gen2 shown in Figure 10-24.

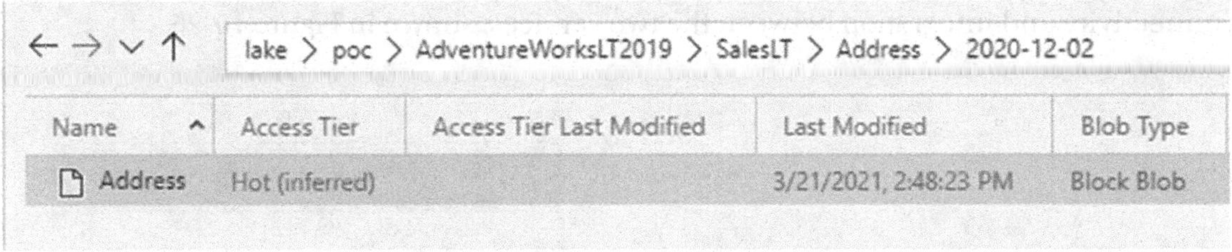

Figure 10-24. *ADLS Gen2 files created from ADF pipeline*

Step 2: Design the Data Lake Storage Gen2 to Snowflake ADF Pipeline

Once the files have been landed into ADLS Gen2 in snappy compressed file format, there are a few available options for loading the parquet files into Snowflake. For the lake to Snowflake ingestion process, the following ADF pipeline options have been evaluated in this chapter, as shown in Figure 10-25.

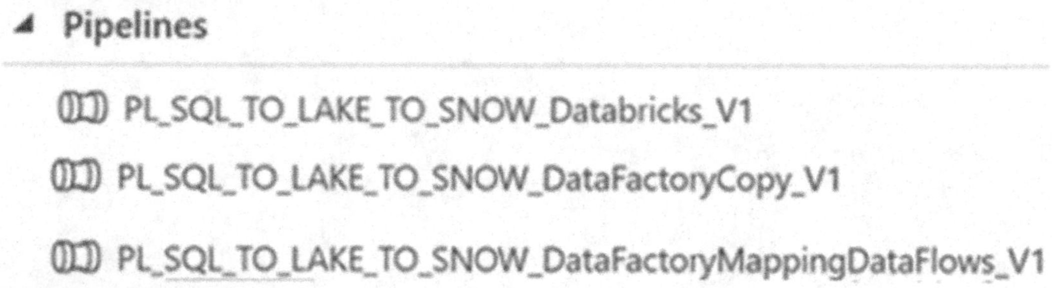

⊿ Pipelines

⬭⬭ PL_SQL_TO_LAKE_TO_SNOW_Databricks_V1

⬭⬭ PL_SQL_TO_LAKE_TO_SNOW_DataFactoryCopy_V1

⬭⬭ PL_SQL_TO_LAKE_TO_SNOW_DataFactoryMappingDataFlows_V1

Figure 10-25. *Options for loading the parquet files into Snowflake*

Option 1: ADF Pipeline to Load ADLS Gen2 to Snowflake Using Azure Databricks

The ADLS Gen2 to Snowflake ADF pipeline that uses Databricks is an option that can pass parameters from Data Factory to a parameterized Databricks notebook and ensures connectivity and integration between the two services, shown in Figure 10-26.

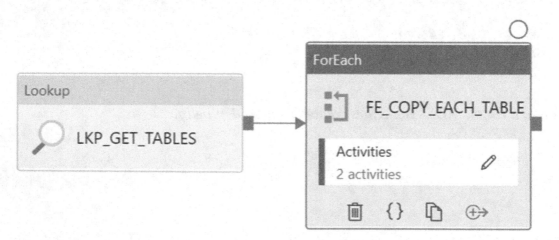

Figure 10-26. *ADF pipeline data flow for option 1*

Within the ForEach loop activity, add the Databricks notebook activity and connect it to the Copy data activity that we ran in a previous section shown in Figure 10-27.

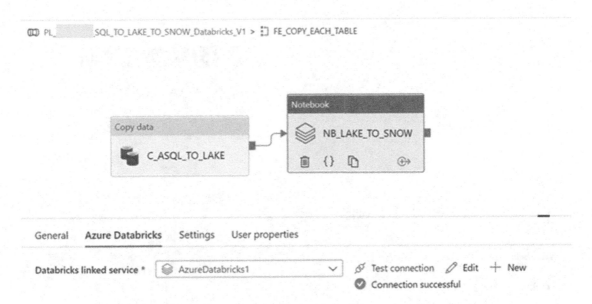

Figure 10-27. *ADF pipeline Azure Databricks connection properties*

Note from the Settings tab in Figure 10-28 that the notebook path references the Databricks notebook containing the code along with the base required parameters for dynamically passing the values to a Databricks notebook for further processing.

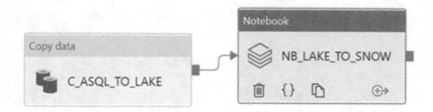

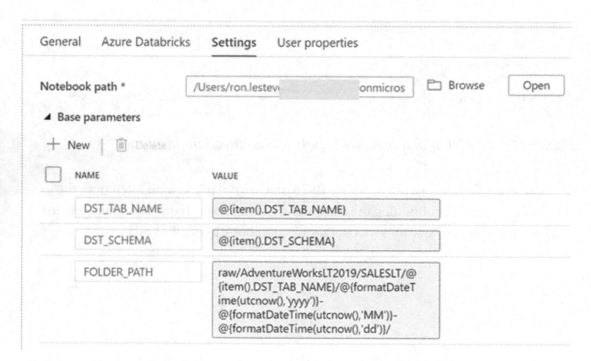

Figure 10-28. *ADF pipeline with Databricks notebook path and parameter settings*

Here is the code that is being used for the base parameters' values in the Settings section of the Databricks notebook shown in Figure 10-28:

```
@item().DST_TAB_NAME}
```

```
@{item().DST_SCHEMA}
```

```
raw/AdventureWorksLT2019/SALESLT/@{item().DST_TAB_NAME}/@{formatDateTime
(utcnow(),'yyyy')}-@{formatDateTime(utcnow(),'MM')}-@{formatDateTime(utcnow
(),'dd')}/
```

Within Databricks, the notebook would contain the following Scala code as shown in Figure 10-29, which accepts the parameters from the ADF Copy activity dynamically and then passes them to a data frame that reads the parquet file based on the dynamic parameters and then writes it to the Snowflake table:

```scala
import org.apache.spark.sql.{SaveMode, SparkSession}
spark.conf.set(
  "fs.azure.account.key.adl001.dfs.core.windows.net",
  "ENTER-ACCOUNT-KEY-HERE"
)

val DST_TAB_NAME = dbutils.widgets.get("DST_TAB_NAME")
val DST_SCHEMA = dbutils.widgets.get("DST_SCHEMA")
val FOLDER_PATH = dbutils.widgets.get("FOLDER_PATH")

var options = Map(
  "sfUrl" -> "px.east-us-2.azure.snowflakecomputing.com",
  "sfUser" -> "USERNAME",
  "sfPassword" -> "PW",
  "sfDatabase" -> "ADVENTUREWORKS",
  "sfSchema" -> DST_SCHEMA,
  "truncate_table" -> "ON",
  "usestagingtable" -> "OFF",
  "sfWarehouse" -> "COMPUTE_WH"
)

val df = spark.read.parquet("abfss://poc@gze2np1adl001.dfs.core.windows.
net/"+FOLDER_PATH+DST_TAB_NAME)

df.write
    .format("snowflake")
    .options(options)
    .option("dbtable", DST_TAB_NAME)
    .mode(SaveMode.Overwrite)
    .save()
```

```
Microsoft Azure | Databricks

lake_to_snow (Scala)

⬧ ● iDFPOC        ⌄ ⬚File ▾   ✏ Edit ▾   ◼ View: Standard ▾   🔒 Permissions   ⊙ Run All   ✦ Clear ▾

Cmd 1

 1  import org.apache.spark.sql.{SaveMode, SparkSession}
 2  spark.conf.set(
 3
 4
 5  )
 6
 7  val DST_TAB_NAME = dbutils.widgets.get("DST_TAB_NAME")
 8  val DST_SCHEMA = dbutils.widgets.get("DST_SCHEMA")
 9  val FOLDER_PATH = dbutils.widgets.get("FOLDER_PATH")
10
11  var options - Map(
12    "sfUrl" -> "   70001.east-us-2.azure.snowflakecomputing.com",
13    "sfUser" -> "     _accountadmin",
14    "sfPassword" -> "
15    "sfDatabase" -> "ADVENTUREWORKS",
16    "sfSchema" -> DST_SCHEMA,
17    "truncate_table" -> "ON",
18    "usestagingtable" -> "OFF ,
19    "sfWarehouse" -> "COMPUTE_WH"
20  )
21
22  val df = spark.read.parquet("abfss:/         poc@gze2npladl001.dfs.core.windows.net/"+FOLDER_PATH+DST_TAB_NAME)
23
24  df.write
25    .format("snowflake")
26    .options(options)
27    .option("dbtable", DST_TAB_NAME)
28    .mode(SaveMode.Overwrite)
29    .save()
30
```

Figure 10-29. *Databricks notebook containing code to load data from ADLS Gen2 to Snowflake, being called from ADF pipeline activity*

Option 2: ADF Pipeline to Load ADLS Gen2 to Snowflake Using ADF Copy Activity

This next ADLS Gen2 to Snowflake ADF pipeline option will use all Data Factory–native tooling using the regular Copy activity. The ADF pipeline's high-level data flow is shown in Figure 10-30.

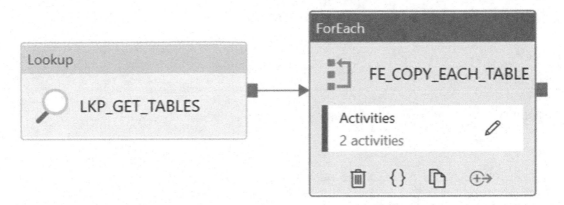

Figure 10-30. *Option 2, which will use Data Factory–native tooling using the regular Copy activity*

Additionally, this option will require a Blob Storage service to be created and configured, and it requires SAS URI authentication only for the Blob linked service connection, for which a sample configuration is shown in Figure 10-31.

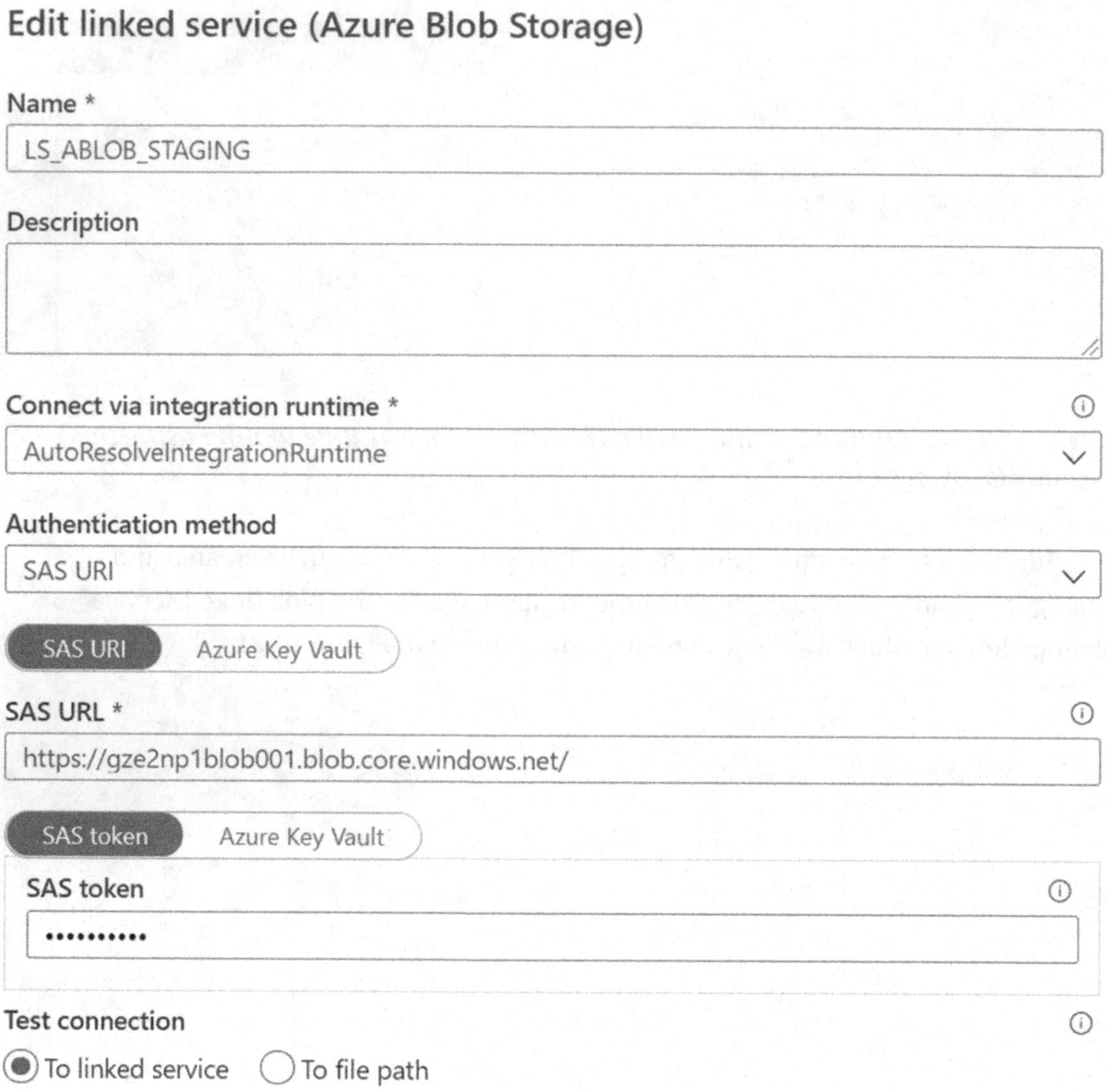

Edit linked service (Azure Blob Storage)

Name *

LS_ABLOB_STAGING

Description

Connect via integration runtime *   ⓘ

AutoResolveIntegrationRuntime   ⌄

Authentication method

SAS URI   ⌄

SAS URI   Azure Key Vault

SAS URL *   ⓘ

https://gze2np1blob001.blob.core.windows.net/

SAS token   Azure Key Vault

SAS token   ⓘ

••••••••••

Test connection   ⓘ

◉ To linked service   ◯ To file path

Figure 10-31. *Blob Storage connection properties*

Ensure that the following Copy data activity source settings have been configured, shown in Figure 10-32.

Figure 10-32. *Copy data source dataset properties*

Figure 10-33 confirms that the following Copy data activity sink settings have been configured correctly.

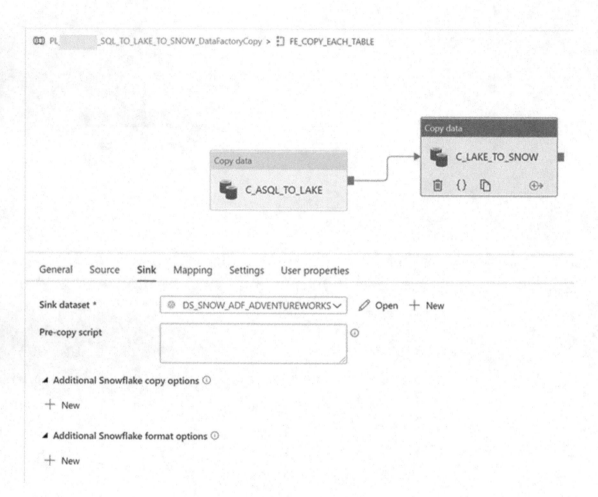

Figure 10-33. *Copy data sink setting details*

Note that in the Settings section in Figure 10-34, staging will need to be enabled and linked to a service.

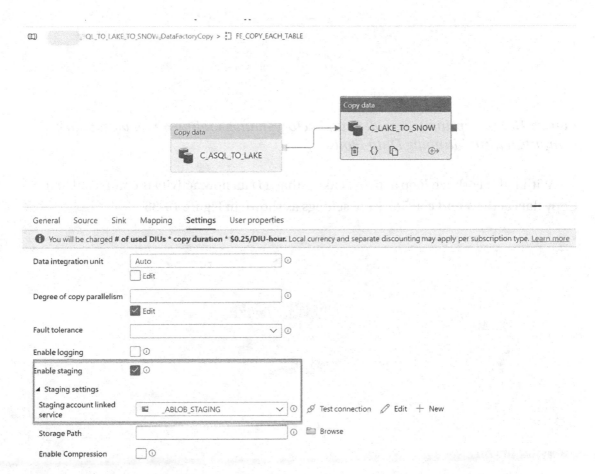

Figure 10-34. *Copy data setting details*

Option 3: ADF Pipeline to Load ADLS Gen2 to Snowflake Using Mapping Data Flows

This ADLS Gen2 to Snowflake ADF pipeline option will use all Data Factory–native tooling along with Spark compute within Mapping Data Flows. Figure 10-35 shows you what this high-level ADF data flow will look like.

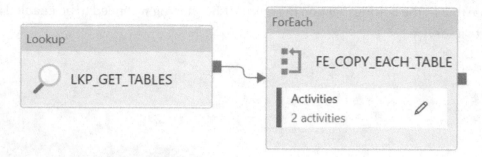

Figure 10-35. *Option to use all Data Factory–native tooling along with Spark compute within Mapping Data Flows*

Within the ForEach loop activity, ensure that a Data flow activity is connected to the Copy data activity and configure the settings as shown in Figure 10-36.

Figure 10-36. *Mapping Data Flows settings*

Within the Mapping Data Flows activity, there is a source connection to the ADLS Gen2 account that will need to be configured, shown in Figure 10-37.

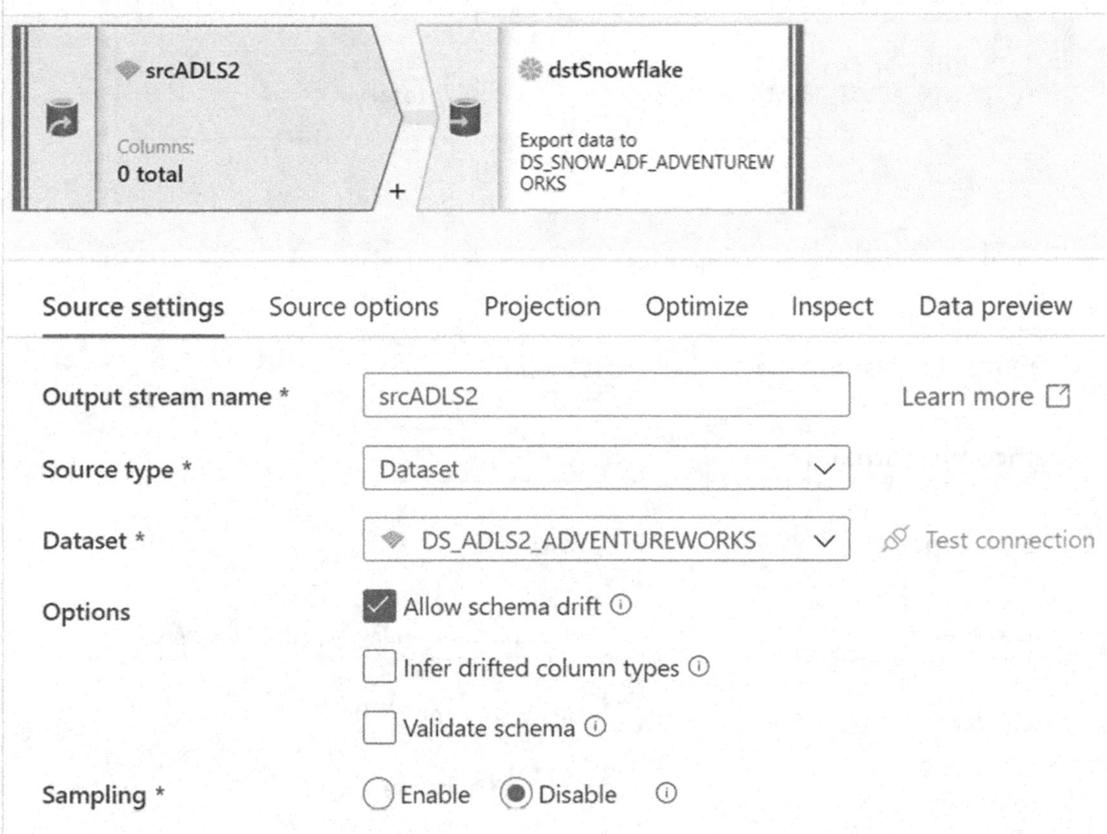

Figure 10-37. *Mapping Data Flows activity source connection to ADLS Gen2*

Also ensure that the destination that references the target Snowflake account is configured as follows in Figure 10-38.

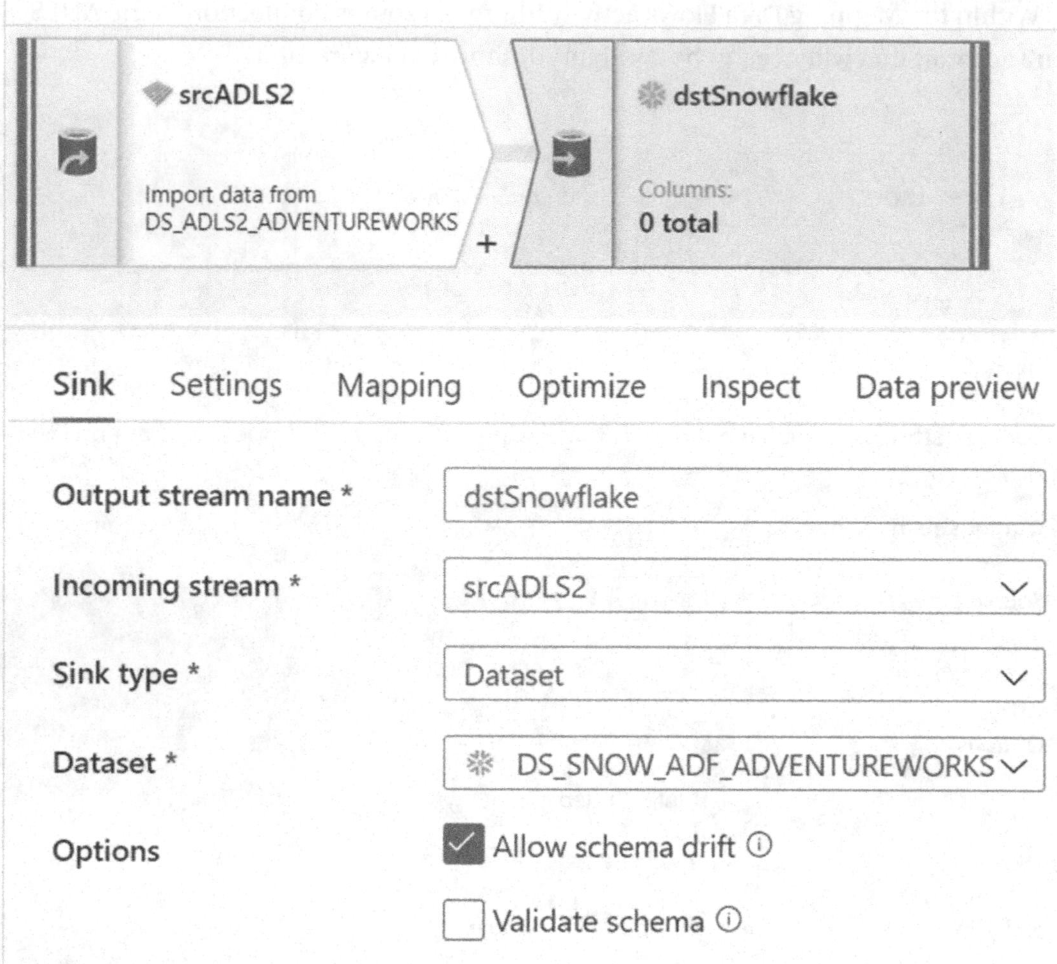

Figure 10-38. *Mapping Data Flows activity sink Snowflake connection properties*

Note that within the destination settings of Figure 10-39, we have specified a "Recreate table" action.

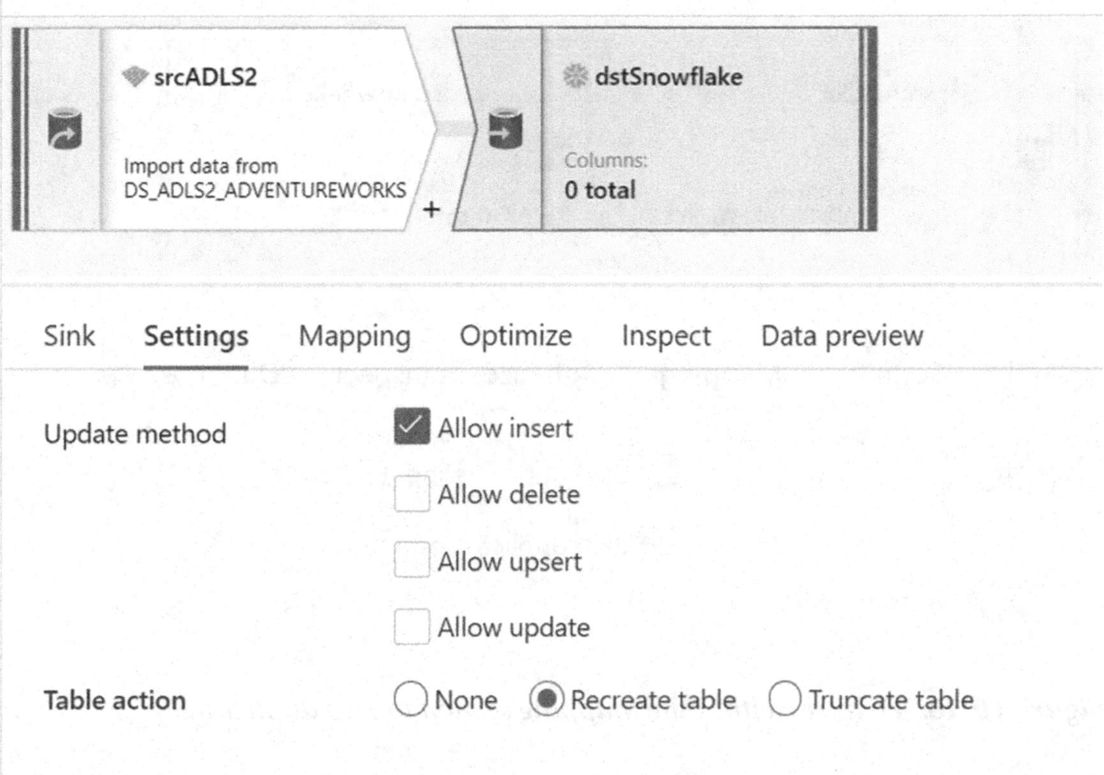

Figure 10-39. *Destination settings specifying "Recreate table" action*

Figure 10-40 shows that there are other valuable features within the Mapping section to skip duplicates.

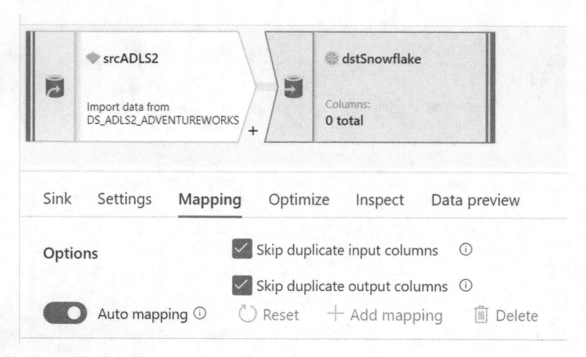

Figure 10-40. *Features within the Mapping section to skip duplicates*

Additionally, Figure 10-41 shows that there are options for optimizing partitioning.

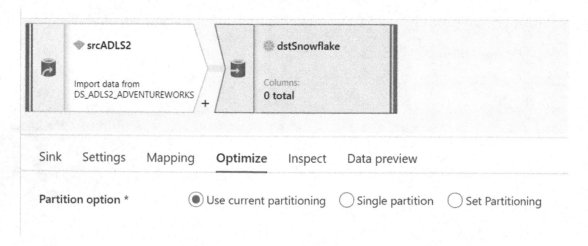

Figure 10-41. *Options for optimizing partitioning*

Comparing the Various ADLS Gen2 to Snowflake Ingestion Options

In the previous sections of this chapter, I have demonstrated a few methods for loading data into Snowflake. In this section, I will compare these various ingestion options to provide a summary of notable pros and cons with these options.

Table 10-1 shows a few noted comparisons of the executed pipelines for all three activities, which compares critical components for a stable, scalable, and successful ingestion pipeline.

Table 10-1. *Comparing the Various ADLS Gen2 to Snowflake Ingestion Options*

	Parameterized Databricks Notebook	**Data Factory Mapping Data Flows**	**Data Factory Copy Activity**
Data type mapping issues	None noted from AdventureWorks sample tables.	Binary data types are causing errors.	Binary data types are causing errors.
Dynamic auto-create table capability	Yes (currently, the auto-create table process is maxing out data types in Snowflake).	Yes (currently, the auto-create table process is maxing out data types in Snowflake).	No (currently, there is no out-of-the-box capability to dynamically auto-create tables).
Dynamic auto-truncate capability	Yes	Yes	Yes
Pipeline activity audit capturing capability	Yes	Yes	Yes
Intermediate blob staging needed	No	No	Yes
SAS URI authentication needed	No	No	Yes

(continued)

211

Table 10-1. (*continued*)

	Parameterized Databricks Notebook	Data Factory Mapping Data Flows	Data Factory Copy Activity
Cluster warm-up time or Running state needed	Yes (approximately 5 minutes for Databricks cluster to warm up; capability to specify core count, compute type, and time to live).	Yes (approximately 5 minutes for Mapping Data Flows cluster to warm up; capability to specify core count, compute type, and time to live).	No
Capability to specify compute and core count	Yes	Yes	N/A (this activity does not utilize Spark compute).
Capability to manage schema drift/evolution	Yes	Yes	N/A
Capability to optimize partitions	Yes	Yes	N/A
Capability to append libraries (jar,egg,wheel)	Yes	Yes	N/A (this activity does not utilize Spark compute, hence no append library capability).
Capability to run parallel activities	Yes	Yes	Yes

Swim Lanes

Since ADF currently has the capability of running 50 tables at a time in parallel through its ForEach loop pipeline activity, if there is a need to run a large number of tables in parallel, you can add a custom attribute to the Snowflake control table called swim_lane and then run the following code that will assign a unique number to a set of tables that can then all be run through one activity set by filtering on the swim_lane column. With this approach, you could have multiple swim lanes using multiple ForEach loops in the ADF pipeline.

The following code will update the table to create the swim_lane column:

```
UPDATE adf_db.etl.pipeline_ctl PL1
SET     swim_lane = 2
FROM    (SELECT src_tab_name,
                Row_number()
                  OVER (
                    ORDER BY src_tab_name ) AS rn
          FROM    adf_db.etl.pipeline_ctl) b
WHERE   PL1.src_tab_name = b.src_tab_name
        AND b.rn > 5
```

Data Validation

It is essential to perform a basic sanity check via validation of the number of rows that were copied from source Azure SQL database server to ADLS Gen2 and from there to the final destination of the Snowflake target set of tables.

This level of detailed information can be captured in a control table within your Snowflake environment and could be named AUDIT_TAB. This table would accept dynamically passed parameters from ADF and the Databricks notebook. Please see Chapters 8 and Chapter 9 for more detail on how to build a robust logging framework in ADF and then combine it with the learnings of this chapter to integrate this logging process with ADF, Databricks and Snowflake.

The following code can be used to create the log table in a Snowflake DW:

```
CREATE OR replace TABLE audit_tab
  (
    pipeline_name    varchar(100),
    src_db_name      varchar(20),
    dest_db_name     varchar(20),
    sch_name         varchar(20),
    table_name       varchar(50),
```

```
  source_count     number(38,0),
  adls_count       number(38,0),
  snowflake_count number(38,0),
  load_time timestamp_ntz(9) DEFAULT CURRENT_TIMESTAMP()
);
```

Summary

In this chapter, I demonstrated a two-step process chained together in a single ADF pipeline to (1) load data from Azure SQL Database to Data Lake Storage Account Gen2 and (2) load data from Data Lake Storage Gen2 into a Snowflake data warehouse. Within part 2 of the process (ADLS Gen2 to Snowflake), I demonstrated three options: Databricks, ADF Copy activity, and Mapping Data Flows. I also introduced the idea of logging ADF pipeline metrics into a Snowflake table by leveraging the learnings from Chapters 8 and 9. Finally, I compared the three options to capture pros and cons of each option.

CHAPTER 11

Mapping Data Flows for Data Warehouse ETL

The advent of Azure's numerous data services has triggered both excitement and confusion in the data warehousing space for several reasons. Customers are interested in the concept of a modern cloud-based data warehouse but may be overwhelmed by the overabundance of cloud-based services available on the market. They are interested to know how they could get started with implementing traditional data warehousing concepts such as dimensional modeling, Slowly Changing Dimensions (SCDs), Star Schemas, and data warehouse ETL or ELT in the cloud. Traditional on-premises Microsoft technologies such as SQL Server Integration Services (SSIS) are mature data integration tools that have been prevalent for over a decade. As more cloud-based tools are becoming available, customers are interested in either migrating their ETL to the cloud or simply getting started with understanding data warehousing in the cloud.

Microsoft continues to expand their service offerings within Azure Data Factory with Mapping Data Flows, which allows for visual and code-free data transformation logic that is executed as activities with Azure Data Factory pipelines using scaled-out Azure Databricks clusters. Essentially this offering brings Azure Data Factory one step closer to Microsoft's traditional SQL Server Integration Services, which has been used for data warehouse ETL for many years. Mapping Data Flows has tremendous potential in the data warehousing space for several data warehousing patterns such as Slowly Changing Dimension Type I and Type II and fact extraction, transformation, and data loading.

In this chapter, I will discuss the typical data warehousing load pattern known as Slowly Changing Dimension Type I to demonstrate a practical example of how Azure Data Factory's Mapping Data Flows can be used to design this data flow pattern.

© Ron C. L'Esteve 2021
R. C. L'Esteve, *The Definitive Guide to Azure Data Engineering*, https://doi.org/10.1007/978-1-4842-7182-7_11

Modern Data Warehouse

Data Factory plays a key role in the modern data warehouse landscape since it integrates well with both structured, unstructured, and on-premises data. More recently, it is beginning to integrate quite well with Data Lake Storage Gen2 and Azure Databricks as well. The diagram in Figure 11-1 does a good job of depicting where Data Factory fits in the modern data warehouse landscape.

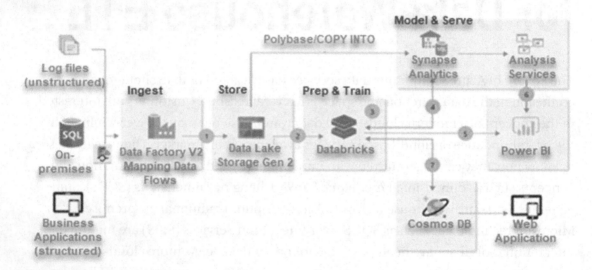

Figure 11-1. *Modern Azure Data Platform data flow architecture diagram*

As you can see from Figure 11-1, Data Factory is a key integrator between source, target, and analytical systems. Additionally, by adding a code-free graphical user-based interface such as Mapping Data Flows that utilizes Spark clusters under the hood, Data Factory is sure to play a key role in the design and development of the modern data warehouse.

Creating the Base Azure Data Resources

For the purposes of this chapter's sample exercise, you will need to create a source OLTP database along with a target OLAP database that contains transformed dimensions. In Chapter 4, I described how to obtain sample data such as the AdventureWorksLT tables and data for free that you can then load to an Azure SQL Database. This exercise will leverage the same AdventureWorksLT database as the OLTP_Source. Once the database is created in Azure, notice in Figure 11-2 that it contains the SalesLT.Customer table that you will work with in this exercise.

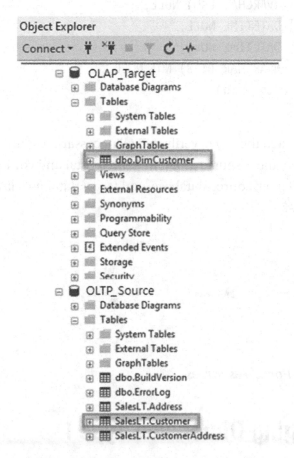

Figure 11-2. *Source OLTP and destination OLAP databases*

Create the following DimCustomer table in the OLAP_Target database. The schema of DimCustomer is as follows:

```
CREATE TABLE dbo.dimcustomer
    (
        [customerid]    INT NOT NULL,
        [title]         NVARCHAR (8) NULL,
        [firstname]     [dbo].[NAME] NOT NULL,
        [middlename]    [dbo].[NAME] NULL,
        [lastname]      [dbo].[NAME] NOT NULL,
```

```
    [suffix]         NVARCHAR (10) NULL,
    [companyname]    NVARCHAR (128) NULL,
    [salesperson]    NVARCHAR (256) NULL,
    [emailaddress]   NVARCHAR (50) NULL,
    [inserteddate]   DATETIME NULL,
    [updateddate]    DATETIME NULL,
    [hashkey]        NVARCHAR (128) NOT NULL,
    PRIMARY KEY (customerid)
);
```

Once you have created the table, verify that all necessary resources have been created by navigating to the resource group in Azure Portal and confirming that a Data Factory, Logical SQL Server, source database, and target database have all been created, as shown in Figure 11-3.

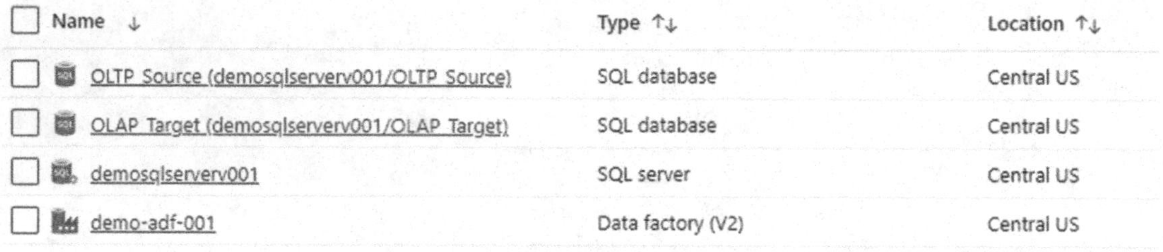

Name ↓	Type ↑↓	Location ↑↓
☐ 📁 OLTP Source (demosqlserverv001/OLTP Source)	SQL database	Central US
☐ 📁 OLAP Target (demosqlserverv001/OLAP Target)	SQL database	Central US
☐ 📁 demosqlserverv001	SQL server	Central US
☐ 📁 demo-adf-001	Data factory (V2)	Central US

Figure 11-3. *Azure Portal resources*

Slowly Changing Dimension Type I

Slowly Changing Dimensions are commonly used advanced techniques for dimensional data warehouses and are used to capture changing data within the dimension over time. The exercise in this chapter will leverage Slowly Changing Dimension Type I within the Data Factory pipeline.

While there are numerous types of Slowly Changing Dimensions, Type I, which I will cover in this chapter, simply overwrites the existing data values with new values. The advantage of this method is that it makes updating the dimension easy and limits the growth to only new records. The disadvantage is that historical data will be lost since the dimension will always only contain current values for each attribute.

Create a Data Factory Pipeline and Datasets

Begin creating the ADF pipeline for a Slowly Changing Dimension Type I ETL pattern by clicking "Author & Monitor" to launch the Data Factory console, as shown in Figure 11-4.

Figure 11-4. *Launch Azure Data Factory by clicking Author & Monitor*

Next, create a new pipeline by clicking "Create pipeline," shown in Figure 11-5.

Figure 11-5. *Create a new ADF pipeline*

Name the new pipeline DW ETL, which will contain the following two datasets:

- **AzureSqlCustomerTable:** This is the OLTP Azure SQL source database that contains the AdventureWorksLT tables. Specifically, use the Customer table for the ETL.

- **AzureSqlDimCustomerTable:** This is the OLAP Azure SQL target database that contains the Dimension Customer table. This dimension table will vary from the OLTP source table in that it contains fewer fields and contains an InsertedDate, UpdatedDate, and HashKey, which you will use for the SCD Type I ETL pattern.

After you create the pipeline and datasets, the Factory Resources section will be populated as shown in Figure 11-6.

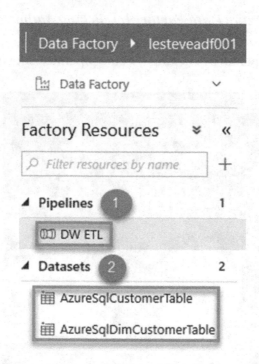

Figure 11-6. *ADF pipeline and datasets*

Create a Data Factory Mapping Data Flow

Now that you have created the ADF pipeline and datasets for both source and target, you're ready to create your data flow using SCD Type I. Start by dragging the data flow from Activities onto the Data Factory canvas shown in Figure 11-7.

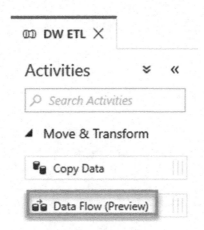

Figure 11-7. *ADF Data Flow activity*

Then give your data flow an intuitive name for what it is. In this case, it would be DimCustomer, shown in Figure 11-8.

Figure 11-8. *ADF Data Flow activity for DimCustomer*

Prior to creating the data flow, be sure to turn on Data Flow Debug mode (https://docs.microsoft.com/en-us/azure/data-factory/concepts-data-flow-debug-mode) since you will be testing out the pipeline prior to deploying it into production. Figure 11-9 shows Data Flow Debug mode, which will be sufficient for the tests. When you toggle on the Debug mode, you will be asked if you wish to proceed. Click OK to continue.

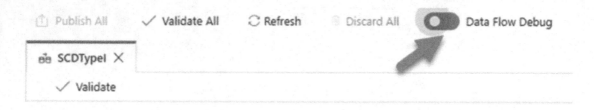

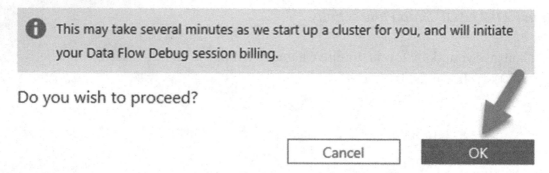

Figure 11-9. *ADF Data Flow Debug mode*

Figure 11-10 shows that the clusters will take a few minutes to get ready, typically anywhere between 5 and 7 minutes to be precise.

ADF introduced a new "Quick re-use" feature in the Azure integration runtime for data flow time to live (TTL). With TTL enabled, ADF can maintain the Spark cluster for the period of time after your last data flow executes in a pipeline, which will provide faster consecutive executions using that same Azure IR in your data flow activities.

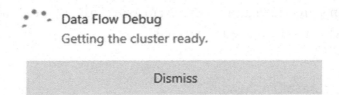

Figure 11-10. ADF Data Flow Debug mode cluster is getting ready

Once Data Flow Debug mode is turned on, a green dot, shown in Figure 11-11, will appear to the right of it to confirm that it is ready to begin a Debug mode pipeline run. At this point, you can start creating the Slowly Changing Dimension Type I data flow.

Figure 11-11. ADF Data Flow Debug mode cluster is ready

Begin the data flow by adding and configuring the two datasets as sources shown in Figure 11-12.

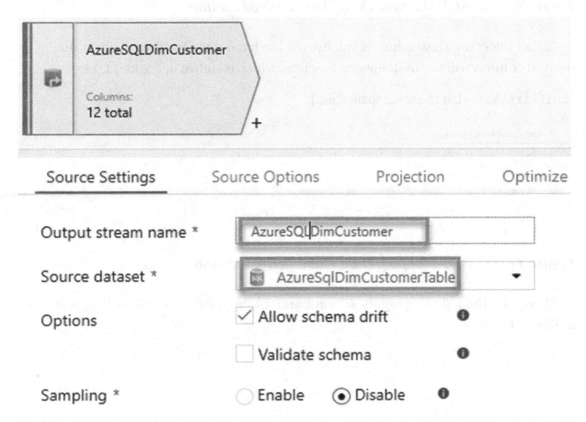

Figure 11-12. ADF Mapping Data Flows Source Settings

Click the + icon shown in Figure 11-13 next to `AzureSqlCustomerTable` to add a derived column, which will be called `CreateHash`.

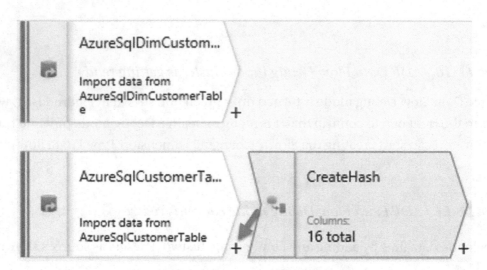

Figure 11-13. *ADF Mapping Data Flows derived column*

Then select the Hash columns and use the Hash function. For this scenario, the derived column will use the following function, which is shown in Figure 11-14:

`sha1(FirstName+LastName+CompanyName)`

Derived Column's Settings	Optimize	Inspect	Data Preview ●
Output stream name *	CreateHash		
Incoming stream *	AzureSqlCustomerTableNoHash		
Columns * ●	HashKey ▼		sha1(FirstName+LastName+CompanyName)

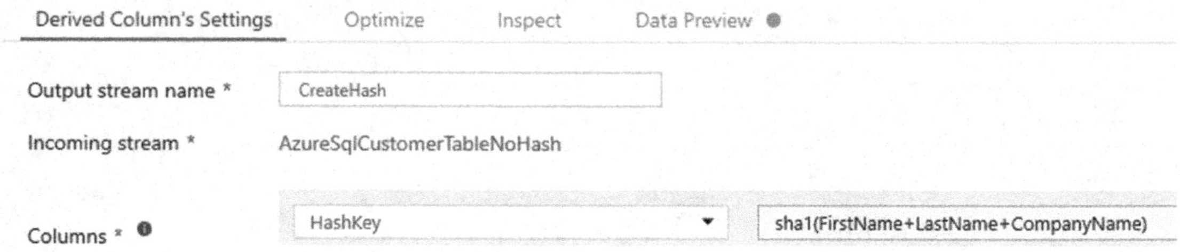

Figure 11-14. *ADF Mapping Data Flows derived column settings*

Then add the following inputs, schema, and row modifiers to the data flow, as shown in Figure 11-15.

Figure 11-15. *ADF Mapping Data Flows activities for SCD Type 1 in DimCustomer*

The following subsections describe each of the steps in Figure 11-15.

Exists

The Exists step will check if the hash key already exists by comparing the newly created source hash key to the target hash key shown in Figure 11-16.

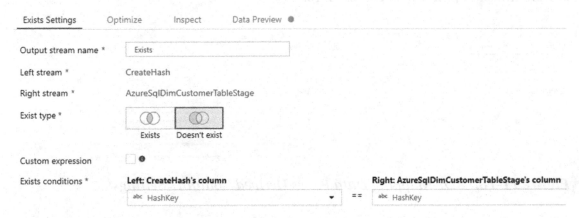

Figure 11-16. *ADF Mapping Data Flows Exists Settings*

LookupDates

Then the LookupDates step will join the CustomerID from the source to the target to ensure that pertinent records are included, as shown in Figure 11-17.

Figure 11-17.   ADF Mapping Data Flows Lookup Settings

SetAttributes

The SetAttributes step will add two derived columns. InsertedDate will insert the current timestamp if it is null, and UpdatedDate will always update the row with the current timestamp, shown in Figure 11-18.

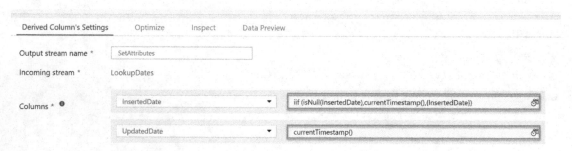

Figure 11-18.   ADF Mapping Data Flows Derived Column's Settings

Here is the code that has been used in the Columns fields for InsertedDate and UpdatedDate in Figure 11-18:

```
iif(isNull(InsertedDate),currentTimestamp(),{InsertedDate})
```

```
currentTimestamp()
```

AlterRows

AlterRows allows for upserts, with a condition set to true(), which will update or insert everything that passes through the flow, as shown in Figure 11-19.

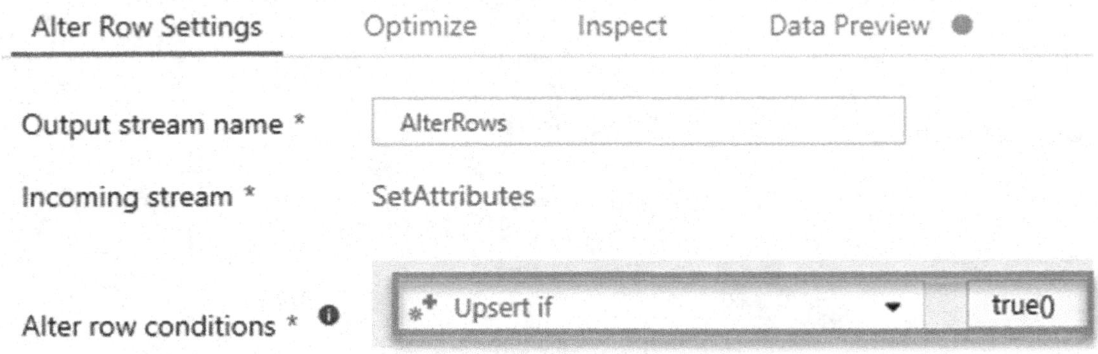

Alter Row Settings   Optimize   Inspect   Data Preview ●

Output stream name *   AlterRows

Incoming stream *   SetAttributes

Alter row conditions * ●   *⁺ Upsert if   ▼   true()

Figure 11-19. *ADF Mapping Data Flows Alter Row Settings*

sink1

Finally, the sink step will write back to the DimCustomer table shown in Figure 11-20.

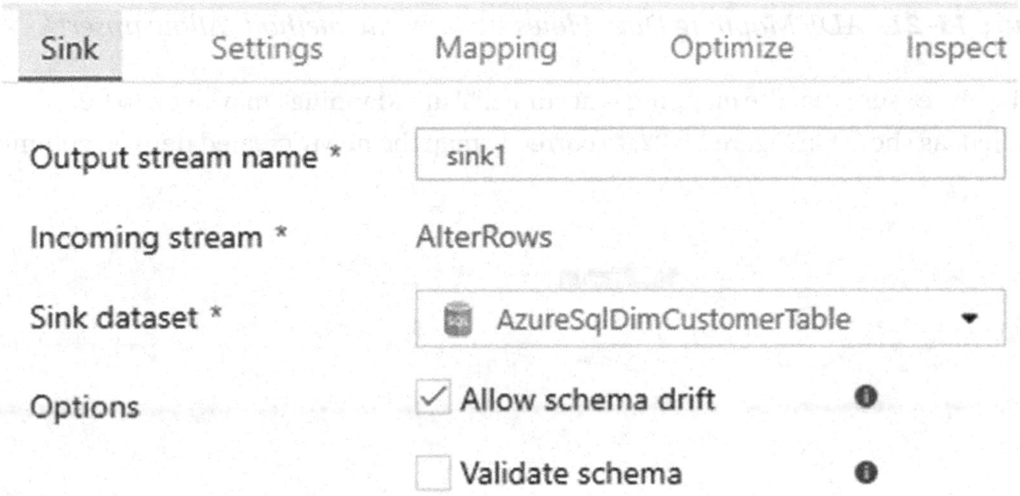

Sink   Settings   Mapping   Optimize   Inspect

Output stream name *   sink1

Incoming stream *   AlterRows

Sink dataset *   ▦ AzureSqlDimCustomerTable   ▼

Options   ☑ Allow schema drift   ●

                 ☐ Validate schema   ●

Figure 11-20. *ADF Mapping Data Flows sink settings*

Within the Settings tab shown in Figure 11-21, ensure that you check "Allow upsert" with CustomerID specified as the key column.

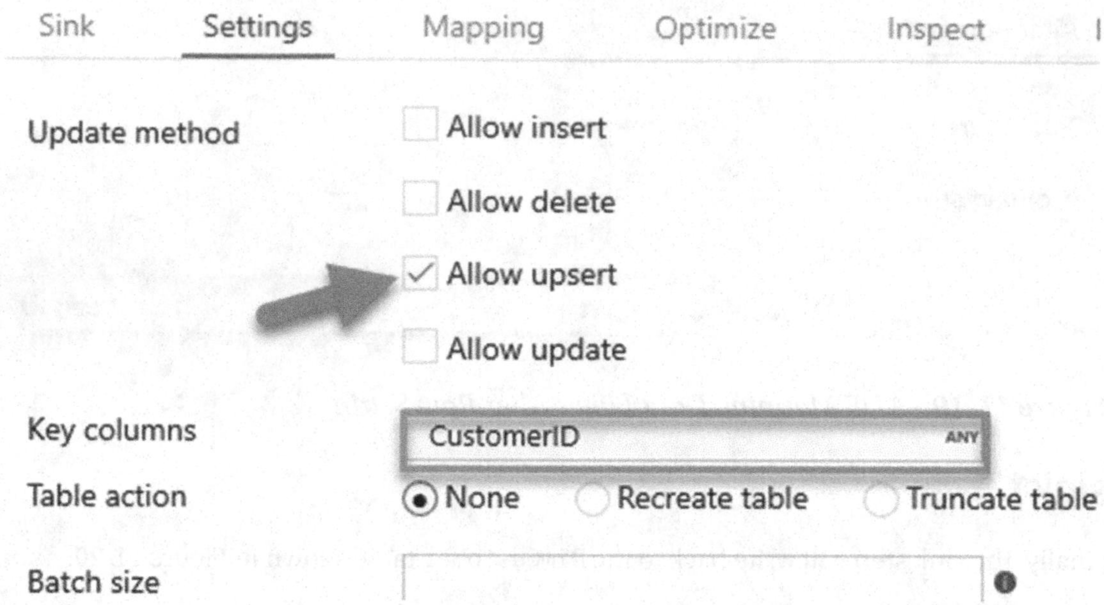

Figure 11-21. *ADF Mapping Data Flows sink update method "Allow upsert"*

Lastly, ensure that the mapping is accurate. "Auto Mapping" may need to be disabled, as shown in Figure 11-22, to correctly map the newly created derived columns.

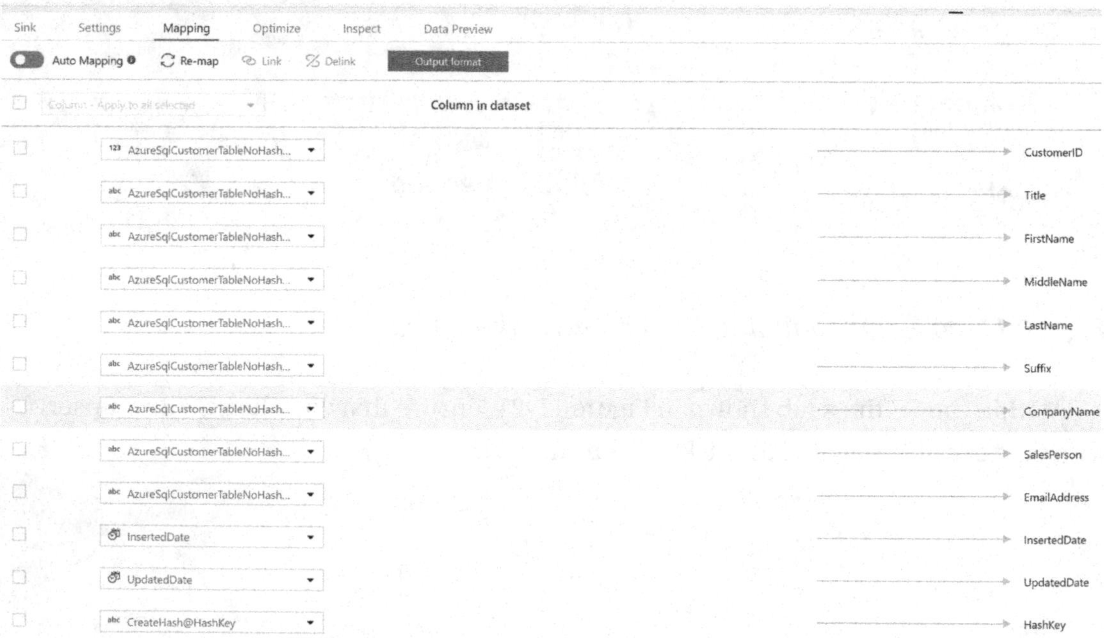

Figure 11-22. *ADF Mapping Data Flows sink mappings*

Once you complete the Slowly Changing Dimension Type I data flow, it will look like the illustration in Figure 11-23.

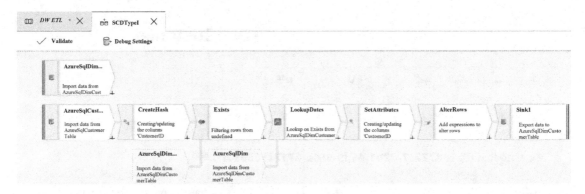

Figure 11-23. *ADF Mapping Data Flows completed SCD Type 1 in DimCustomer MDF pipeline*

The process of designing and configuring an SCD Type I data flow for `DimCustomer` is now complete. Since Debug mode has been switched on, simply click "Debug" in Figure 11-24 within the pipeline view and wait for the pipeline to complete running.

Figure 11-24. *ADF Mapping Data Flows pipeline run in Debug mode*

Once the pipeline completes running, notice a green check circle to the right along with the pipeline success status. Since there was no data in the `DimCustomer` table, this pipeline run loaded all records to the `DimCustomer` table, as shown in Figure 11-25.

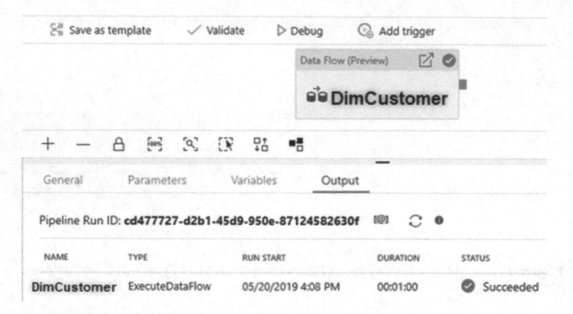

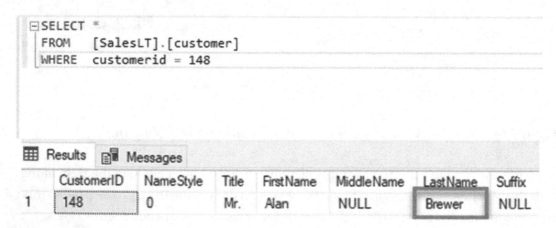

Figure 11-25. *ADF Mapping Data Flows Debug mode pipeline success output*

Updating a Record

To test the functionality of your pipeline, pick a record from the OLTP system as shown in Figure 11-26. It can be any record, and for example purposes, let's assume that you pick Alan Brewer's record.

```
SELECT  *
FROM    [SalesLT].[customer]
WHERE   customerid = 148
```

	CustomerID	NameStyle	Title	FirstName	MiddleName	LastName	Suffix
1	148	0	Mr.	Alan	NULL	Brewer	NULL

Figure 11-26. *Selecting a record from the OLTP source customer table*

Here is the SQL code that has been executed in Figure 11-26:

```
SELECT *
FROM    [SalesLT].[customer]
WHERE   customerid = 148
```

Then update the LastName for CustomerID 148 from Brewer to Update as shown in Figure 11-27.

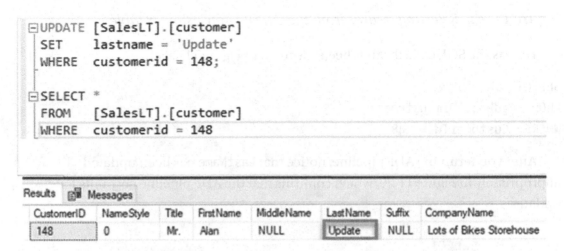

Figure 11-27. *Updating a record from the OLTP source customer table*

Here is the SQL code that has been executed in Figure 11-27:

```
UPDATE [SalesLT].[customer]
SET     lastname = 'Update'
WHERE   customerid = 148;

SELECT *
FROM    [SalesLT].[customer]
WHERE   customerid = 148
```

Prior to rerunning the ADF pipeline, run a select * query on the DimCustomer table and notice that the LastName is still "Brewer," shown in Figure 11-28.

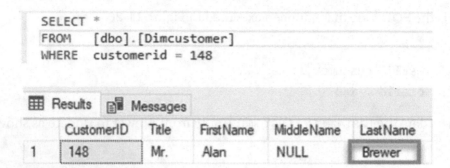

Figure 11-28. *Selecting records from the OLAP DimCustomer table*

Here is the SQL code that has been executed in Figure 11-28:

```
SELECT *
FROM    [dbo].[Dimcustomer]
WHERE   CustomerId = 148
```

After you rerun the ADF pipeline, notice that LastName has been updated appropriately in Figure 11-29, which confirms that the ADF pipeline accounts for updates.

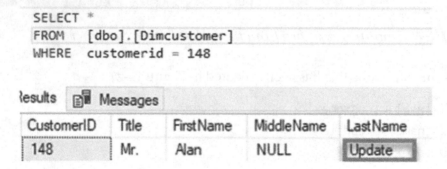

Figure 11-29. *Updating records in the OLAP DimCustomer table*

Here is the SQL code that has been executed in Figure 11-29:

```
SELECT *
FROM    [dbo].[DimCustomer]
```

Inserting a Record

Also test inserting a new record into the OLTP system and then rerun the ADF pipeline to see if the insert is picked up. To that end, execute the USE and INSERT statements in the following example. Execute the following query to insert a new record into the [SalesLT].[Customer] table:

```
USE [OLTP_Source]

go

INSERT INTO [SalesLT].[customer]
           ([namestyle],
            [title],
            [firstname],
            [middlename],
            [lastname],
            [suffix],
            [companyname],
            [salesperson],
            [emailaddress],
            [phone],
            [passwordhash],
            [passwordsalt],
            [rowguid],
            [modifieddate])
VALUES     (1,
            'Mr.',
            'John',
            'B',
            'Doe',
            NULL,
            'MSSQLTips',
            NULL,
            NULL,
            NULL,
            'L/Rlwxzp4w7RWmEgXX+/A7cXaePEPcp+KwQhl2fJL7d=',
```

```
          '1KjXYs4=',
          '2005-08-01 00:00:00:000')
```

go

Then check to make sure that you see results like those in Figure 11-30. You should
see the record that you have just inserted.

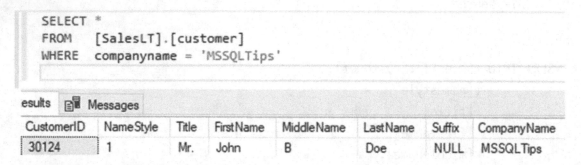

Figure 11-30. *Selecting inserted records in the OLTP customer table*

Here is the SQL code that has been executed in Figure 11-30:

```
SELECT *
FROM    [SalesLT].[customer]
WHERE   companyname = 'MSSQLTips'
```

After you rerun the ADF pipeline, notice in Figure 11-31 that the new OLTP record
has also been inserted into the [dbo].[DimCustomer] table, which confirms that the
ADF pipeline accounts for inserts.

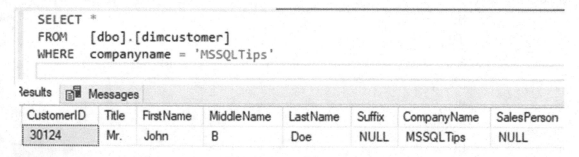

Figure 11-31. *Selecting inserted records in the OLAP DimCustomer table*

Here is the SQL code that has been executed in Figure 11-31:

```
SELECT *
FROM   [dbo].[dimcustomer]
WHERE  companyname = 'MSSQLTips'
```

Remember to turn off "Data Flow Debug" mode when finished to prevent unnecessary costs for an unused feature within Data Factory. Also, remember to clean up and delete any unused resources in your resource group as needed, as shown in Figure 11-32.

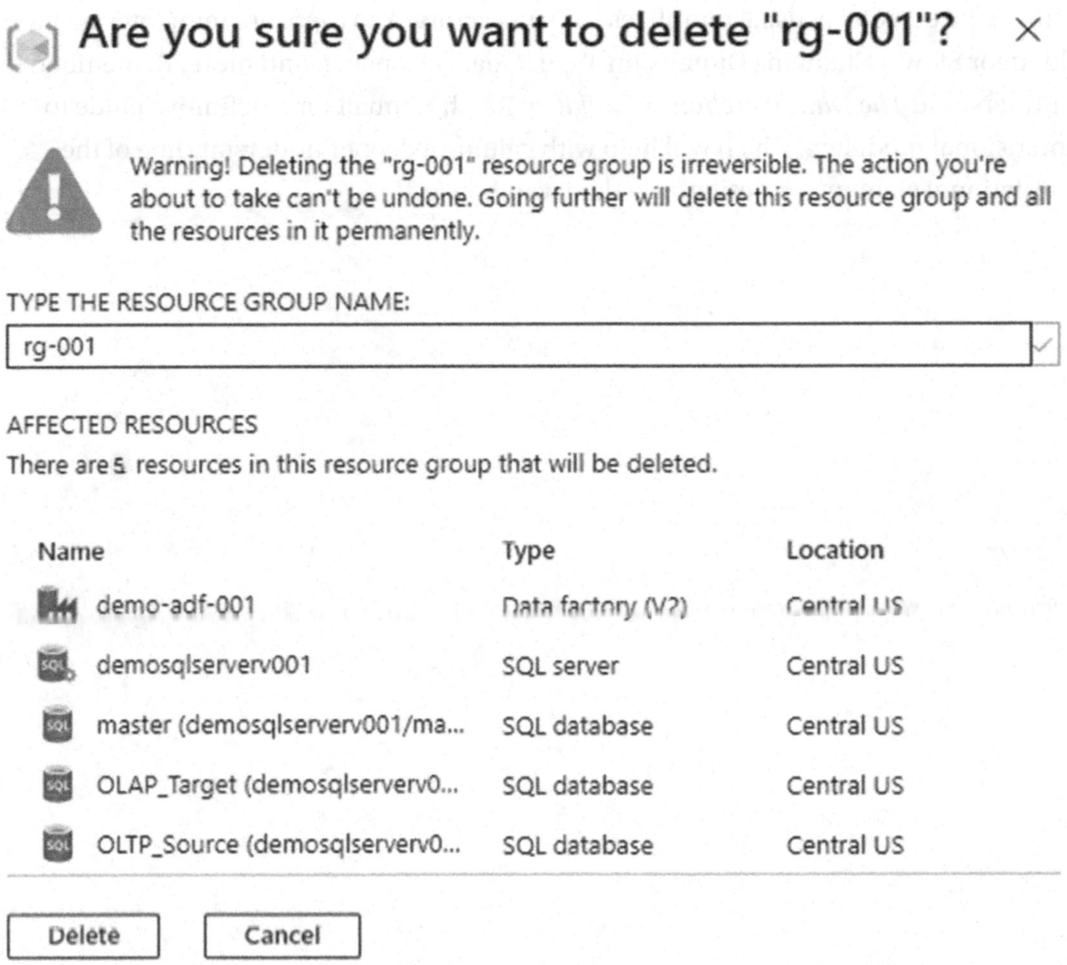

 # Are you sure you want to delete "rg-001"?   ×

 Warning! Deleting the "rg-001" resource group is irreversible. The action you're about to take can't be undone. Going further will delete this resource group and all the resources in it permanently.

TYPE THE RESOURCE GROUP NAME:

rg-001

AFFECTED RESOURCES

There are 5 resources in this resource group that will be deleted.

Name	Type	Location
demo-adf-001	Data factory (V2)	Central US
demosqlserverv001	SQL server	Central US
master (demosqlserverv001/ma...	SQL database	Central US
OLAP_Target (demosqlserverv0...	SQL database	Central US
OLTP_Source (demosqlserverv0...	SQL database	Central US

Delete   Cancel

Figure 11-32. *Delete unused resources and/or resource group as needed*

Summary

In this chapter, I discussed the modern data warehouse along with Azure Data Factory's Mapping Data Flows and its role in this landscape. You learned how to set up your source, target, and Data Factory resources to prepare for designing a Slowly Changing Dimension Type I ETL pattern by using Mapping Data Flows. Additionally, you learned how to design and test a Slowly Changing Dimension Type I data flow and pipeline within Azure Data Factory.

There are a few different types of Slowly Changing Dimensions that can be used for different purposes. Microsoft's Azure Data Factory offers a number of methods for getting started quickly using templates in Data Factory, and there are templates being added for Slowly Changing Dimension Type II, data cleansing, and more. Remember to understand *The Data Warehouse Toolkit* by Ralph Kimball for a definitive guide to dimensional modeling, which will help with gaining a deeper understanding of the foundational elements to designing a data warehouse.

## CHAPTER 12

Aggregate and Transform Big Data Using Mapping Data Flows

The process of cleansing and transforming big datasets in the data lake has become an increasingly popular and critical step in a modern enterprise's data architecture. Microsoft has introduced several big data analytics and orchestration tools to serve the need for big data lake Extract-Load-Transform (ELT). Customers are seeking cloud-based services that can cleanse, transform, and aggregate extremely big datasets with ease, coupled with a low learning curve. They are seeking to understand what tools and technologies could potentially fit the bill for big data lake cleansing and transformations.

Azure Data Factory's Mapping Data Flows has become a promising solution for big data lake cleansing and transformations. In Chapter 11, I discussed the concept of a modern data warehouse and demonstrated a practical example of Mapping Data Flows for enterprise data warehouse transformations. In this chapter, I will continue to demonstrate additional data cleansing and aggregation features of Mapping Data Flows, specifically to process big data files stored in Azure Data Lake Storage Gen2 as hierarchical files.

Add Files and Folders to Azure Data Lake Storage Gen2

Structuring a Data Lake Storage correctly by using best practices is key. When data is stored in Data Lake Storage Gen2, the file size, number of files, and folder structure have an impact on performance.

© Ron C. L'Esteve 2021
R. C. L'Esteve, *The Definitive Guide to Azure Data Engineering*, https://doi.org/10.1007/978-1-4842-7182-7_12

File Size

Depending on what services and workloads are using the data, a good size to consider for files is 256 MB or greater. If the file sizes cannot be batched when landing in Data Lake Storage, you can have a separate compaction job that combines these files into larger ones.

Folder Structure

The folder and file organizational structure can help some queries read only a subset of the data, which improves performance by optimizing for the larger file sizes and a reasonable number of files in each folder. Be cognizant of performance tuning and optimization techniques, along with folder and file structure recommendations.

For this exercise, you will learn how to create an ADLS Gen2 container named **lake**, along with a few additional folders that will organize the data by the year 2016. To get started, ensure that you have an ADLS Gen2 account and ADF account set up. Next, you'll need to upload the 2016 sales files from the following GitHub account (`https://github.com/ronlesteve/sales-datasets`) into the ADLS Gen2 folder structure shown in Figure 12-1.

As we can see from Figure 12-1, within the 2016 sales folder, the additional folders are organized by month number.

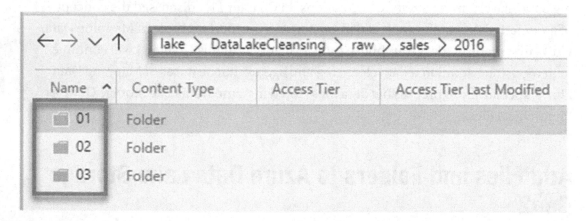

Figure 12-1. *2016 sales folders organized by month number*

Within each month, there are **.txt** files organized and saved by day, as shown in Figure 12-2.

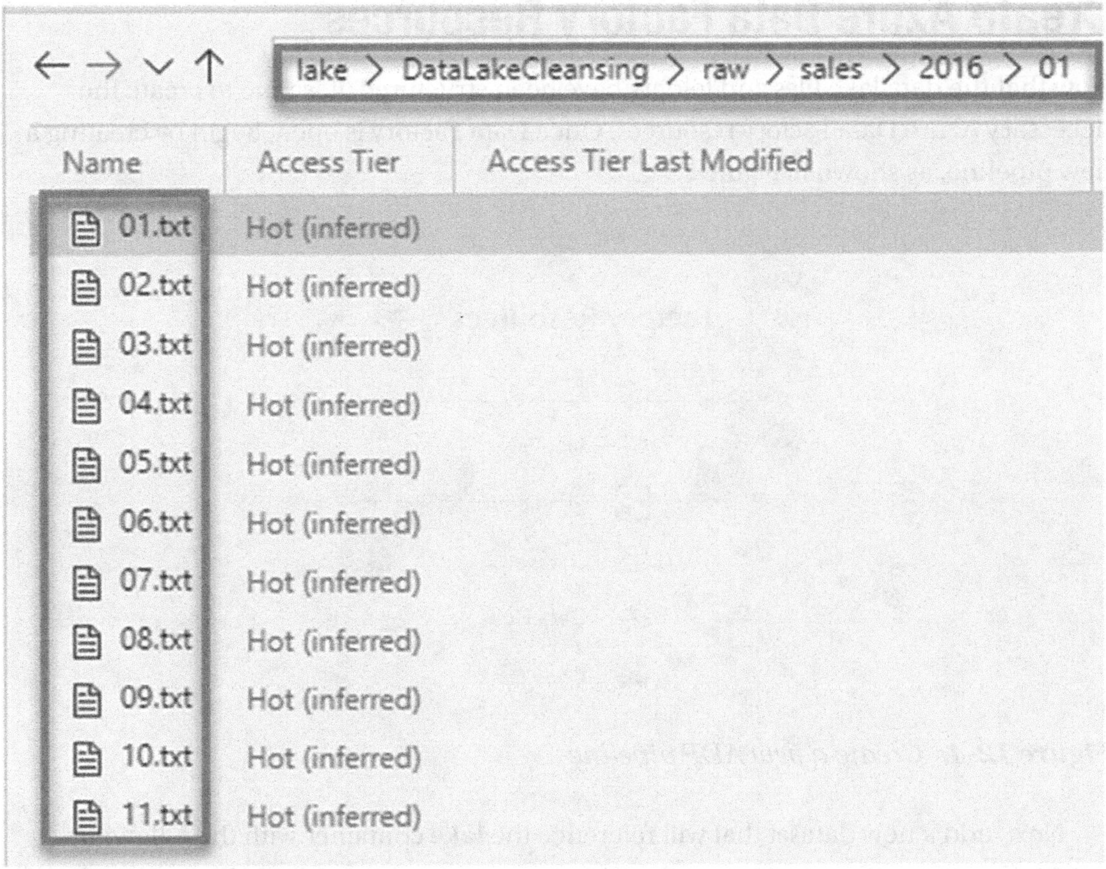

Figure 12-2. *2016 sales folders organized by month day text files*

And finally, upon opening one of the text files, notice the structure of the data in Figure 12-3 consisting of the following columns selected.

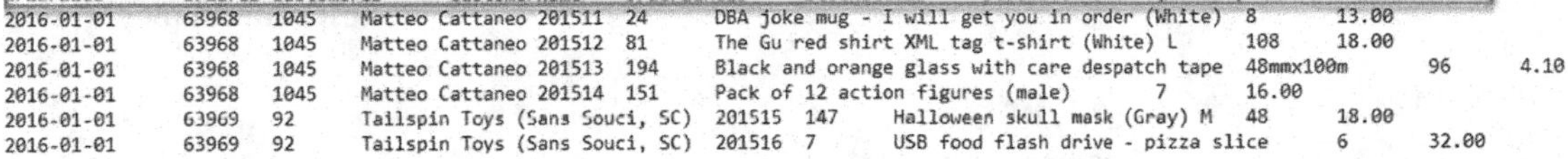

Figure 12-3. *Structure of the text files*

Create Azure Data Factory Resources

Now that the data lake files and folders have been structured, it is time to create the necessary Azure Data Factory resources. Once Data Factory is open, begin by creating a new pipeline, as shown in Figure 12-4.

Figure 12-4. *Create a new ADF pipeline*

Next, add a new dataset that will reference the **lake** container with the following connection properties. Notice that the Directory and File properties in Figure 12-5 have been left empty as this can be dynamically set in the Mapping Data Flows properties. Also, set the column delimiter to **Tab(\)**.

| General | Connection ① | Schema | Parameters |

Linked service * ②	AzureDataLakeStorageGen2 ▼		Test connection
File path * ③	datalake	/ Directory	/ File
	Add dynamic content [Alt+P]		
Compression type	none ▼		
Column delimiter ④	Tab (\t) ▼		

Figure 12-5. *ADLS Gen2 dataset connection properties*

After you publish the resources, the Factory Resources section in Figure 12-6 will be available. These resources consist of pipelines and datasets.

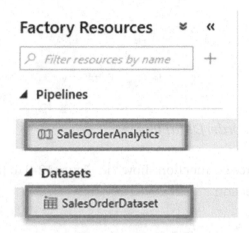

Figure 12-6. *ADF pipeline and dataset resources*

Within the newly created pipeline, expand **Move & Transform** from Activities and then drag the **Data Flow activity**, shown in Figure 12-7, onto the canvas.

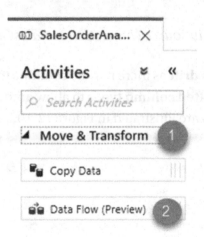

Figure 12-7. *ADF Data Flow activity*

Create the Mapping Data Flow

Now you are ready to create the Data Flow activity shown in Figure 12-8.

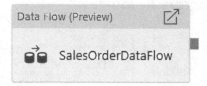

Figure 12-8. *ADF SalesOrderDataFlow*

Start by adding a source connection shown in Figure 12-9 to the **SalesOrderDataset** created in the previous section.

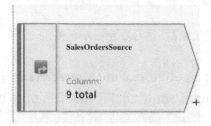

Figure 12-9. *ADF data flow source connection to SalesOrderDataset*

Be sure to **Allow schema drift** as there may be columns that change in the files. Additionally, select **Infer drifted column types** to allow auto-detection of drifted column types. These options are illustrated in Figure 12-10.

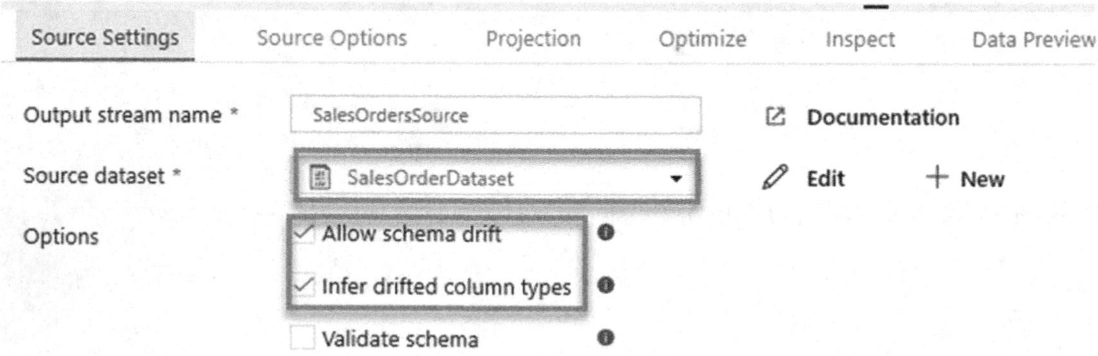

Source Settings	Source Options	Projection	Optimize	Inspect	Data Preview

Output stream name *   SalesOrdersSource   ⬀ Documentation

Source dataset *   ▦ SalesOrderDataset   ▼   ✎ Edit   + New

Options   ✓ Allow schema drift   ❶

✓ Infer drifted column types   ❶

☐ Validate schema   ❶

Figure 12-10. *ADF source data flow settings*

Parametrize the year 2016 so that you can maintain these values outside of the hard-coded path. To add parameters, click the white space in the Mapping Data Flows canvas, and then select and add the desired parameters shown in Figure 12-11.

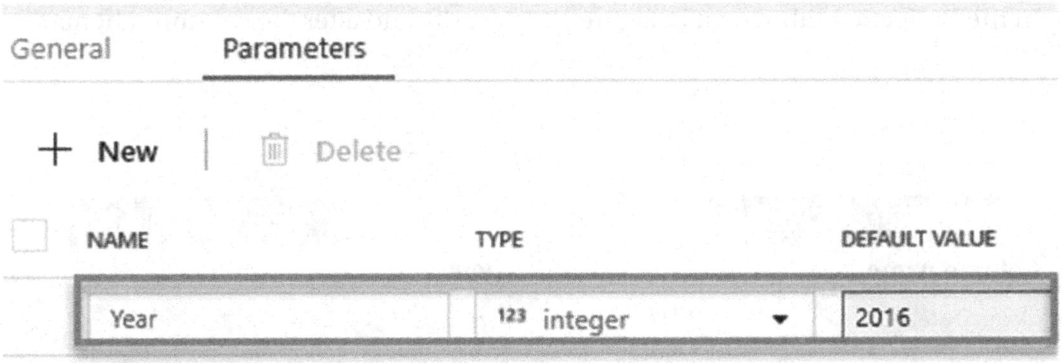

Figure 12-11. *ADF source dataset parameters*

After the parameter is added, return to the Source Options tab. Under Source Options, add the path to your 2016 sales folder in **Wildcard paths**. This setting will override the folder path set in the dataset, starting at the container root. The ****** will allow for recursive directory nesting. Lastly, specify that all text files will be needed by using ***.txt.** These settings for the Source Options tab have been listed in Figure 12-12.

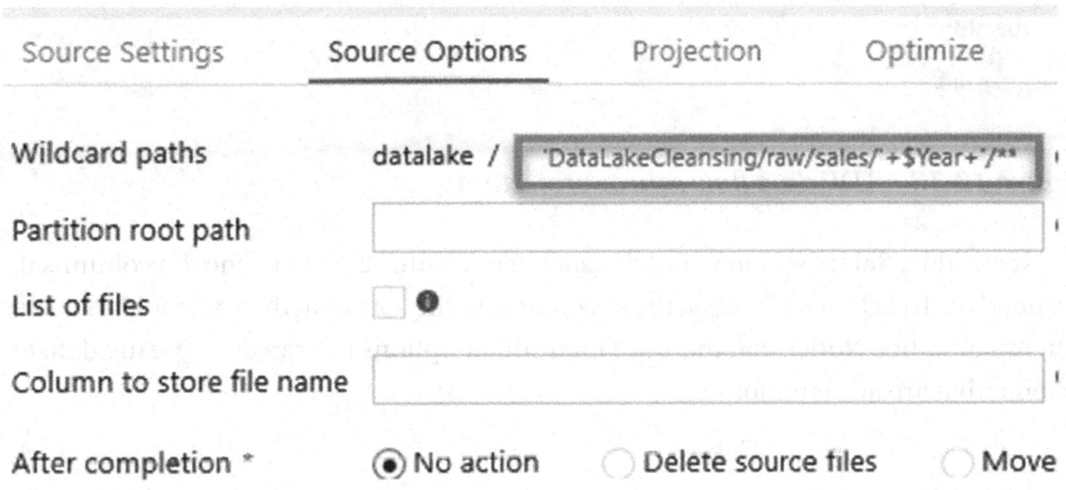

Figure 12-12. *ADF source options*

Here is the code that has been added to the Wildcard paths field within the Source Options tab in Figure 12-12:

```
'DataLakeCleansing/raw/sales/'+$Year+'/**/*.txt'
```

In the Projection tab shown in Figure 12-13, verify and alter the column schema.

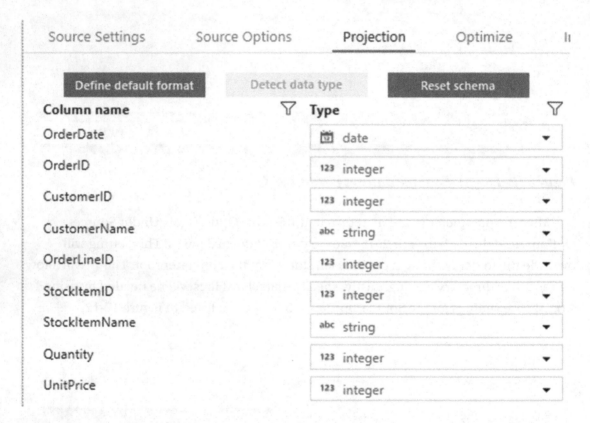

Figure 12-13. *ADF data flow source projection*

Next, add a **Select** schema modifier shown in Figure 12-14 to prune the columns that are needed. To achieve this, click the + icon next to the source activity and add the Select schema modifier. Notice that there are many other options for transforming the data and schema that are available for use.

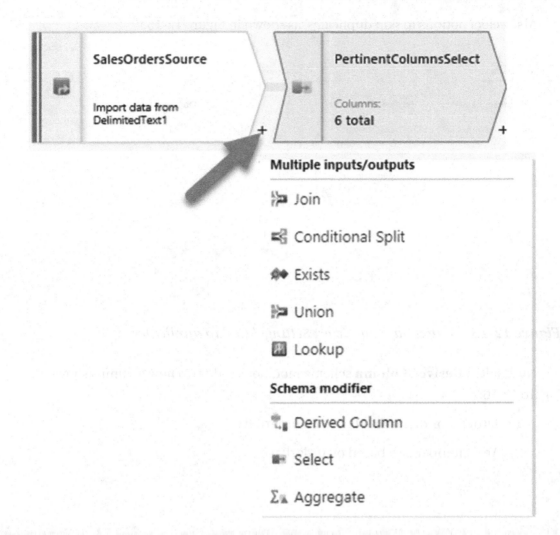

Figure 12-14. *ADF Mapping Data Flows "Select" modifier activity*

Also select options to skip duplicates, as shown in Figure 12-15.

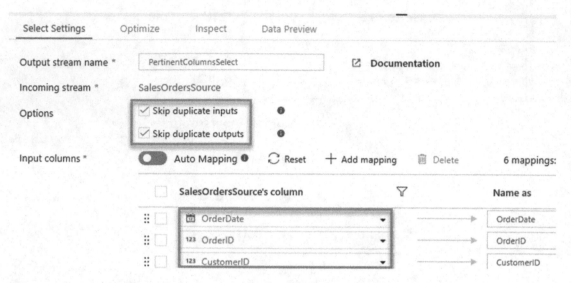

Figure 12-15. *ADF data flow "Select Settings" to skip duplicates*

Next, add a **Derived Column** schema modifier to add two new columns shown in Figure 12-16:

1. Order month number based on OrderDate

2. Year number also based on OrderDate

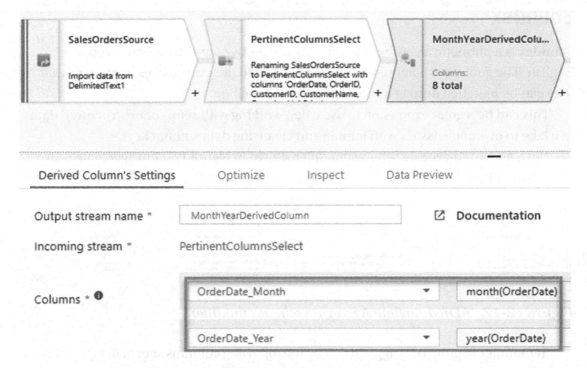

Figure 12-16. *ADF data flow derived columns*

Derived columns are great for data cleansing through the power of expressions.

Regular Expressions (Regex)

A regular expression is a sequence of characters that specifies a search pattern. Usually, such patterns are used by string-searching algorithms for "find" or "find and replace" operations on strings or for input validation. It is a technique developed in theoretical computer science and formal language theory. Regex functions are compatible with Mapping Data Flows and can be added to activities within the ADP pipelines. Here are some sample regular expressions and their intended purpose:

- RegexReplace(Address,`^a-zA-Z\d\s:`,''): Removes all non-alphanumeric characters

- RegexReplace(Address,`[ ]{2}|\.`,' '): Takes the Address field, which contains street address strings, and replaces any occurrence of two spaces or dots "." with a single space.

- Regex_extract(Address, `^(\d+)`, 1): Uses the street address to extract just the house number.

Soundex

Soundex is a phonetic algorithm for indexing names by sound, as pronounced in English. The goal is for homophones to be encoded to the same representation so that they can be matched despite minor differences in spelling.

This can be a great expression to use when working with semi- or unstructured data in a lake to overcome issues with joining and cleansing data without keys.

Next, add an **Aggregate** schema modifier shown in Figure 12-17 to aggregate unit price * quantity.

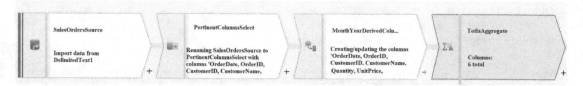

Figure 12-17. *ADF data flow Aggregate schema modifier*

Remember to group this aggregation by the following columns shown in Figure 12-18.

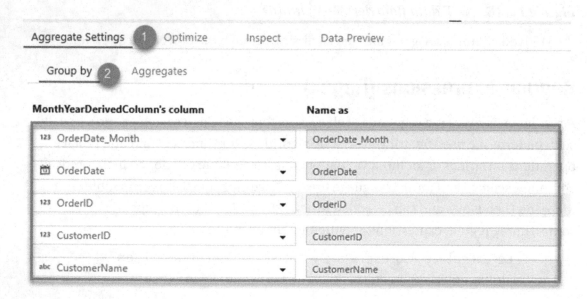

Figure 12-18. *ADF data flow Group by aggregate settings*

Figure 12-19 shows where you would need to enter the total calculation.

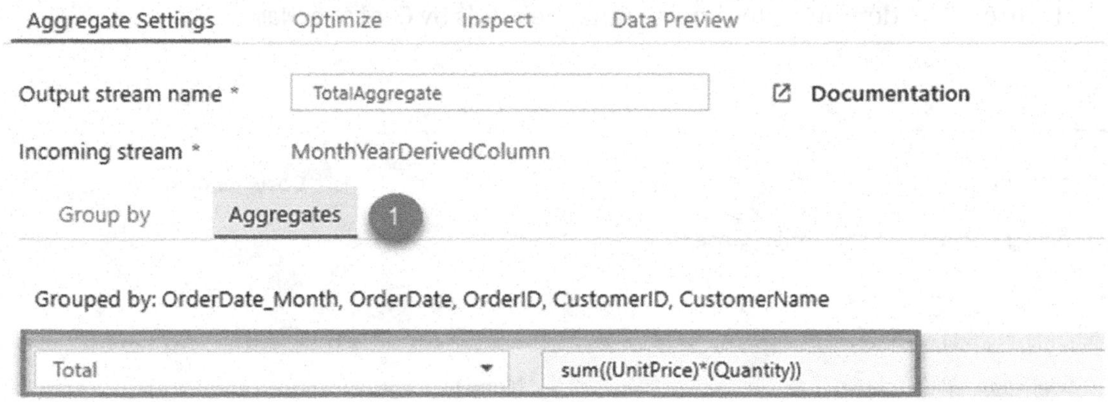
Aggregate Settings
Optimize
Inspect
Data Preview
Output stream name *
TotalAggregate
Documentation
Incoming stream *
MonthYearDerivedColumn
Group by
Aggregates
1
Grouped by: OrderDate_Month, OrderDate, OrderID, CustomerID, CustomerName
Total
sum((UnitPrice)*(Quantity))

***Figure 12-19.** ADF data flow Aggregates*

Now that the aggregation is complete, add a Windows schema modifier, as shown in Figure 12-20.

The window transformation is where you will define window-based aggregations of columns in your data streams. In the Expression Builder, you can define different types of aggregations that are based on data or time windows (SQL OVER clause such as LEAD, LAG, NTILE, CUMEDIST, RANK, etc.). A new field will be generated in your output that includes these aggregations. You can also include optional group-by fields.

SalesOrdersSource
Import data from DelimitedText1
PertinentColumnsSelect
SalesOrdersSource to PertinentColumnsSelect with columns 'OrderDate, OrderID,
MonthYearDerivedCol...
Creating/updating the columns 'OrderDate, OrderID, CustomerID, CustomerName, Quantity, UnitPrice,
TotalAggregate
'OrderDate_Month, OrderDate, OrderID, CustomerID, CustomerName'
RankbyCustomerWind...
Columns:
7 total

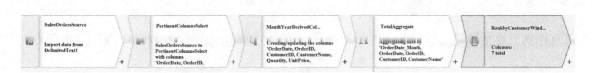

***Figure 12-20.** ADF data flow Windows schema modifier*

Figure 12-21 demonstrates how to rank the totals by `CustomerName`.

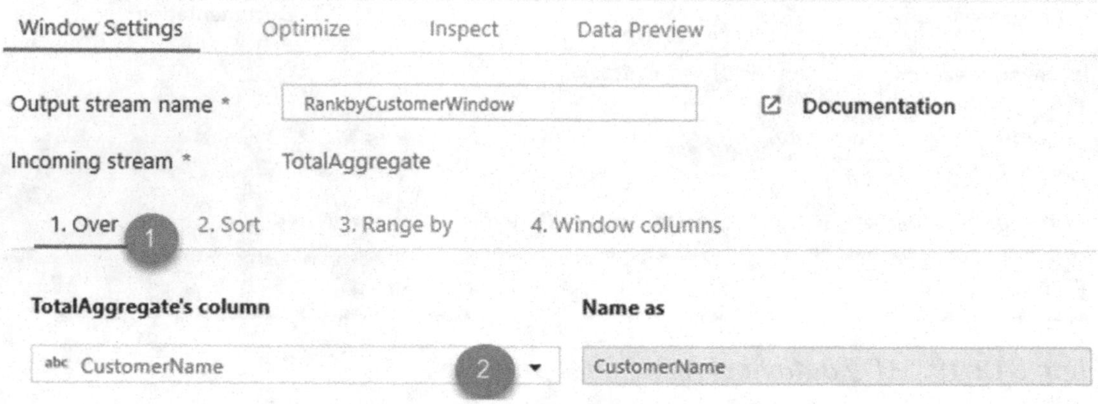

Figure 12-21. *ADF Windows Settings for "Over"*

Figure 12-22 shows how to sort the totals in descending order to sort and rank the totals from highest to lowest.

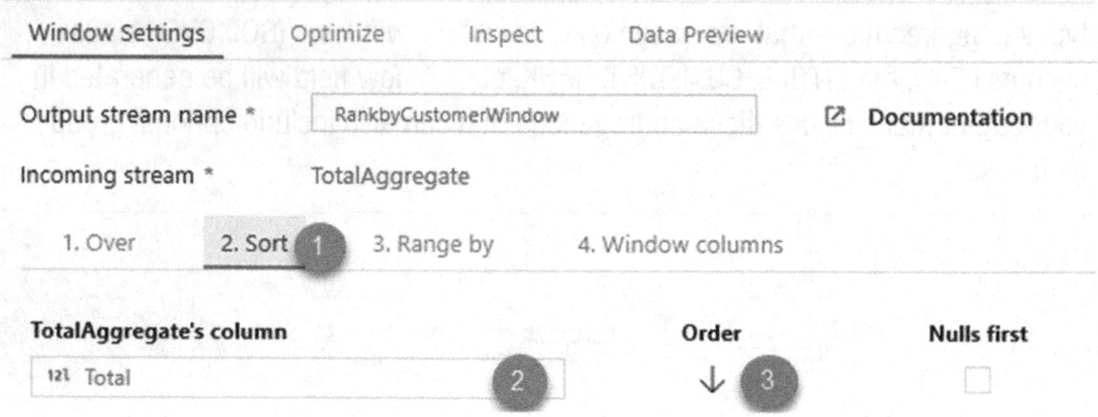

Figure 12-22. *ADF Windows Settings for "Sort"*

Leave the Range by as **Unbounded**, as shown in Figure 12-23.

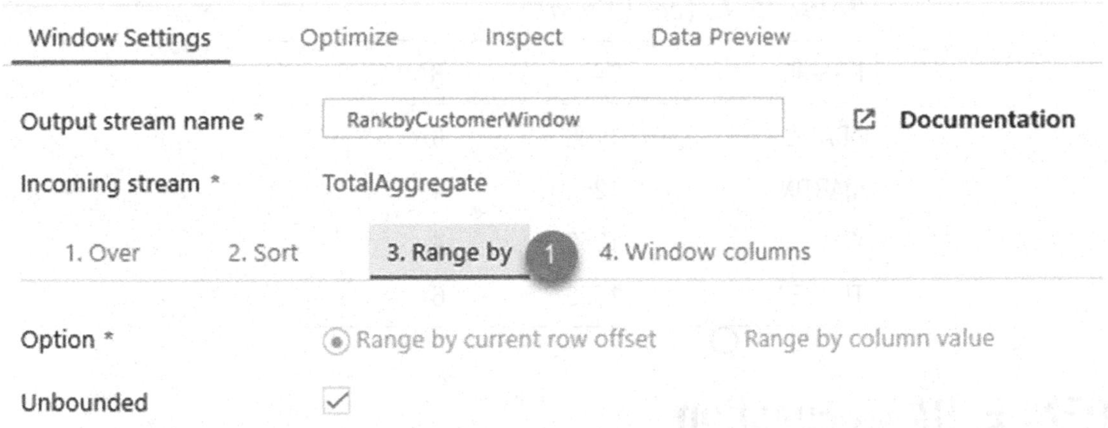

Window Settings	Optimize	Inspect	Data Preview

Output stream name *   RankbyCustomerWindow   ⬈ **Documentation**

Incoming stream *   TotalAggregate

1. Over	2. Sort	3. Range by ①	4. Window columns

Option *   ⦿ Range by current row offset   ○ Range by column value

Unbounded   ☑

Figure 12-23. *ADF Windows Settings for "Range by"*

Next, add a dense rank function to the total. Note that there are a few rank and row number functions that fit specific needs and use cases.

RANK Function

The RANK function is used to retrieve ranked rows based on the condition of the ORDER BY clause.

Here is a sample query containing the RANK function:

```
SELECT ename,
       sal,
       Rank()
         OVER (
           ORDER BY sal) RANK
FROM   emp;
```

Table 12-1 shows a sample output for the RANK function which has been run on a sample dataset.

Table 12-1. *RANK function sample output*

ENAME	SAL	RANK
SMITH	800	1
JAMES	950	2

(continued)

Table 12-1. (*continued*)

ENAME	SAL	RANK
ADAMS	1100	3
MARTIN	1250	4
WARD	1250	4
TURNER	1500	6

DENSE_RANK Function

The DENSE_RANK function is like the RANK function. However, the DENSE_RANK function does not skip any ranks if there is a tie between the ranks of the preceding records.

Here is a sample query containing the DENSE_RANK function:

```
SELECT ename,
       sal,
       Dense_rank()
         OVER (
           ORDER BY sal) DEN_RANK
FROM   emp;
```

Table 12-2 shows a sample output for the DENSE_RANK function which has been run on a sample dataset.

Table 12-2. *DENSE_RANK function sample output*

ENAME	SAL	RANK
SMITH	800	1
JAMES	950	2
ADAMS	1100	3
MARTIN	1250	4
WARD	1250	4
TURNER	1500	5

ROW_NUMBER Function

Unlike the RANK and DENSE_RANK functions, the ROW_NUMBER function simply returns the row numbers of the sorted records starting with 1. The Window Settings with the Mapping Data Flows transformation task in ADF can be seen in Figure 12-24.

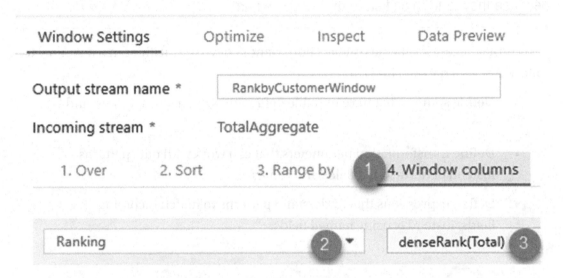

Figure 12-24. *ADF Windows Settings for "Window columns"*

Once the window function is complete, add a sink to store the enriched results in the data lake, as shown in Figure 12-25.

Figure 12-25. *ADF data flow sink to store enriched data*

Use the following sink dataset settings shown in Figure 12-26 and remember to check the option Allow schema drift. Azure Data Factory natively supports flexible schemas that change from execution to execution so that you can build generic data transformation logic without the need to recompile your data flows.

Schema drift is the case where your sources often change metadata. Fields, columns, and types can be added, removed, or changed on the fly. Without handling schema drift, your data flow becomes vulnerable to upstream data source changes. Typical ETL patterns fail when incoming columns and fields change because they tend to be tied to those source names.

To protect against schema drift, it's important to have the facilities in a data flow tool to allow you, as a data engineer, to

- Define sources that have mutable field names, data types, values, and sizes.

- Define transformation parameters that can work with data patterns instead of hard-coded fields and values.

- Define expressions that understand patterns to match incoming fields, instead of using named fields.

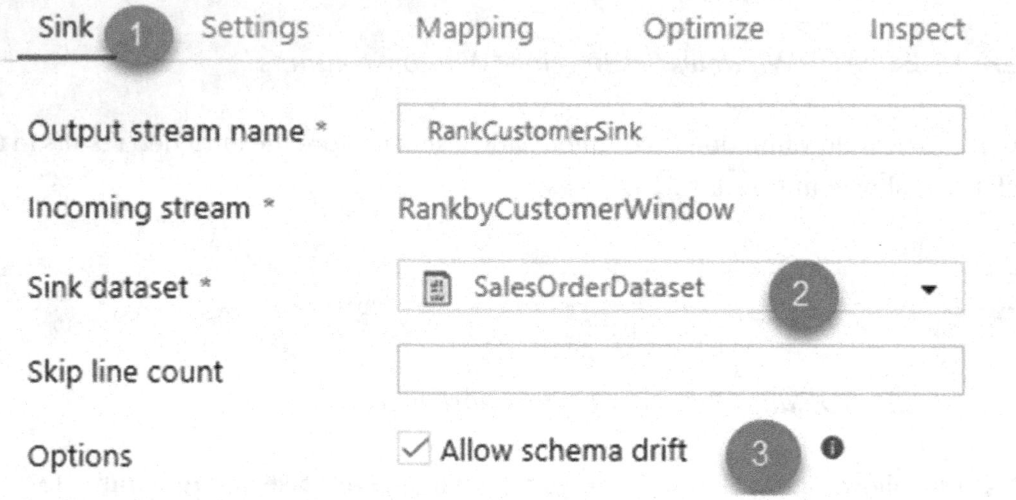

Figure 12-26. *ADF data flow sink dataset connection properties*

Also, configure the settings to output to a single file and specify the file name, as shown in Figure 12-27.

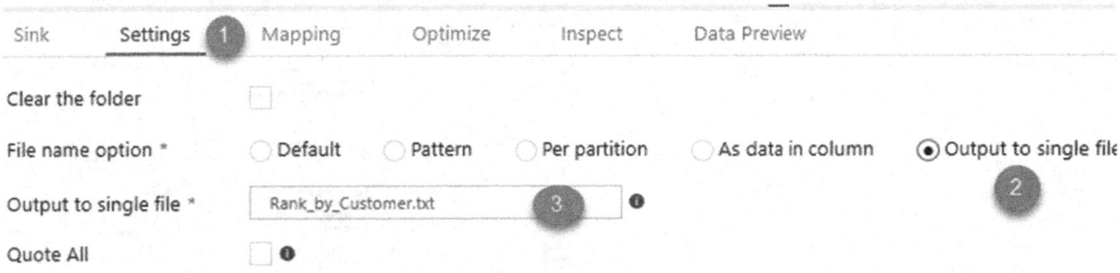

Sink	Settings ①	Mapping	Optimize	Inspect	Data Preview

Clear the folder ☐

File name option *   ○ Default   ○ Pattern   ○ Per partition   ○ As data in column   ⦿ Output to single file ②

Output to single file *   [Rank_by_Customer.txt   ③] ❶

Quote All   ☐ ❶

Figure 12-27. *ADF data flow sink settings*

One of the benefits of Mapping Data Flows is the **Data Flow Debug** mode, which allows for a preview of the transformed data without having to manually create clusters and run the pipeline.

Remember to turn on Debug mode in order to preview the data as shown in Figure 12-28, and then turn it off before logging out of Azure Data Factory. Note that the Debug mode will auto-terminate after a period.

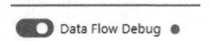 Data Flow Debug ●

Figure 12-28. *ADF data flow Debug trigger*

The ranked total results by customer will look similar to the results displayed in Figure 12-29.

¹²³ Typecast ∨   ⚙ Modify ∨   📊 Statistics   ✕ Remove

OrderID ¹²³	CustomerID ¹²³	CustomerName abc	Total ¹²¹	Ranking ¹²¹
67157	905	Sara Huiting	2880	1
64488	905	Sara Huiting	2208	2
69588	905	Sara Huiting	1944	3
65853	905	Sara Huiting	1944	3
66541	905	Sara Huiting	1842	4

Figure 12-29. *Ranked total results by customer*

This next exercise will show you how to split the Total Aggregate into a new branch to create a new file with a different window function, this time ranking the totals by month and outputting the results to a different file in the data lake, as shown in Figure 12-30.

Figure 12-30. *Split Total Aggregate into new branch*

Ensure that the new window function's settings are configured as shown in Figure 12-31.

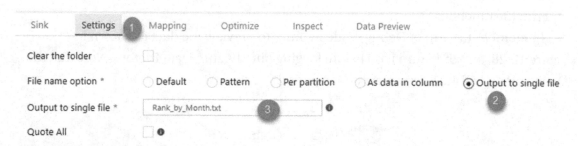

Figure 12-31. *Settings to rank the totals by month and output the results to file*

Once the Mapping Data Flow is complete, it will look like what is being displayed in Figure 12-32.

Figure 12-32. *Full end-to-end ADF Mapping Data Flow*

Summary

In this chapter, I demonstrated how to create a dataset pointing to a data lake container. Next, I showed you how to add a parameterized wildcard path to all text files in the 2016 sales folder. I then showed you how to select the pertinent columns, add a few key derived columns, perform aggregations, add window functions, split branches, and export the desired results to enriched text files in the data lake.

The output dataset contains aggregated and descending ranked totals (unit price * quantity) by customer name and by month. All of this was done by utilizing Azure Data Factory's Mapping Data Flows feature and tested with the Data Flow Debug functionality.

CHAPTER 13

Incrementally Upsert Data

If you recall from Chapters 4 through 6, I demonstrated how to do the following:

- Fully load Azure Data Lake Storage Gen2 from a SQL database.

- Fully load Azure Synapse Analytics DW with the Data Lake Storage Gen2 parquet files using Azure Data Factory's Copy activity.

This ELT process was driven by a metadata approach by using pipeline parameters populated in a SQL table. While this is a useful ELT method, this chapter will focus on advancing this process to address the following:

- Incrementally load Azure Data Lake Storage from a SQL source.

- Update and insert (upsert) the incremental records into an Azure Synapse Analytics DW destination.

In this chapter, I will demonstrate the capabilities of the built-in upsert feature of Azure Data Factory's Mapping Data Flows to update and insert data from Azure Data Lake Storage Gen2 parquet files into Azure Synapse Analytics DW. It is important to note that Mapping Data Flows does not currently support on-premises data sources and sinks. Therefore, this exercise will utilize an Azure SQL Database as the source to create the dataset. Additionally, I will demonstrate how to use a custom method of incrementally populating Data Lake Storage Gen2 with parquet files based on the created date within the Azure SQL Database. This chapter assumes that you are familiar with the previous chapters in this book, which discuss the metadata-driven ETL approach using Azure Data Factory.

© Ron C. L'Esteve 2021
R. C. L'Esteve, *The Definitive Guide to Azure Data Engineering*, https://doi.org/10.1007/978-1-4842-7182-7_13

Create a Parameter Table

Let's begin the process by creating a `pipeline_parameter` table in an Azure SQL Database. You've already created this in previous chapters, and you'll simply need to add the additional columns listed in the following. As we recall from the previous chapters, this table will contain the parameter values that will be configured to control the ADF pipeline:

- **upsert_key_column**: This is the key column that must be used by Mapping Data Flows for the upsert process. It is typically an ID column.

- **incremental_watermark_value**: This must be populated with the source SQL table's value to drive the incremental process. This is typically either a primary key ID or created/last updated date column. It can be updated by a stored procedure.

- **incremental_watermark_column**: This is simply the column name for the value populated in the `incremental_watermark_value` column. This is typically either a primary key ID or created/last updated date column. It can be updated by a stored procedure.

- **process_type**: This must be set to incremental for ADF to know which records within this table are incremental.

The following is the SQL code to reproduce this updated `pipeline_parameter` table:

```
SET ansi_nulls ON

go

SET quoted_identifier ON

go

CREATE TABLE [dbo].[pipeline_parameter]
  (
      [parameter_id]                    [INT] IDENTITY(1, 1) NOT NULL,
      [server_name]                     [NVARCHAR](500) NULL,
      [src_type]                        [NVARCHAR](500) NULL,
      [src_schema]                      [NVARCHAR](500) NULL,
```

```
    [src_db]                                 [NVARCHAR](500) NULL,
    [src_name]                               [NVARCHAR](500) NULL,
    [dst_type]                               [NVARCHAR](500) NULL,
    [dst_name]                               [NVARCHAR](500) NULL,
    [include_pipeline_flag]                  [NVARCHAR](500) NULL,
    [partition_field]                        [NVARCHAR](500) NULL,
    [process_type]                           [NVARCHAR](500) NULL,
    [priority_lane]                          [NVARCHAR](500) NULL,
    [pipeline_date]                          [NVARCHAR](500) NULL,
    [pipeline_status]                        [NVARCHAR](500) NULL,
    [load_synapse]                           [NVARCHAR](500) NULL,
    [load_frequency]                         [NVARCHAR](500) NULL,
    [dst_folder]                             [NVARCHAR](500) NULL,
    [file_type]                              [NVARCHAR](500) NULL,
    [lake_dst_folder]                        [NVARCHAR](500) NULL,
    [spark_flag]                             [NVARCHAR](500) NULL,
    [dst_schema]                             [NVARCHAR](500) NULL,
    [distribution_type]                      [NVARCHAR](500) NULL,
    [load_sqldw_etl_pipeline_date]           [DATETIME] NULL,
    [load_sqldw_etl_pipeline_status]         [NVARCHAR](500) NULL,
    [load_sqldw_curated_pipeline_date]       [DATETIME] NULL,
    [load_sqldw_curated_pipeline_status]     [NVARCHAR](500) NULL,
    [load_delta_pipeline_date]               [DATETIME] NULL,
    [load_delta_pipeline_status]             [NVARCHAR](500) NULL,
    [upsert_key_column]                      [NVARCHAR](500) NULL,
    [incremental_watermark_column]           [NVARCHAR](500) NULL,
    [incremental_watermark_value]            [DATETIME] NULL,
    PRIMARY KEY CLUSTERED ( [parameter_id] ASC )WITH
    (statistics_norecompute =
    OFF, ignore_dup_key = OFF) ON [PRIMARY]
  )
ON [PRIMARY]

go
```

Create a Source Query for the ADF Pipeline

Now that you have created the pipeline_parameter, write a custom SQL query that will be used as the source of the ADF pipeline. Notice the addition of SQLCommand and WhereValue that could be used to dynamically create a custom SQL statement and where clause based on whether the process type is full or incremental. For the purposes of this exercise, only apply a filter for the incremental values, but the query demonstrates the flexibility to incorporate full loads into the same source query. Here's our query:

```sql
SELECT src_schema,
       src_db,
       src_name,
       dst_schema,
       dst_type,
       dst_name,
       dst_folder,
       process_type,
       file_type,
       load_synapse,
       distribution_type,
       upsert_key_column,
       incremental_watermark_column,
       CASE
         WHEN process_type = 'FULL' THEN 'select * from ' + src_schema
         + '.' +
                                       src_name
                                       + ' where  1 = '
         WHEN process_type = 'Incremental' THEN
         'select * from ' + src_schema + '.' + src_name
         + ' where  ' + incremental_watermark_column
         + ' > '
       END                            AS SQLCommand,
       CASE
         WHEN process_type = 'FULL' THEN '1'
         WHEN process_type = 'incremental' THEN Cast(
         Isnull(incremental_watermark_value, 'yyyy-MM-dd') AS VARCHAR(50))
```

```
        END                                 AS WhereValue,
        dst_folder + '/' + dst_name + '/' + file_type + '/'
        + Format(Getdate(), 'yyyy-MM-dd') AS FolderName,
        dst_name + '.' + file_type          AS FileName
FROM    dbo.pipeline_parameter
WHERE   load_synapse = 1
        AND process_type = 'incremental'
```

Please note that in this exercise, the source query will be embedded into the ADF pipeline. As a best practice, consider converting the queries to stored procedures for easier maintenance of the SQL code in a centralized SQL database outside of the ADF pipelines.

Figure 13-1 shows the results in SQL Server Management Studio (SSMS) from executing the query after populating the `pipeline_parameter` with one incremental record that you'd want to run through the ADF pipeline. Notice that the `SQLCommand` column builds the incremental source query that will be used by the ADF pipeline.

process_type	upsert_key_column	incremental_watermark_column	SQLCommand	WhereValue	FolderName	FileName
incremental	LogID	CreatedDt	select * from db.Log where CreatedDt >	Mar 26 2021 12:00AM	lake:/raw/AdventureWorksLT//Log/parquet/2021-03-26	Log.parquet

Figure 13-1. *Custom SQL query that will be used as the source of the ADF pipeline*

Here is the SQL query that is contained in the `SQLCommand` column in Figure 13-1. This `SQLCommand` when combined with `WhereValue` dynamically forms the source SQL query that can be integrated and called as parameters within the ADF pipelines. The purpose of creating this process is to demonstrate the vast capabilities of building source SQL queries dynamically and then integrating them into an ADF pipeline:

```
select * from db.Log where CreatedDt >
```

Add the ADF Datasets

Next, head over to ADF and create the datasets described in the following sections. There are three datasets that you'll need to create. You'll want a source dataset in Azure SQL Database, an Azure Data Lake Storage Gen2 dataset for use in Mapping Data Flows, and finally an Azure Synapse Analytics dataset to be the destination.

Azure SQL Database

An Azure SQL Database will be needed as the source for datasets within the ADF pipeline. For the purposes of this exercise, you'll need to use a log table that resides on this Azure SQL Database. Figure 13-2 does a good job of illustrating how this source connection would look in the dataset view after it has been created.

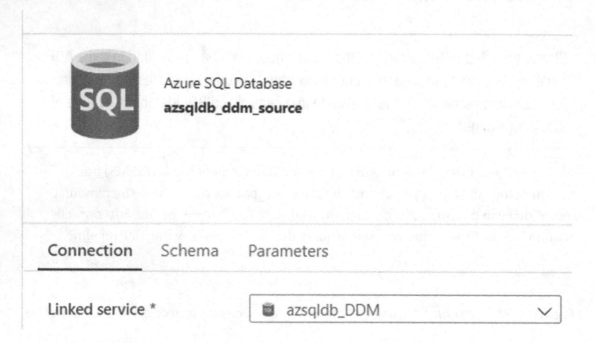

Figure 13-2. *ADF Azure SQL DB source dataset*

Azure Data Lake Storage Gen2

Also, an Azure Data Lake Storage Gen2 dataset will be needed to use within Mapping Data Flows to create the parquet files. This ADLS Gen2 account will serve as the landing zone for the parquet files that are sourced from the SQL database. Figure 13-3 illustrates how this connection and file path setting would appear once successfully created. Notice the parameterized folder name, which is from the FolderName column in the source query.

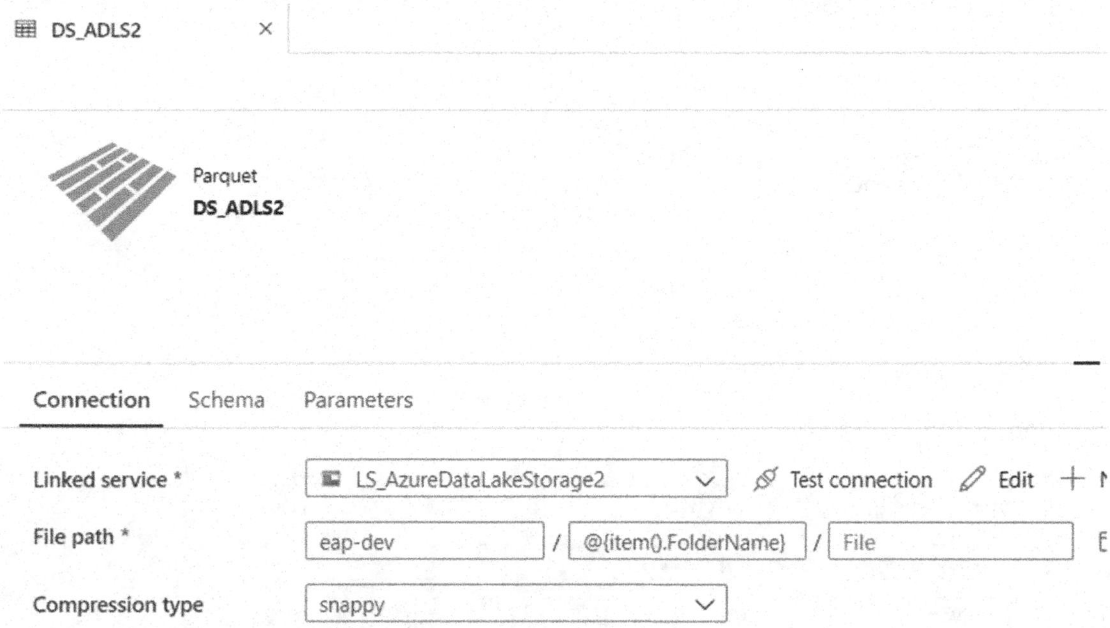

▦ DS_ADLS2	×	

Parquet
DS_ADLS2

Connection   Schema   Parameters

Linked service *   ▣ LS_AzureDataLakeStorage2   ⌄   ✐ Test connection   ✎ Edit   ＋ ℕ

File path *   eap-dev   /   @{item().FolderName}   /   File   E

Compression type   snappy   ⌄

Figure 13-3. *Azure Data Lake Storage Gen2 dataset*

Also add the following parameters to the parquet configuration section, shown in Figure 13-4. Both FolderName and FileName come from the source SQL query that was listed in the preceding sections.

Figure 13-4. *Parameters for parquet configuration*

Azure Synapse Analytics DW

Finally, you will need an Azure Synapse Analytics DW destination dataset to store both initial and incremental data that has been identified from the source systems. Mapping Data Flows in ADF will be used to build this transformation step. Figure 13-5 shows what this successful connection to the Synapse Analytics dataset would look like. Notice that there is a dynamic definition for the table name, and the schema will be defined as "etl" consistently.

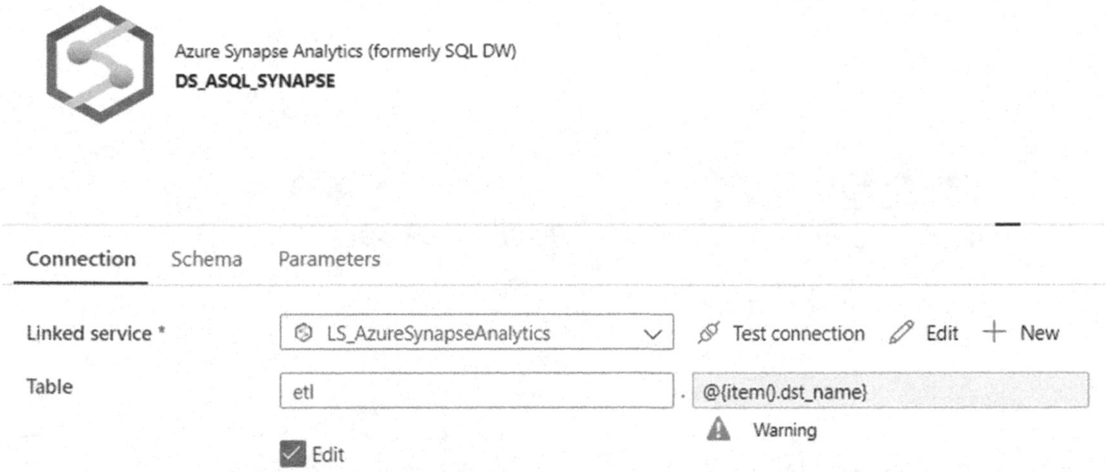

Azure Synapse Analytics (formerly SQL DW)
DS_ASQL_SYNAPSE

Connection   Schema   Parameters

Linked service *   ⊙ LS_AzureSynapseAnalytics   ⌄   ⌀ Test connection   ✎ Edit   + New

Table   etl   . @{item().dst_name}
                                                                 ⚠ Warning
                     ☑ Edit

Figure 13-5. *Azure Synapse Analytics DW destination dataset*

Create the ADF Pipeline

Now that you have created the required datasets, it's time to begin configuring the ADF pipeline activities by adding a Lookup activity to get the list of tables, a ForEach loop activity to copy each table, and a Mapping Data Flow activity to incrementally copy data from ADLS Gen2 to Synapse Analytics DW. This pipeline will demonstrate the end-to-end capabilities of upserting data from source to sink using ADF's Mapping Data Flows.

Add a Lookup Activity to Get the List of Tables

Begin by adding a Lookup activity to the ADF pipeline canvas to get the tables needed for the ADF pipeline. Figure 13-6 shows where you will need to add the query that was created in the previous steps.

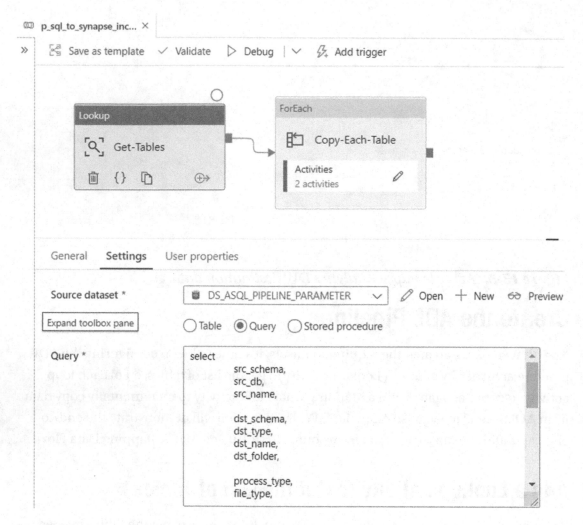

Figure 13-6. *ADF source dataset settings*

Add a ForEach Activity to Iterate and Copy Each Table

The ForEach loop activity, when connected to the Lookup activity, will iterate over each table and copy the table from ADLS Gen2 to Synapse Analytics DW. Figure 13-7 shows how to configure the ForEach loop activity settings within the ADF pipeline.

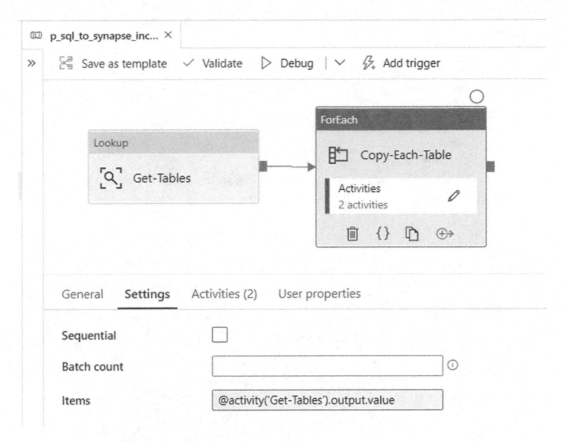

Figure 13-7. *ADF ForEach loop settings*

Mapping Data Flow for SQL to Lake Incremental ADF Pipeline

Now you can get started with building the Mapping Data Flow for the incremental loads from the source Azure SQL Database to the sink Data Lake Store Gen2 parquet folders and files. The FolderName and FileName were created in the source ADLS Gen2 parquet dataset and used as a source in the Mapping Data Flow, as shown in Figure 13-8.

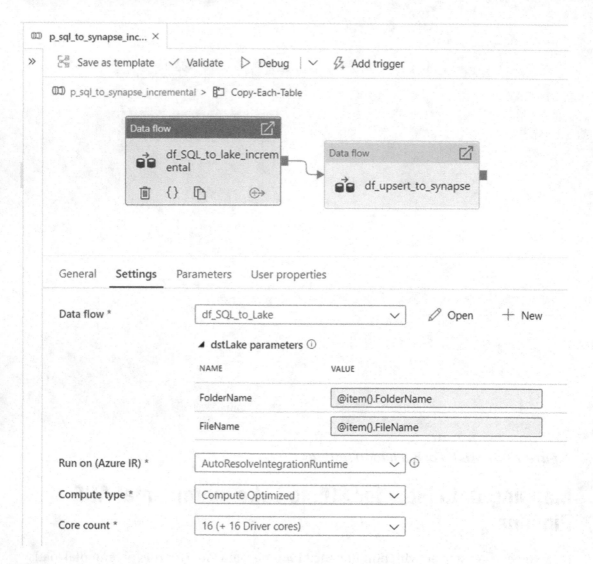

Figure 13-8. *ADF SQL to Lake data flow settings*

The parameters are the SQLCommand, WhereValue, and FileName, as shown in Figure 13-9. These parameters will be used to generate the source incremental query string.

Figure 13-9. *ADF SQL to Lake data flow parameters*

The source settings of the Mapping Data Flow are configured as shown in Figurc 13-10:

- **Allow schema drift**: Select Allow schema drift if the source columns will change often. This setting will allow all incoming fields from your source to flow through the transformations to the sink.

- **Infer drifted column types**: Allows auto-detection of drifted column types.

- **Validate schema**: Setting this option will cause the data flow to fail if any column and type defined in the Projection tab does not match the incoming data.

- **Sampling**: Setting this option will cause the data flow to fail if any column and type defined in the Projection tab does not match the incoming data.

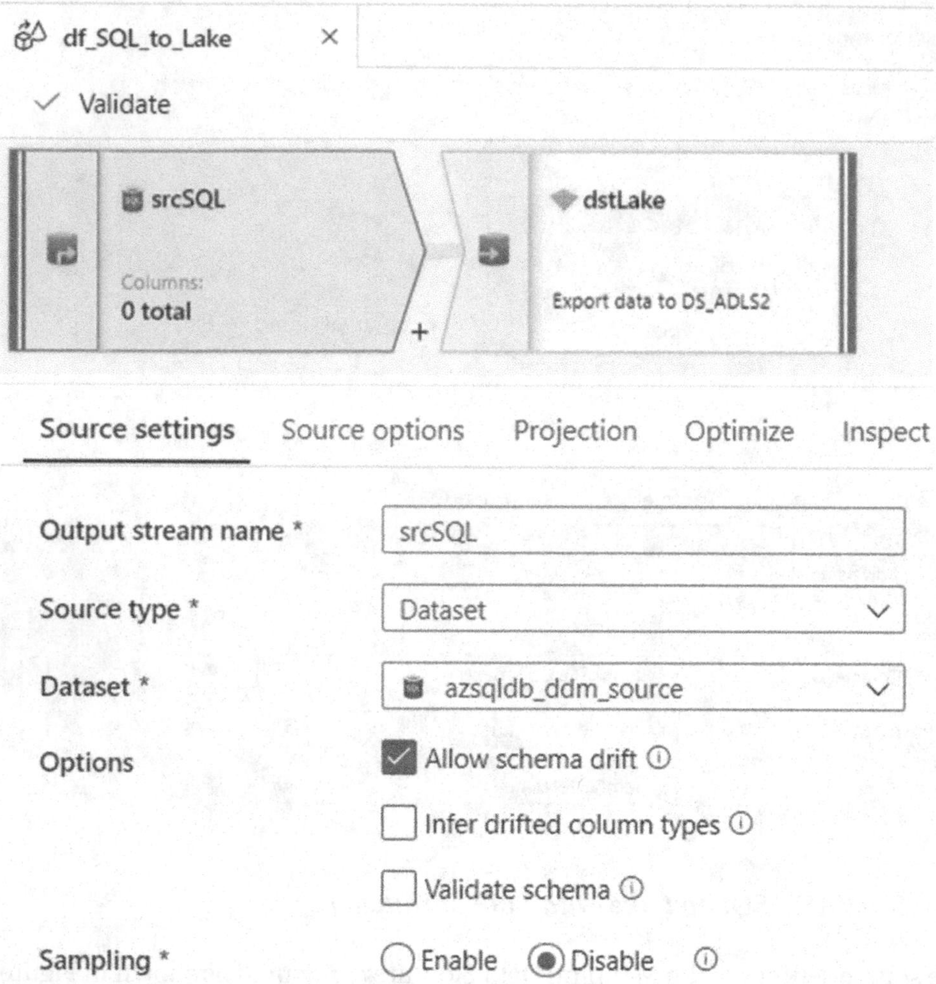

Figure 13-10. *ADF SQL to Lake data flow source connection settings*

The source options of the data flow are configured as shown in Figure 13-11 and use the string interpolation expression features within Mapping Data Flows (https://docs.microsoft.com/en-us/azure/data-factory/concepts-data-flow-expression-builder). Notice that the SQLCommand and WhereValue fields are concatenated to form a dynamic SQL query.

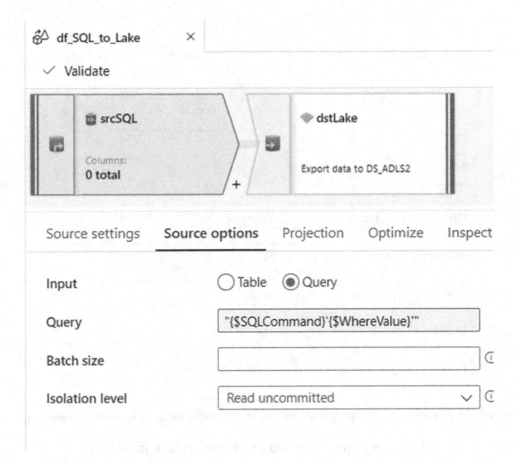

Figure 13-11. *ADF SQL to Lake data flow source options*

Next, configure the destination sink of the Mapping Data Flow as shown in Figure 13-12.

Figure 13-12. *ADF data flow destination sink connection properties*

The settings shown in Figure 13-13 will indicate the single file output of FileName, which is generated from the SQL query listed in the previous sections of this chapter.

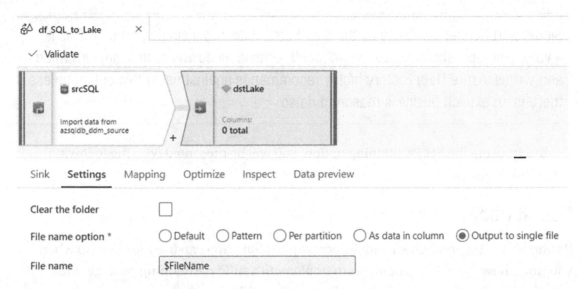

Figure 13-13. *ADF data flow destination settings*

Use single partitioning in the Optimize tab as shown in Figure 13-14.

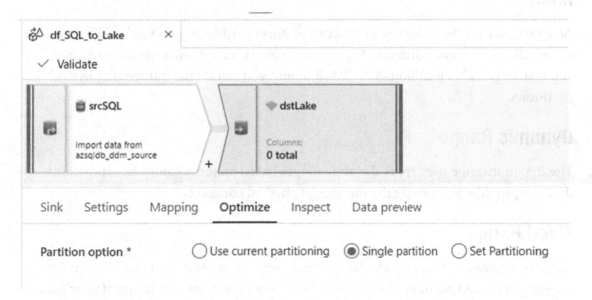

Figure 13-14. *ADF data flow destination partition optimization properties*

Single partitioning combines all the distributed data into a single partition. This is a very slow operation that also significantly affects all downstream transformation and writes. Azure Data Factory highly recommends against using this option unless there is an explicit business reason to do so.

By choosing the Set Partitioning option, you will be presented with the following partitioning options.

Round Robin

Round robin distributes data equally across partitions. You can use round robin when you don't have good key candidates to implement a solid partitioning strategy since round robin distributes the data equally across partitions and allows you to set the number of physical partitions.

Hash

ADF creates a hash of columns to produce uniform partitions such that rows with similar values fall in the same partition. When you use the Hash option, it will be important that you test for possible partition skew. Additionally, you can set the number of physical partitions.

Dynamic Range

The dynamic range uses Spark dynamic ranges based on the columns or expressions that you provide. You can set the number of physical partitions.

Fixed Range

Build an expression that provides a fixed range for values within your partitioned data columns. To avoid partition skew, you should have a good understanding of your data before you use this option. The values you enter for the expression are used as part of a partition function. You can set the number of physical partitions.

Key

If you have a good understanding of the cardinality of your data, key partitioning might be a good strategy. Key partitioning creates partitions for each unique value in your column. You can't set the number of partitions because the number is based on unique values in the data.

By default, Use current partitioning is selected that instructs Azure Data Factory to keep the current output partitioning of the transformation. As repartitioning data takes time, Use current partitioning is recommended in most scenarios.

By clicking the blank space of the Mapping Data Flow canvas, the Mapping Data Flow Parameters tab will appear. Add the following parameters shown in Figure 13-15.

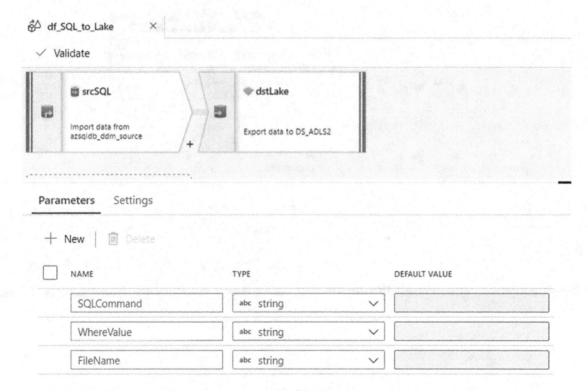

Figure 13-15. *ADF Mapping Data Flow parameters*

Mapping Data Flow to Incrementally Upsert from Lake to Synapse Analytics DW

Now that you have created and configured the SQL to lake incremental Mapping Data Flow pipeline, it's time to create and configure the lake to Synapse incremental upsert pipeline, as shown in Figure 13-16.

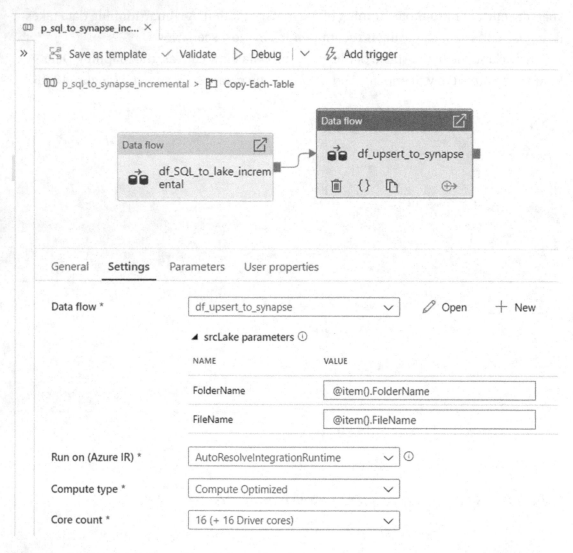

Figure 13-16. *ADF data flow lake to Synapse incremental upsert pipeline settings*

Be sure to configure the following data flow parameters for FolderName and upsert_key_column, as shown in Figure 13-17. These parameters will be used for the incremental data ingestion process.

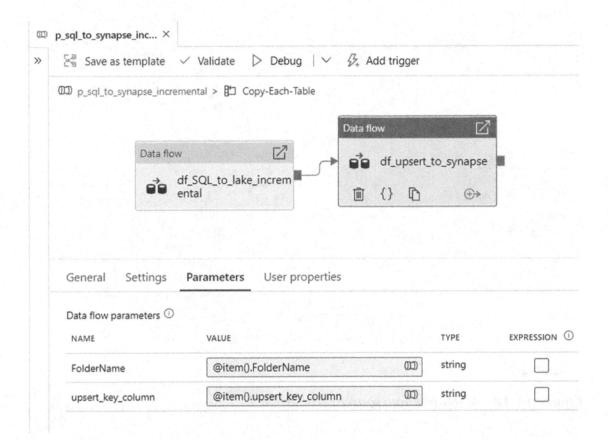

Figure 13-17. *ADF data flow lake to Synapse incremental upsert pipeline parameters*

This data flow will contain the following three activities:

1) **SrcLake:** This is the source connection to the ADLS Gen2 account.

2) **AlterRow:** This is the step that will begin the upsert process by identifying which rows need to be upserted into the sink.

3) **dstSynapse:** This is the sink connection to Synapse Analytics DW.

Begin by configuring the settings of the lake source, as shown in Figure 13-18.

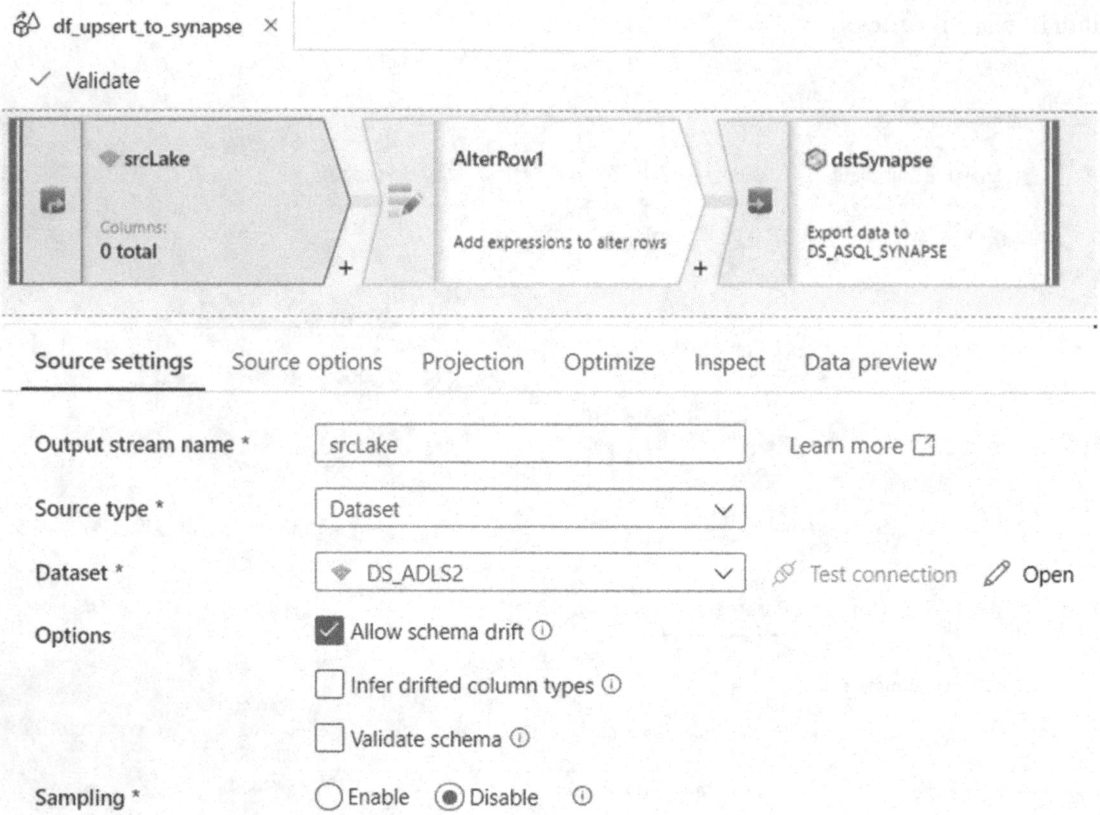

Figure 13-18. *ADF data flow source settings*

Next, ensure that the Source options tab contains the parameterized `FolderName`, as shown in Figure 13-19.

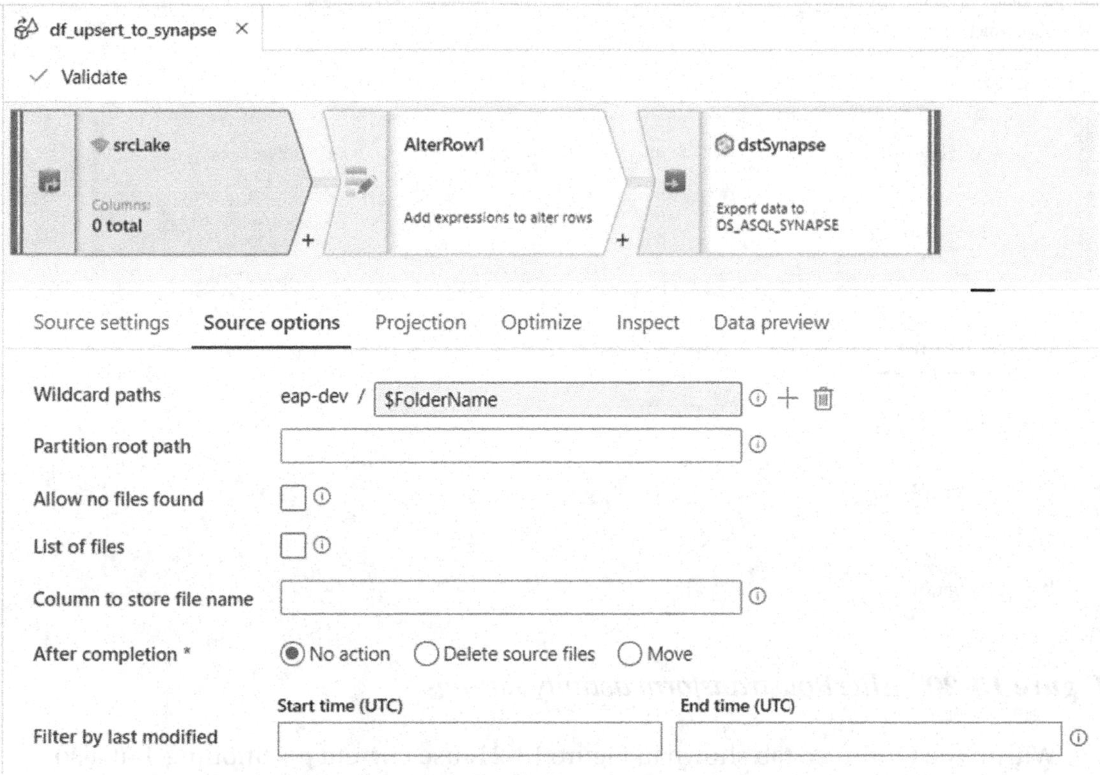

Figure 13-19. *ADF data flow source options*

Figure 13-20 shows how to add an AlterRow transform activity and then set the alter row condition to Upsert if row condition equals `true()`.

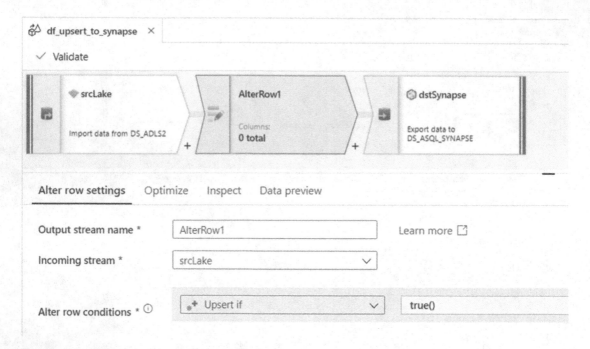

Figure 13-20. *AlterRow transform activity settings*

Within the Optimize tab shown in Figure 13-21, use current partitioning, but also be aware that there may be opportunities to explore setting the partitioning for big data workloads and to improve performance.

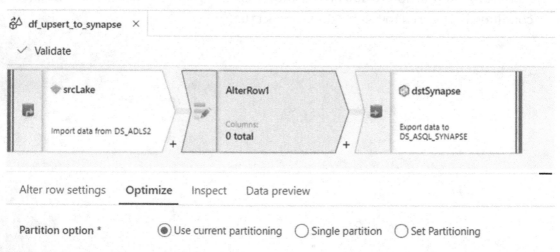

Figure 13-21. *AlterRow transform activity partition optimization options*

Finally, configure the destination Synapse Analytics DW dataset, shown in Figure 13-22.

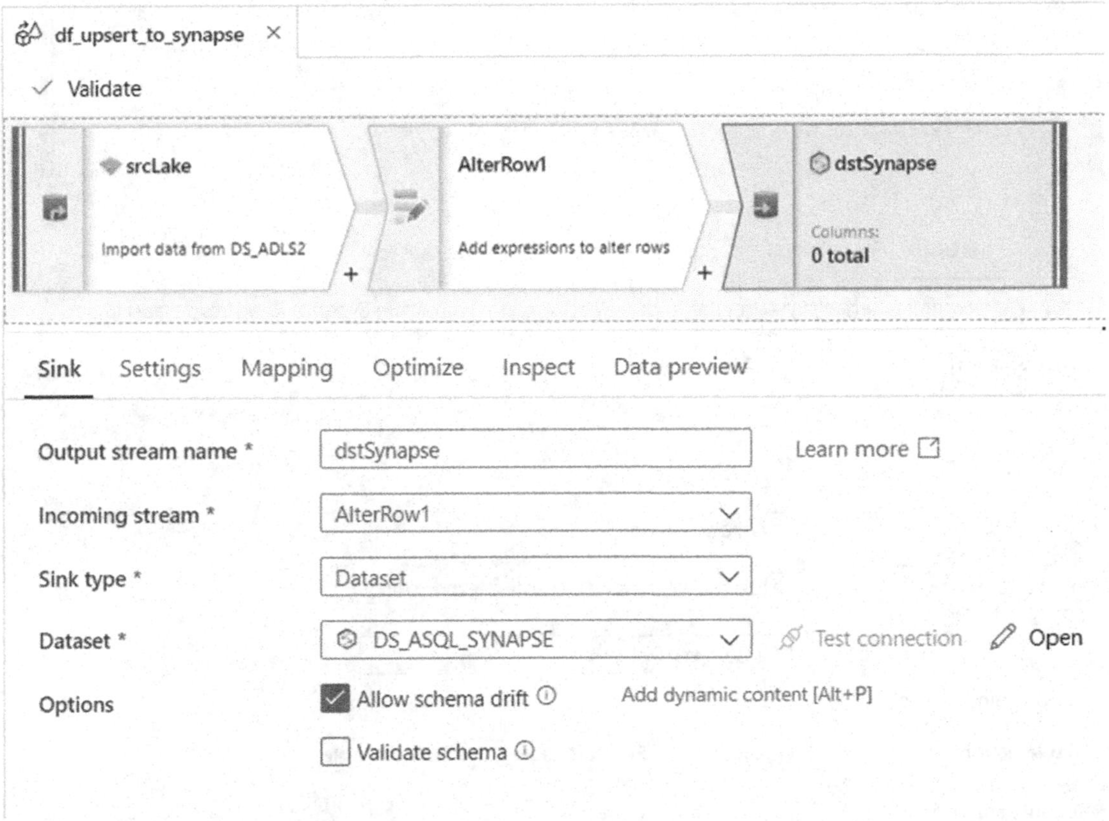

Figure 13-22. *ADF data flow sink connection properties*

Within the Settings tab, shown in Figure 13-23, choose "Allow upsert" for the update method and add the upsert_key_column that you created and populated in the pipeline parameter table. You could choose to not enable staging for this exercise, but this may be a good option for performance optimization purposes.

Figure 13-23. *ADF data flow sink settings*

Finally, ensure that the Mapping Data Flow parameters contain `FolderName` and `upsert_key_column,` shown in Figure 13-24. To get to these parameters, remember to click the white space of the Mapping Data Flow canvas.

Figure 13-24. *ADF data flow parameters*

Run the ADF Pipeline

Now that the ADF pipelines have been built, it's time to run them in order to test
and verify the results of the incremental Azure SQL to ADLS Gen2 pipeline and the
incremental ADLS Gen2 to Synapse Analytics DW pipeline. Additionally, as a verification
step, this section will run SQL counts against the source and sink databases using SSMS
to further test, verify, and validate the accuracy of the results.

Verify Incremental SQL to Lake Pipeline Results

After running this pipeline, notice in Figure 13-25 that the end-to-end pipeline
succeeded and copied over one table since there was only one record in the `pipeline_
parameter` table.

p_sql_to_synapse_incremental

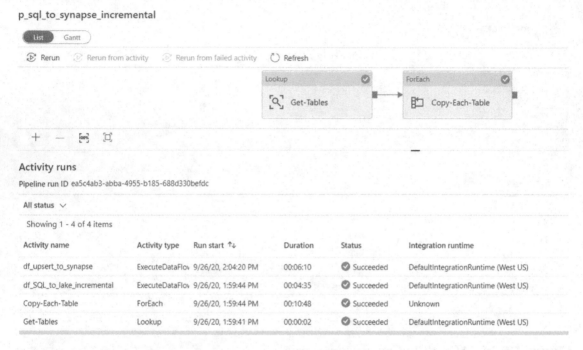

Activity runs

Pipeline run ID ea5c4ab3-abba-4955-b185-688d330befdc

All status ∨

Showing 1 - 4 of 4 items

Activity name	Activity type	Run start ↑↓	Duration	Status	Integration runtime
df_upsert_to_synapse	ExecuteDataFlov	9/26/20, 2:04:20 PM	00:06:10	✅ Succeeded	DefaultIntegrationRuntime (West US)
df_SQL_to_lake_incremental	ExecuteDataFlov	9/26/20, 1:59:44 PM	00:04:35	✅ Succeeded	DefaultIntegrationRuntime (West US)
Copy-Each-Table	ForEach	9/26/20, 1:59:44 PM	00:10:48	✅ Succeeded	Unknown
Get-Tables	Lookup	9/26/20, 1:59:41 PM	00:00:02	✅ Succeeded	DefaultIntegrationRuntime (West US)

Figure 13-25. *ADF incremental pipeline status of "Succeeded"*

Figure 13-26 shows that the pipeline copied 493 rows from the source Azure SQL Database table to a parquet file in ADLS Gen2.

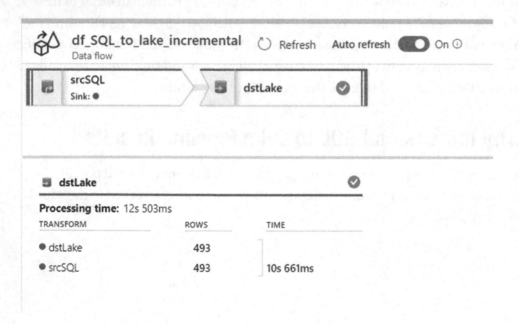

Figure 13-26. *ADF data flow incremental pipeline processing details*

Verify Incremental Upsert Lake to Synapse ADF Pipeline Results

It's time to verify that the incremental ADF pipeline has moved the pertinent data from ADLS Gen2 to Synapse Analytics DW. Figure 13-27 shows that the incremental upsert from ADLS Gen2 copied over 493 rows to Synapse Analytics DW.

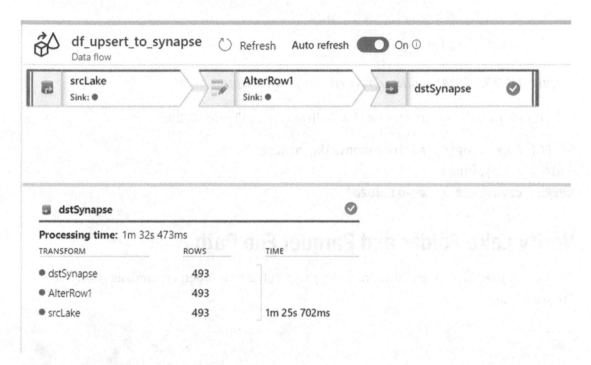

Figure 13-27. *ADF data flow incremental upsert pipeline processing details*

Verify Source SQL Record Count

The reason there are 493 rows, as shown in Figure 13-28, is because the source contains 493 rows with a created date greater than April 1, 2020, and since this was your incremental_watermark_value defined in the pipeline_parameter table, that is how many records the ADF pipeline is expected to incrementally load.

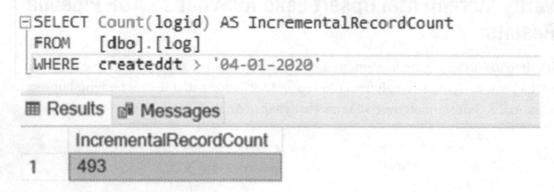

```
SELECT Count(logid) AS IncrementalRecordCount
FROM    [dbo].[log]
WHERE   createddt > '04-01-2020'
```

Results   Messages

	IncrementalRecordCount
1	493

Figure 13-28. *SSMS count of records for verification*

Here is the SQL query that has been displayed in Figure 13-28:

```
SELECT Count(logid) AS IncrementalRecordCount
FROM    [dbo].[log]
WHERE   createddt > '04-01-2020'
```

Verify Lake Folder and Parquet File Path

Also verify that ADLS Gen2 folders and a parquet file have been created, as shown in Figure 13-29.

	lake > rl-sql-001 > AdventureWorksLT > SalesLT > Log > parquet > 2021-03-27				
Name ^	Access Tier	Access Tier Last Modified	Last Modified	Blob Type	Content Type
📄 Log.parquet	Hot (inferred)		3/27/2021, 12:26:41 PM	Block Blob	application/octet-stream

Figure 13-29. *ADLS Gen2 folders containing parquet files from ADF pipeline*

Verify Destination Synapse Record Count

Finally, after running a SQL count statement on the destination Synapse Analytics DW table, the count shown in Figure 13-30 confirms that there are 493 records, proving that the incremental pipeline worked as expected.

```
SELECT Count(logid) AS IncrementalRecordCount
FROM    [etl].[log]
WHERE   createddt > '04-01-2020'
```

	IncrementalRecordCount
1	493

Figure 13-30. *SSMS count of incremental pipeline records in sink*

Here is the SQL query that has been displayed in Figure 13-30:

```
SELECT Count(logid) AS IncrementalRecordCount
FROM    [etl].[log]
WHERE   createddt > '04-01-2020'
```

Insert a Source SQL Record

Now that you have confirmed that the incremental SQL to Synapse Analytics DW pipeline worked as expected, also verify that the insert portion of the Upsert command works as expected by adding an additional record to the source SQL table where the created date is greater than April 1, 2020.

After adding the record and running a count, notice that the query now returns 494 records rather than 493, as shown in Figure 13-31.

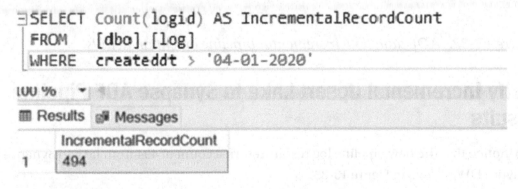

Figure 13-31. *SQL count for inserting a source record*

Here is the SQL query that has been displayed in Figure 13-31:

```
SELECT  Count(logid) AS IncrementalRecordCount
FROM    [dbo].[log]
WHERE   createddt > '04-01-2020'
```

Verify Incremental SQL to Lake ADF Pipeline Results

Once again, run the pipeline and notice that the new pipeline log results return a count of 494 from SQL to lake, shown in Figure 13-32.

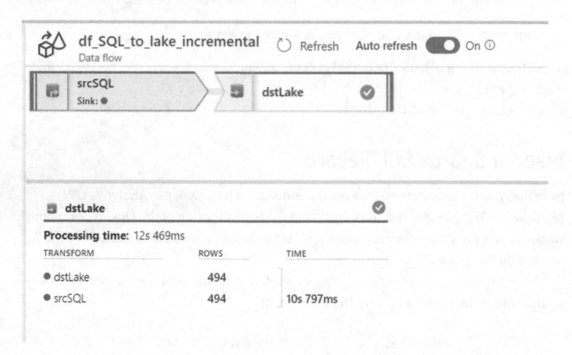

Figure 13-32. *ADF data flow incremental pipeline execution details*

Verify Incremental Upsert Lake to Synapse ADF Pipeline Results

Also, notice that the new pipeline log results return a count of 494 from lake to Synapse Analytics DW, shown in Figure 13-33.

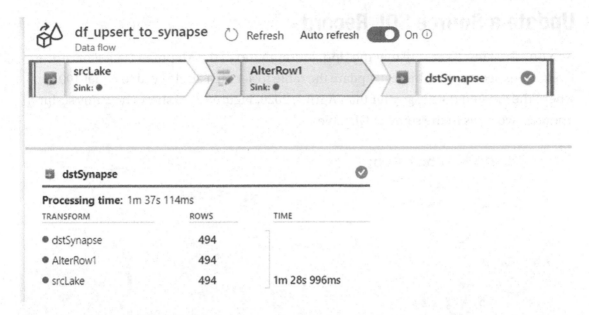

Figure 13-33. *ADF data flow incremental upsert pipeline execution details*

Verify Destination Synapse Analytics DW Record Count

Finally, the destination Synapse Analytics DW table shown in Figure 13-34 also contains 494 records, which confirms that the insert worked as expected.

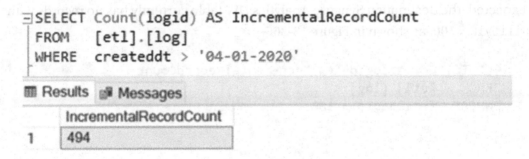

Figure 13-34. *SQL count of destination Synapse Analytics DW records*

Here is the SQL query that has been displayed in Figure 13-34:

```
SELECT Count(logid) AS IncrementalRecordCount
FROM   [etl].[log]
WHERE  createddt > '04-01-2020'
```

Update a Source SQL Record

Run one final test to ensure that the Update command of the Mapping Data Flow upsert works as expected. For this test, update the table to set the FacilityId to equal 100 where the created date is greater than April 1, 2020. Figure 13-35 shows how this script updates two rows in the source SQL table.

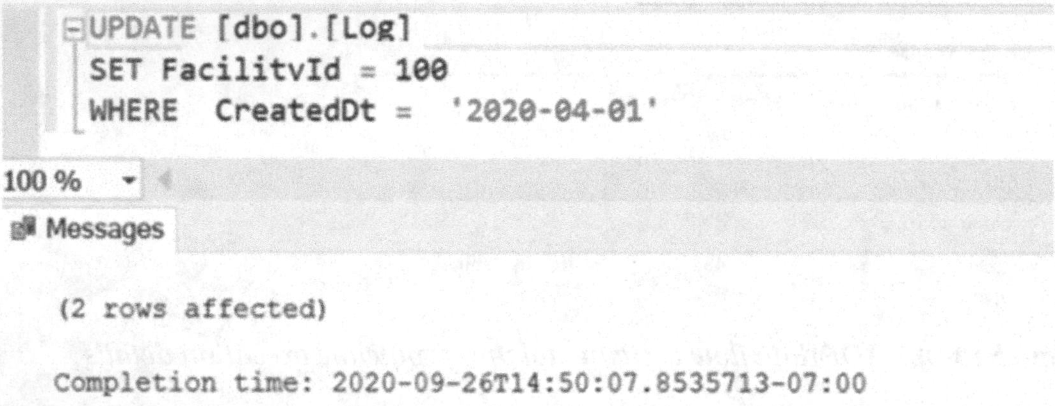

```
UPDATE [dbo].[Log]
   SET FacilityId = 100
   WHERE  CreatedDt =  '2020-04-01'
```

100 %   ▾

🔲 Messages

```
(2 rows affected)

Completion time: 2020-09-26T14:50:07.8535713-07:00
```

Figure 13-35. *SQL to update a source record*

Verify Destination Synapse Analytics DW Record Count

As expected, the destination Synapse Analytics DW table currently has no records with FacilityID = 100, as shown in Figure 13-36.

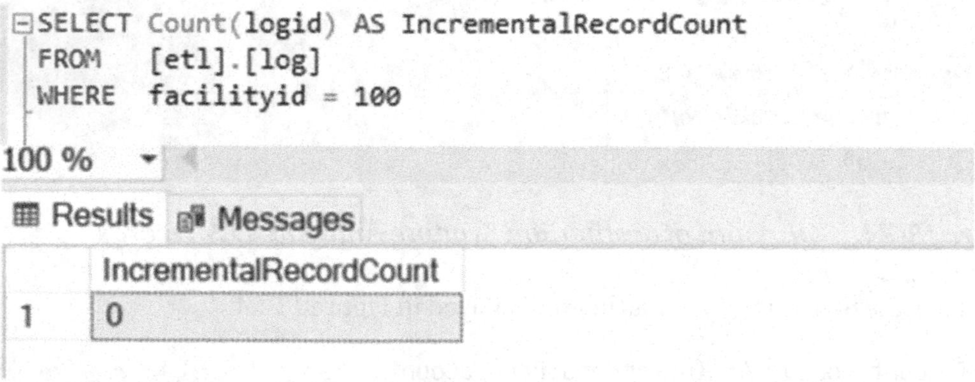

```
SELECT Count(logid) AS IncrementalRecordCount
  FROM   [etl].[log]
  WHERE  facilityid = 100
```

100 %   ▾

⊞ Results   🔲 Messages

	IncrementalRecordCount
1	0

Figure 13-36. *SQL to verify pre-update on the destination table*

Here is the SQL query that has been displayed in Figure 13-36:

```
SELECT Count(logid) AS IncrementalRecordCount
FROM    [etl].[log]
WHERE   facilityid = 100
```

After running the pipeline again, notice in Figure 13-37 that the destination Synapse Analytics DW table still contains 494 rows and has two records that have been updated with a `FacilityId` of 100. This finally confirms that both the Update and Insert commands along with incremental loads work as expected from the source SQL table to ADLS Gen2 and finally to Synapse Analytics DW.

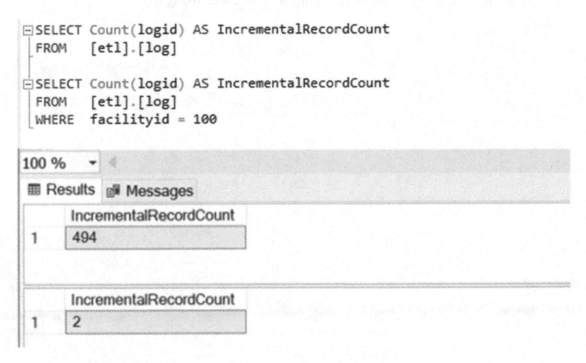

Figure 13-37. *SQL to verify destination table's updated records*

Here are the SQL queries that are displayed in Figure 13-37:

```
SELECT Count(logid) AS IncrementalRecordCount
FROM    [etl].[log]
```

```
SELECT Count(logid) AS IncrementalRecordCount
FROM    [etl].[log]
WHERE   facilityid = 100
```

Summary

In this chapter, I have demonstrated some of the capabilities of Azure Data Factory's Mapping Data Flows upsert feature to update and insert data from Azure Data Lake Storage Gen2 parquet files into Azure Synapse Analytics DW. I showed you how to build a custom method of incrementally populating Data Lake Storage Gen2 with parquet files based on the created date within the Azure SQL Database and test and verify the results. It is important to mention that this custom method of incrementally loading data from source to sink is one of many possible methods that could be used. Microsoft, the ADF product team, and a variety of community contributors frequently blog about alternative methods for incrementally loading data from source to sink using ADF.

## CHAPTER 14

Load Excel Sheets into Azure SQL Database Tables

The need to load data from Excel spreadsheets into SQL databases has been a long-standing requirement for many organizations over the course of many years. Previously, tools such as VBA, SSIS, C#, and more have been used to perform this data ingestion orchestration process. Microsoft Azure's Excel connector (`https://docs.microsoft.com/en-us/azure/data-factory/format-excel`) for Azure Data Factory (ADF) is a promising cloud-based connector that allows for an easy-to-use, low-code experience for working with Excel, much like an on-premises SSIS Excel connector.

The Excel connector in Azure Data Factory provides the capability of leveraging dynamic and parameterized pipelines to load Excel spreadsheets into Azure SQL Database tables using ADF. In this chapter, I will demonstrate how to dynamically load an Excel spreadsheet residing in an ADLS Gen2 account that contains multiple sheets into a single Azure SQL table and also into multiple tables for every sheet.

Prerequisites

There are a few setup steps that are required for successfully creating an end-to-end ADF pipeline to load Excel files to a relational SQL database. First, you'll need to do some prep work to create the desired Excel spreadsheet. Then you'll need to upload it to an ADLS Gen2 account. Finally, you will need to create the corresponding linked services and datasets in ADF to prepare for the creation of the ADF pipeline.

© Ron C. L'Esteve 2021
R. C. L'Esteve, *The Definitive Guide to Azure Data Engineering*, https://doi.org/10.1007/978-1-4842-7182-7_14

Create an Excel Spreadsheet

The image in Figure 14-1 shows a sample Excel spreadsheet containing four sheets, each containing the same headers and schema that you will need to use in your ADF pipelines to load data in Azure SQL tables. The expectation is that each of the four sheets will have the same schema, column order, and headers. This process does not test corner cases such as accounting for schema drift, embedded images in the Excel file, and more. You can find a sample Excel file in my GitHub repo, SampleExcelFiles, in the following: https://github.com/ronlesteve/SampleExcelFiles.

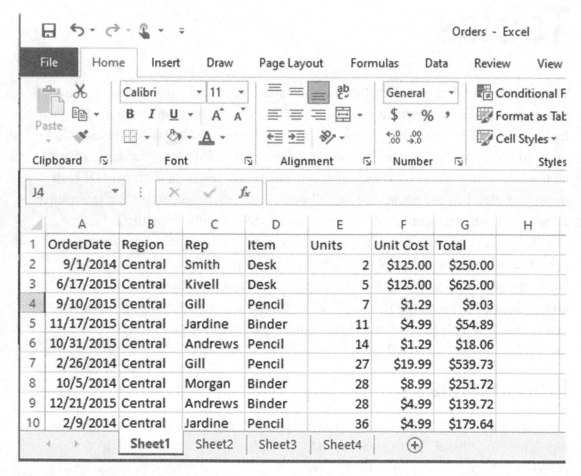

***Figure 14-1.** Sample Excel spreadsheet containing four sheets with data*

Upload to Azure Data Lake Storage Gen2

Once you have created the Excel spreadsheet, upload the spreadsheet to your ADLS Gen2 account, as shown in Figure 14-2.

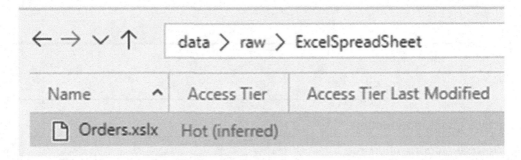

Figure 14-2. *Excel file loaded to ADLS Gen2*

Create Linked Services and Datasets

Within Data Factory, add an ADLS Gen2 linked service for the file path of the Excel spreadsheet, which has been added to the data/raw/ExcelSpreadSheet folder path, as shown in Figure 14-3. As a reminder, since ADLS Gen2 is a hierarchical file system, the ADF pipeline will easily be able to iterate through multiple folder hierarchies to locate the file.

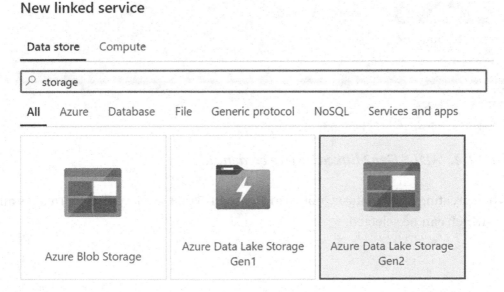

Figure 14-3. *ADLS Gen2 linked service for the location of the Excel spreadsheet*

297

Figure 14-4 shows the proper configuration properties of the ADLS Gen2 connection. Ensure that the ADLS Gen2 linked service credentials are configured accurately.

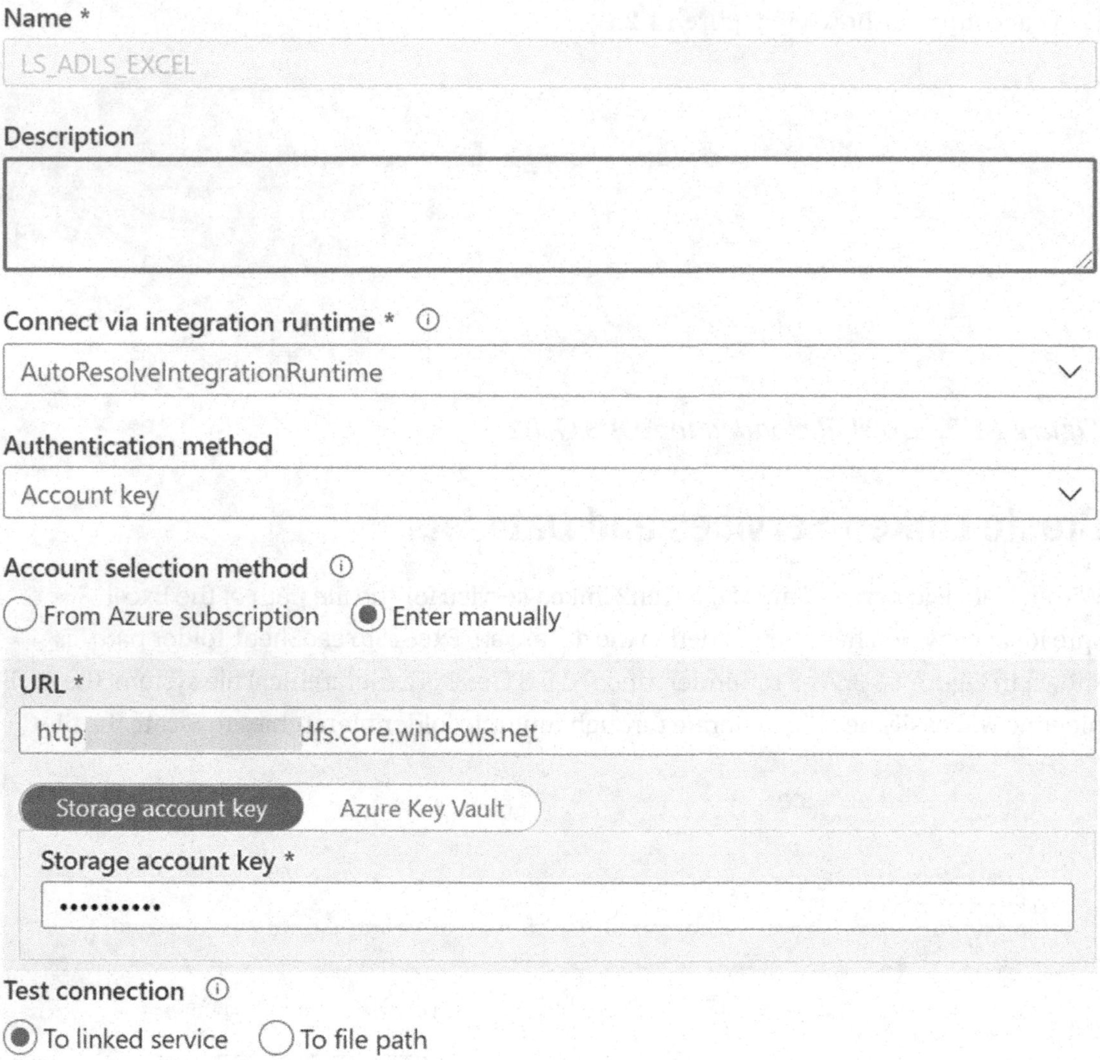

Name *

LS_ADLS_EXCEL

Description

Connect via integration runtime * ⓘ

AutoResolveIntegrationRuntime   ⌄

Authentication method

Account key   ⌄

Account selection method ⓘ

◯ From Azure subscription   ⦿ Enter manually

URL *

http⬛⬛⬛⬛⬛⬛⬛⬛dfs.core.windows.net

(Storage account key)   Azure Key Vault

Storage account key *

••••••••••

Test connection ⓘ

⦿ To linked service   ◯ To file path

Figure 14-4. *ADLS Gen2 linked service credentials*

When creating a new dataset, notice in Figure 14-5 that we have Excel format as an option, which can be selected.

Select format

Choose the format type of your data

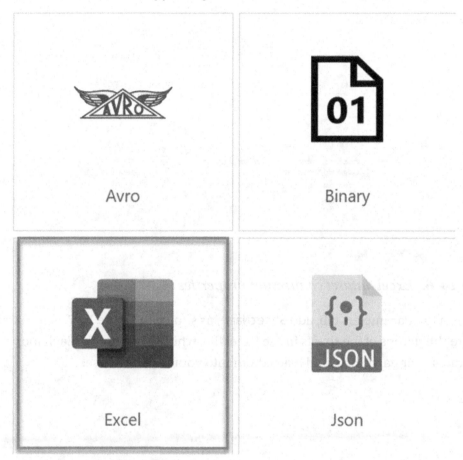

Figure 14-5.   Excel dataset format

The connection configuration properties for the Excel dataset can be found in Figure 14-6. The sheet name property will need to be configured with the dynamic parameterized `@dataset().SheetName` value. Also, since headers exist in the file, check "First row as header." Again, this exercise assumes that there are no floating images within the spreadsheet. Additionally, schema drift along with additional headers is not tested in this exercise.

Excel
Orders

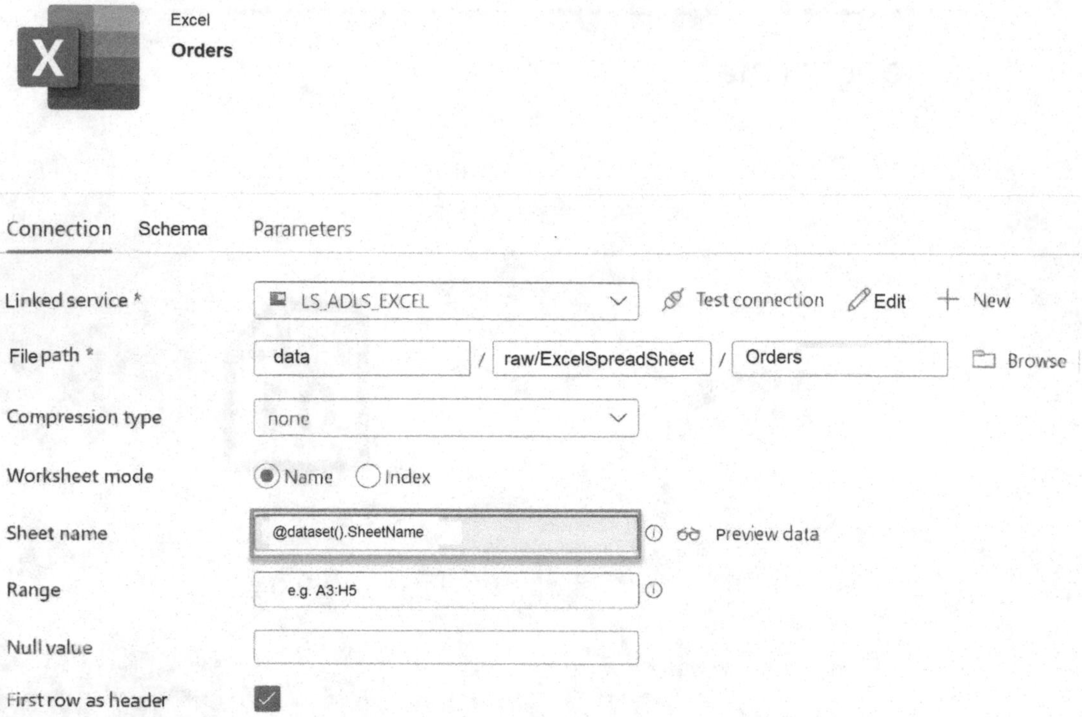

| Connection | Schema | Parameters |

Linked service *   [▣ LS_ADLS_EXCEL ▾]   ⌖ Test connection   ✎ Edit   + New

File path *   [data] / [raw/ExcelSpreadSheet] / [Orders]   🗀 Browse

Compression type   [none ▾]

Worksheet mode   ◉ Name   ◯ Index

Sheet name   [@dataset().SheetName] ⓘ 👁 Preview data

Range   [e.g. A3:H5] ⓘ

Null value   [ ]

First row as header   ☑

Figure 14-6. *Excel dataset connection properties*

Within the Parameters tab, add SheetName, as shown in Figure 14-7. This parameter will store the names of the sheets in the Excel file, which feeds the ForEach loop activity that iterates over each sheet and loads them into your SQL database.

Figure 14-7. *Excel dataset parameter properties*

Next, add a sink dataset to the target Azure SQL table with a connection to the appropriate linked service as shown in Figure 14-8. Please note that although you have specified a hard-coded table path including schema and table name, you do not need to worry about creating this corresponding table in your SQL database. Within ADF, when the sink dataset properties are configured in the pipeline, there is a property that you will need to enable to "auto-create" the table based on the incoming schema of the source Excel file, which you will learn about later in this chapter.

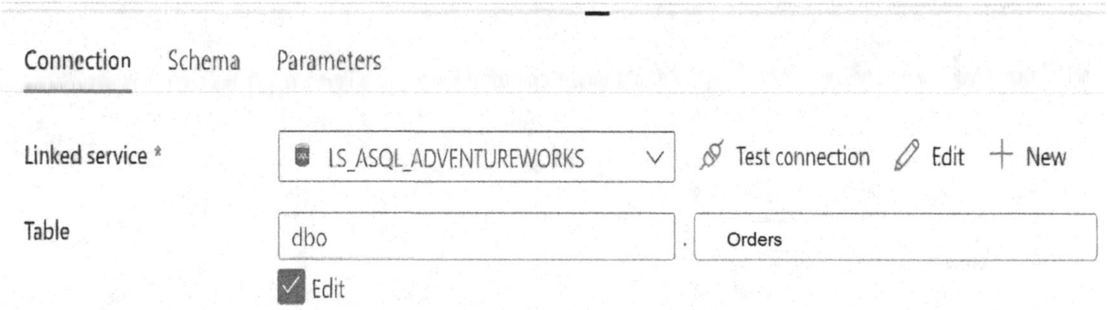

Figure 14-8. *Azure SQL sink dataset connection properties*

To recap, in this section, you have created the necessary prerequisites including creation and upload of Excel spreadsheets to ADLS Gen2 and creation of datasets and linked services. At this point, you are ready to proceed to the next section where you will learn how to create the actual ADF pipeline to load the Excel spreadsheets into Azure SQL Database tables.

Create a Pipeline to Load Multiple Excel Sheets in a Spreadsheet into a Single Azure SQL Table

Now let's create a pipeline to load multiple Excel sheets from a single spreadsheet file into a single Azure SQL table. Within the ADF pane, create a new pipeline and then add a ForEach loop activity to the pipeline canvas. Next, click the white space of the canvas within the pipeline to add a new Array variable called SheetName containing default values of all the sheets in the spreadsheet from Sheet1 through Sheet4, as depicted in Figure 14-9.

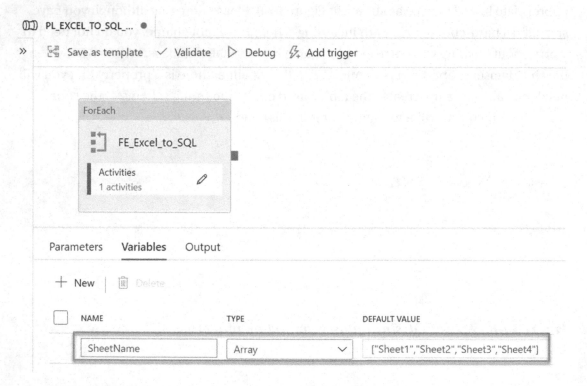

Figure 14-9. *ADF ForEach loop connection variables*

Next, click the ForEach loop activity to enable it and then add @variables('SheetName') to the Items property of the ForEach loop activity's Settings tab, as shown in Figure 14-10.

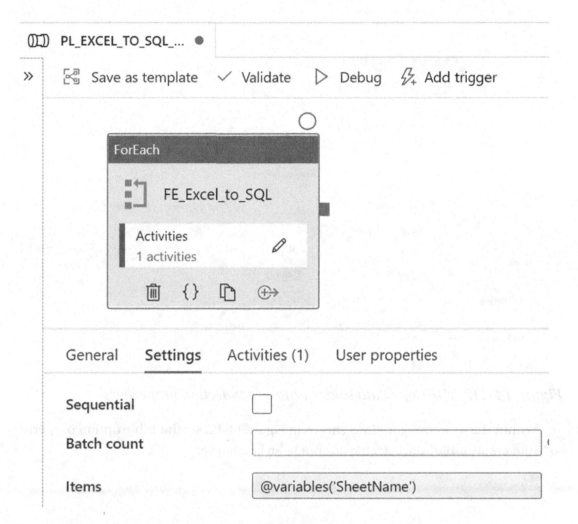

Figure 14-10. *ADF ForEach connection settings*

Next, navigate into the ForEach loop activity by clicking the Activities tab in Figure 14-10. Alternatively, you could get to the activities by clicking the pencil icon shown in the ForEach loop activity. Add a Copy data activity with the source configurations, as shown in Figure 14-11.

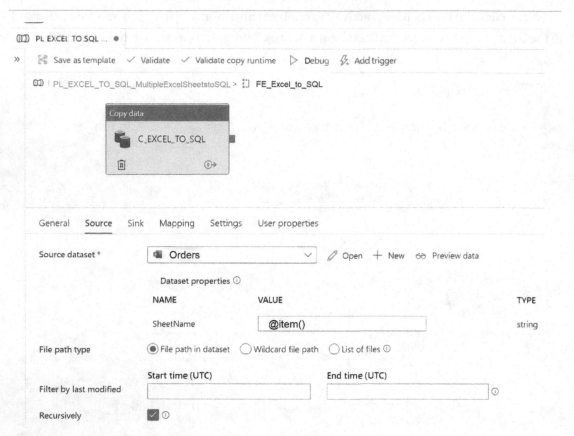

Figure 14-11. *ADF Copy data source dataset connection properties*

Within the sink configurations shown in Figure 14-12, set the Table option property to "Auto create table" since a table has not been created yet.

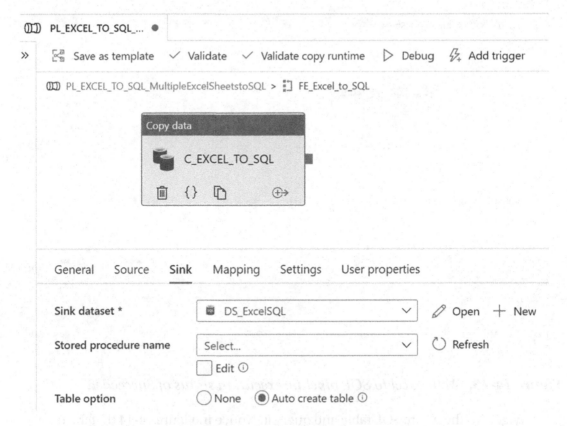

Figure 14-12. ADF Copy activity sink connection properties

After executing the pipeline, notice in Figure 14-13 that the four sheets have been loaded into the Azure SQL table.

PL_EXCEL_TO_SQL_MultipleExcelSheetstoSQL

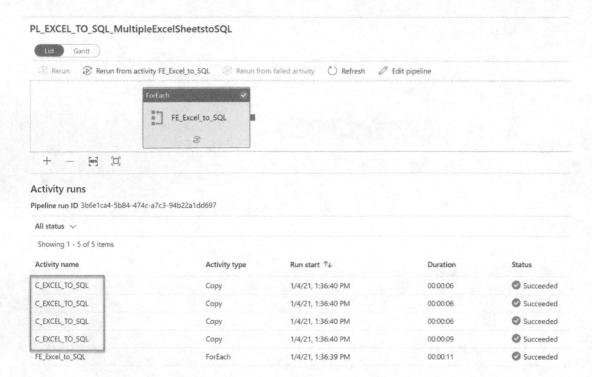

Figure 14-13. *ADF Excel to SQL pipeline execution status of Succeeded*

Navigate to the Azure SQL table and query it. Notice in Figure 14-14 that the data from all the Excel sheets are loaded into the single Azure SQL table.

Here is the SQL query that was executed in Figure 14-14:

```
SELECT [orderdate],
       [region],
       [rep],
       [item],
       [units],
       [unit cost],
       [total]
FROM   [dbo].[orders]
```

```
SELECT [OrderDate]
      ,[Region]
      ,[Rep]
      ,[Item]
      ,[Units]
      ,[Unit Cost]
      ,[Total]
  FROM [dbo].[ExcelTable]
```

100 % ▾ ◀

⊞ Results ⬛ Messages

	OrderDate	Region	Rep	Item	Units	Unit Cost	Total
1	2015-08-07	Central	Kivell	Pen Set	42	23.95	1005.9
2	2015-01-15	Central	Gill	Binder	46	8.99	413.54
3	2014-01-23	Central	Kivell	Binder	50	19.99	999.5
4	2015-03-24	Central	Jardine	Pen Set	50	4.99	249.5
5	2015-05-14	Central	Gill	Pencil	53	1.29	68.37
6	2015-07-21	Central	Morgan	Pen Set	55	12.49	686.95
7	2015-04-10	Central	Andrews	Pencil	66	1.99	131.34
8	2014-12-12	Central	Smith	Pencil	67	1.29	86.43
9	2014-04-18	Central	Andrews	Pencil	75	1.99	149.25
10	2015-05-31	Central	Gill	Binder	80	8.99	719.2
11	2015-07-04	East	Jones	Pen Set	62	4.99	309.38
12	2014-10-22	East	Jones	Pen	64	8.99	575.36
13	2014-12-29	East	Parent	Pen Set	74	15.99	1183.26
14	2014-07-29	East	Parent	Binder	81	19.99	1619.19
15	2014-01-06	East	Jones	Pencil	95	1.99	189.05
16	2015-04-27	East	Howard	Pen	96	4.99	479.04
17	2015-08-24	West	Sorvino	Desk	3	275	825
18	2015-03-07	West	Sorvino	Binder	7	19.99	139.93
19	2014-05-22	West	Thompson	Pencil	32	1.99	63.68
20	2014-03-15	West	Sorvino	Pencil	56	2.99	167.44
21	2015-10-14	West	Thompson	Binder	57	19.99	1139.43
22	2015-09-27	West	Sorvino	Pen	76	1.99	151.24
23	2014-09-01	Central	Smith	Desk	2	125	250
24	2015-06-17	Central	Kivell	Desk	5	125	625
25	2015-09-10	Central	Gill	Pencil	7	1.29	9.03
26	2015-11-17	Central	Jardine	Binder	11	4.99	54.89
27	2015-10-31	Central	Andrews	Pencil	14	1.29	18.06
28	2014-02-26	Central	Gill	Pen	27	19.99	539.73
29	2014-10-05	Central	Morgan	Binder	28	8.00	251.72
30	2015-12-21	Central	Andrews	Binder	28	4.99	139.72
31	2014-02-09	Central	Jardine	Pencil	36	4.99	179.64
32	2015-02-01	Central	Smith	Binder	87	15	1305
33	2014-05-05	Central	Jardine	Pencil	90	4.99	449.1
34	2014-06-25	Central	Morgan	Pencil	90	4.99	449.1
35	2015-12-04	Central	Jardine	Binder	94	19.99	1879.06
36	2014-11-25	Central	Kivell	Pen Set	96	4.99	479.04
37	2015-02-18	East	Jones	Binder	4	4.99	19.96
38	2014-11-08	East	Parent	Pen	15	19.99	299.85
39	2014-09-18	East	Jones	Pen Set	16	15.99	255.84
40	2014-07-12	East	Howard	Binder	29	1.99	57.71
41	2014-08-15	East	Jones	Pencil	35	4.99	174.65
42	2014-04-01	East	Jones	Binder	60	4.99	299.4
43	2014-06-08	East	Jones	Binder	60	8.99	539.4

Figure 14-14. *Query Azure SQL table from SSMS to confirm data is available*

Create a Pipeline to Load Multiple Excel Sheets in a Spreadsheet into Multiple Azure SQL Tables

In this next example, let's test loading multiple Excel sheets from an Excel spreadsheet into multiple Azure SQL tables within the same database. To begin, create a new Excel lookup table, as shown in Figure 14-15, containing the SheetName and TableName, which will be used by the dynamic ADF pipeline parameters.

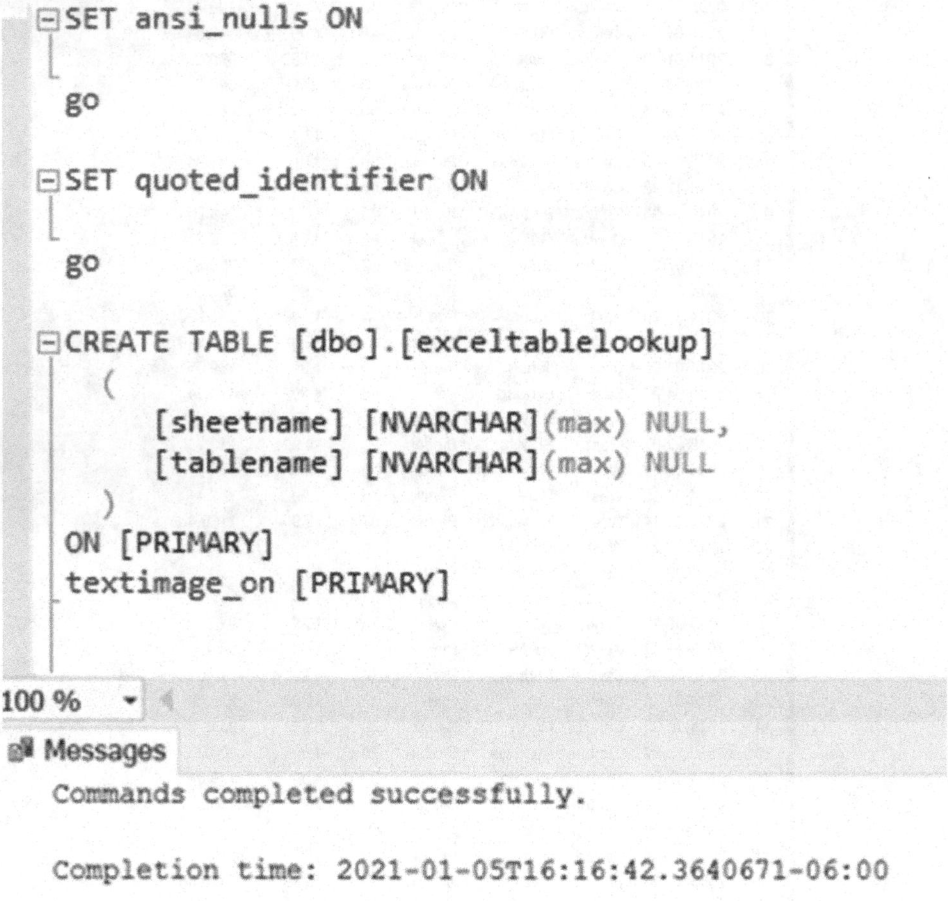

```
SET ansi_nulls ON

  go

SET quoted_identifier ON

  go

CREATE TABLE [dbo].[exceltablelookup]
  (
      [sheetname] [NVARCHAR](max) NULL,
      [tablename] [NVARCHAR](max) NULL
  )
ON [PRIMARY]
textimage_on [PRIMARY]
```

100 %

Messages

Commands completed successfully.

Completion time: 2021-01-05T16:16:42.3640671-06:00

Figure 14-15. *Create Excel lookup table containing SheetName and TableName*

The following script is used to create this lookup table in Figure 14-15:

```
SET ansi_nulls ON

go

SET quoted_identifier ON

go

CREATE TABLE [dbo].[exceltablelookup]
  (
     [sheetname] [NVARCHAR](max) NULL,
     [tablename] [NVARCHAR](max) NULL
  )
ON [PRIMARY]
textimage_on [PRIMARY]
```

Once the table is created, insert the SheetNames and corresponding TableNames into the table, as shown in Figure 14-16.

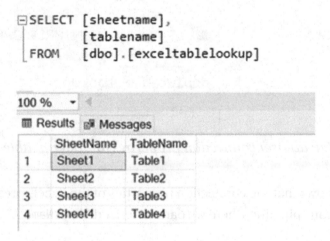

Figure 14-16. Insert the SheetNames and corresponding TableNames into the table

Here is the SQL query that has been executed in Figure 14-16:

```
SELECT [sheetname],
       [tablename]
FROM   [dbo].[exceltablelookup]
```

Next, add a new dataset with a connection to the Excel lookup table, as shown in Figure 14-17.

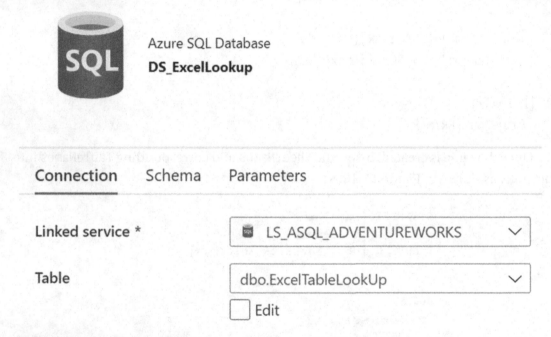

Figure 14-17. *ADF dataset connection properties for lookup table*

Figure 14-18 shows that the connection properties of the Excel spreadsheet will be similar to the previous pipeline where we parameterized SheetName.

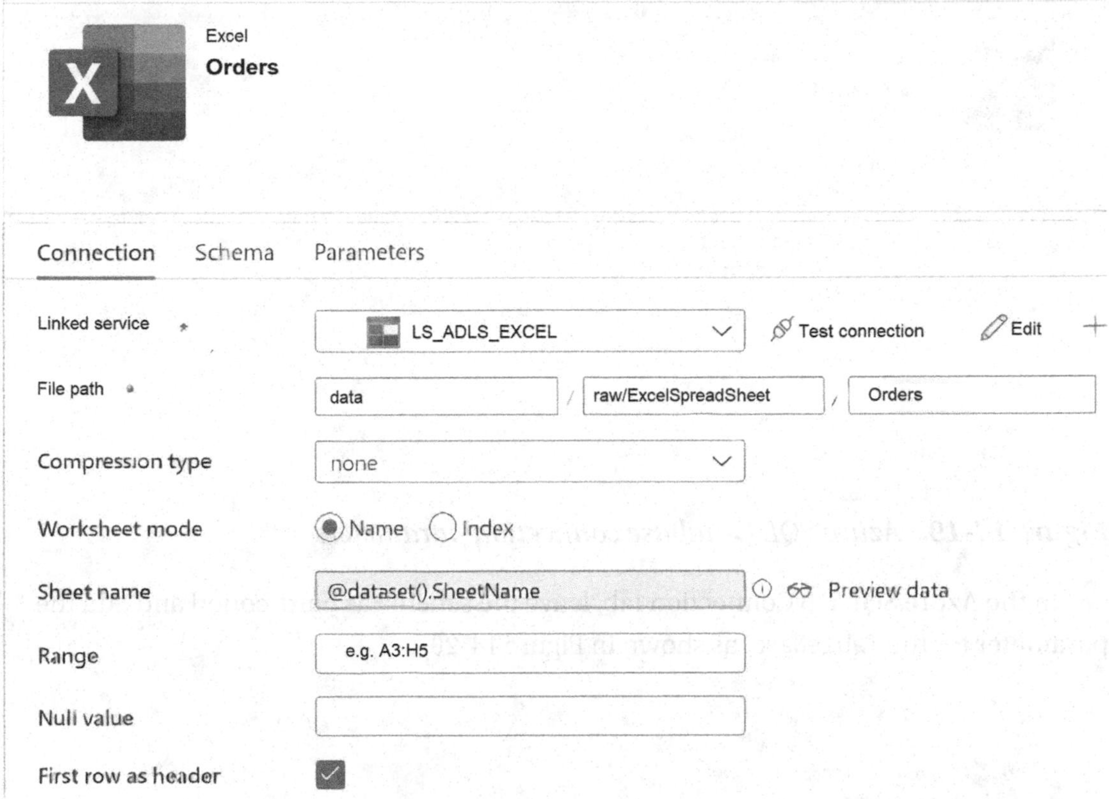

Excel
Orders

Connection	Schema	Parameters

| Linked service | * | LS_ADLS_EXCEL ∨ | 🖉 Test connection | 🖊 Edit | + |

| File path | * | data | / | raw/ExcelSpreadSheet | , | Orders |

| Compression type | none ∨ |

| Worksheet mode | ⦿ Name ◯ Index |

| Sheet name | @dataset().SheetName | ⓘ 👓 Preview data |

| Range | e.g. A3:H5 |

| Null value | |

| First row as header | ✓ |

Figure 14-18. ADF Excel dataset connection properties

In this scenario, add a parameter for the TableName in the Azure SQL Database dataset connection, as shown in Figure 14-19.

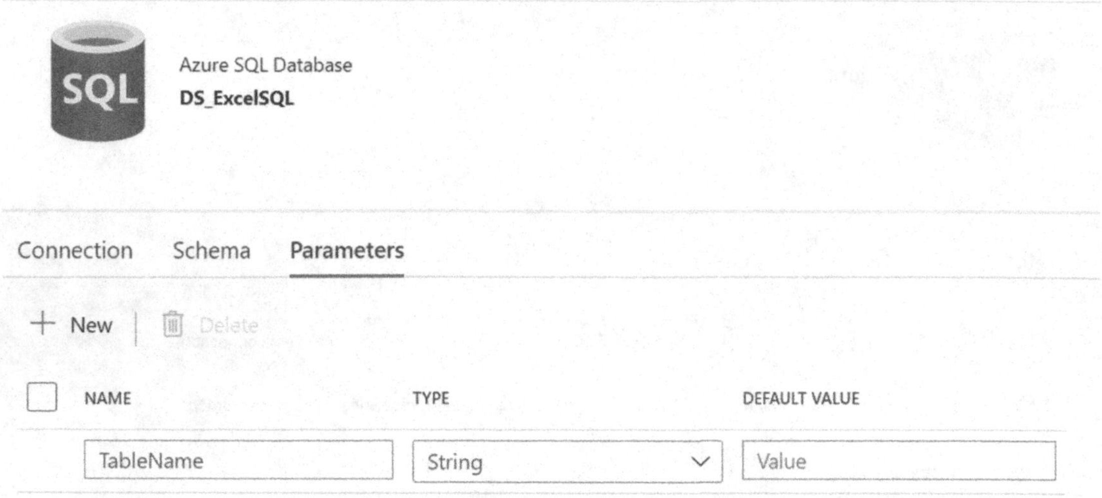

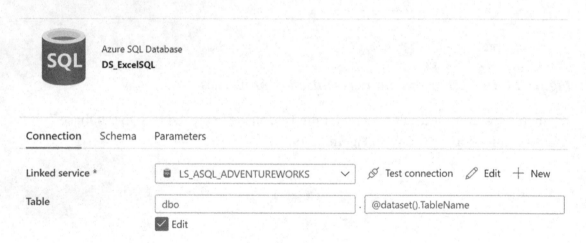

Figure 14-19. *Azure SQL Database connection parameters*

In the Azure SQL DB Connection tab, leave the schema as hard-coded and add the parameter for the TableName, as shown in Figure 14-20.

Figure 14-20. *Azure SQL Database parameterized connections*

In this pipeline, a lookup table will be needed that will serve the purpose of looking up the values in the SQL lookup table through a select * lookup on the table, as shown in Figure 14-21.

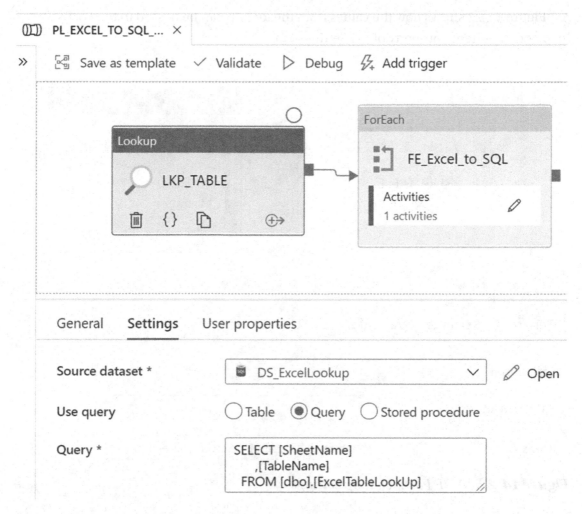

Figure 14-21. *ADF lookup settings*

Note that for the purpose of this exercise, the Query option has been selected, and the source SQL query is embedded. As an alternative option for low embedded SQL code maintenance within the ADF pipeline, consider using either the Table or Stored procedure query option type.

Figure 14-22 shows how the values from the lookup can be passed to the ForEach loop activity's Items property of the Settings tab.

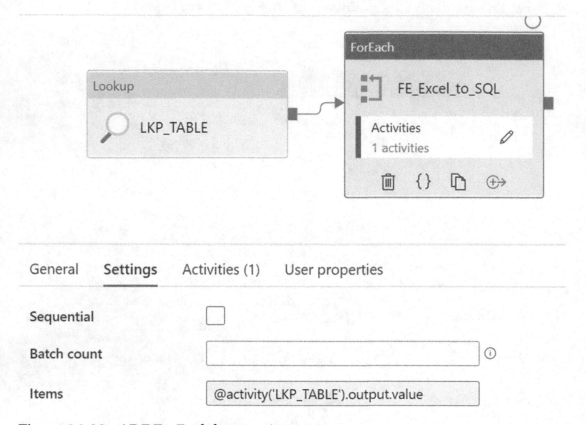

Figure 14-22. *ADF ForEach loop settings*

Next, within the ForEach loop activity, add a Copy data activity with the source dataset properties containing the parameterized SheetName value, as shown in Figure 14-23.

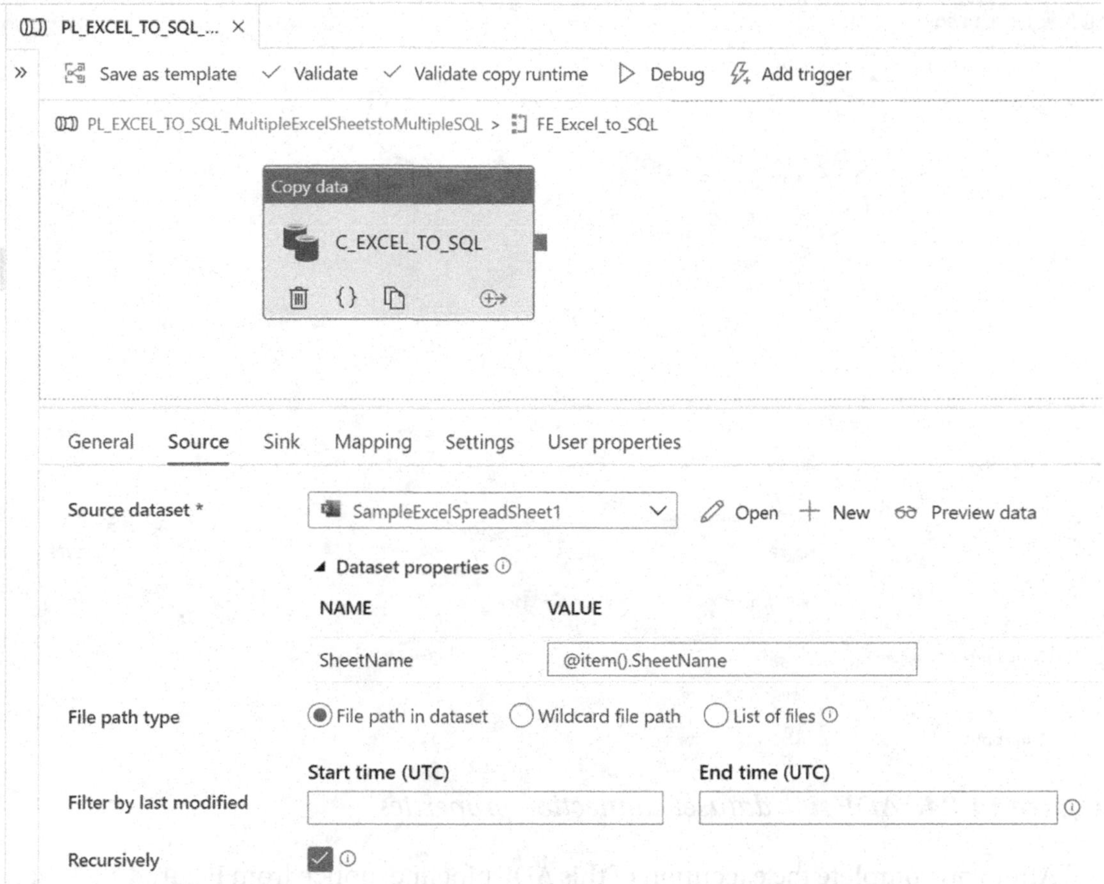

Figure 14-23. *ADF Copy data source dataset connection properties*

Next, the sink dataset properties will also need to contain the parameterized
TableName value, as shown in Figure 14-24. Note that the table option is once again set
to "Auto create table," which will create the table based on the incoming schema of the
source Excel file.

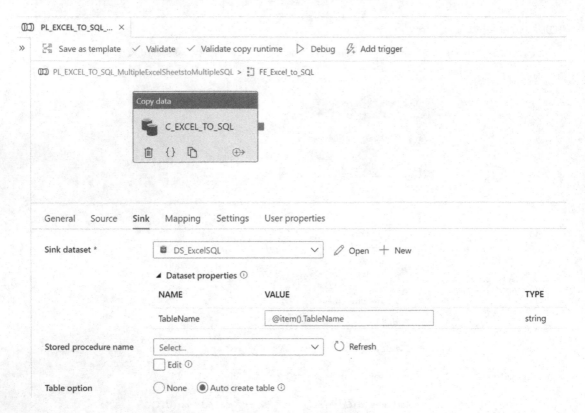

Figure 14-24. *ADF sink dataset connection properties*

After you complete the execution of this ADF pipeline, notice from Figure 14-25 that the pipeline succeeded and four tables were created in your Azure SQL Database.

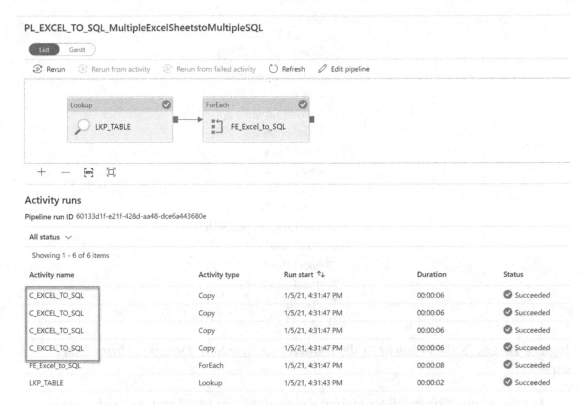

Figure 14-25. *ADF Excel to SQL pipeline status of Succeeded*

Navigate to the Azure SQL Database and notice in Figure 14-26 that all four tables were created with the appropriate names based on the `TableName` values you had defined in the SQL lookup table.

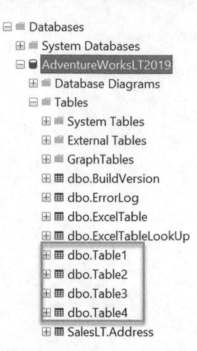

Figure 14-26. *SSMS view of tables loaded to Azure SQL Database from Excel sheets*

As a final check, query all four tables shown in Figure 14-27 and notice that they all contain the data from the Excel sheets, which confirms that the pipeline executed successfully, with the correct mappings of sheets to multiple tables, which were defined in the lookup table.

```
SELECT [OrderDate]
      ,[Region]
      ,[Rep]
      ,[Item]
      ,[Units]
      ,[Unit Cost]
      ,[Total]
  FROM [dbo].[Table1]
SELECT * FROM [dbo].[Table2]
SELECT * FROM [dbo].[Table3]
```

100 %   ▾

⊞ Results ▨ Messages

	OrderDate	Region	Rep	Item	Units	Unit Cost	Total
1	2014-09-01	Central	Smith	Desk	2	125	250
2	2015-06-17	Central	Kivell	Desk	5	125	625
3	2015-09-10	Central	Gill	Pencil	7	1.29	9.03
4	2015-11-17	Central	Jardine	Binder	11	4.99	54.89
5	2015-10-31	Central	Andrews	Pencil	14	1.29	18.06
6	2014-02-26	Central	Gill	Pen	27	19.99	539.73
7	2014-10-05	Central	Morgan	Binder	28	8.99	251.72
8	2015-12-21	Central	Andrews	Binder	28	4.99	139.72

	OrderDate	Region	Rep	Item	Units	Unit Cost	Total
1	2015-08-07	Central	Kivell	Pen Set	42	23.95	1005.9
2	2015-01-15	Central	Gill	Binder	46	8.99	413.54
3	2014-01-23	Central	Kivell	Binder	50	19.99	999.5
4	2015-03-24	Central	Jardine	Pen Set	50	4.99	249.5
5	2015-05-14	Central	Gill	Pencil	53	1.29	68.37
6	2015-07-21	Central	Morgan	Pen Set	55	12.49	686.95
7	2015-04-10	Central	Andrews	Pencil	66	1.99	131.34
8	2014-12-12	Central	Smith	Pencil	67	1.29	86.43

	OrderDate	Region	Rep	Item	Units	Unit Cost	Total
1	2015-02-01	Central	Smith	Binder	87	15	1305
2	2014-05-05	Central	Jardine	Pencil	90	4.99	449.1
3	2014-06-25	Central	Morgan	Pencil	90	4.99	449.1
4	2015-12-04	Central	Jardine	Binder	94	19.99	1879.06
5	2014-11-25	Central	Kivell	Pen Set	96	4.99	479.04
6	2015-02-18	East	Jones	Binder	4	4.99	19.96
7	2014-11-08	East	Parent	Pen	15	19.99	299.85
8	2014-09-18	East	Jones	Pen Set	16	15.99	255.84

	OrderDate	Region	Rep	Item	Units	Unit Cost	Total
1	2015-07-04	East	Jones	Pen Set	62	4.99	309.38
2	2014-10-22	East	Jones	Pen	64	8.99	575.36
3	2014-12-29	East	Parent	Pen Set	74	15.99	1183.26
4	2014-07-29	East	Parent	Binder	81	19.99	1619.19
5	2014-01-06	East	Jones	Pencil	95	1.99	189.05
6	2015-04-27	East	How...	Pen	96	4.99	479.04
7	2015-08-24	West	Sorvi...	Desk	3	275	825
8	2015-03-07	West	Sorvi...	Binder	7	19.99	139.93
9	2014-05-22	West	Tho...	Pencil	32	1.99	63.68
10	2014-03-15	West	Sorvi...	Pencil	56	2.99	167.44
11	2015-10-14	West	Tho...	Binder	57	19.99	1139.43
12	2015-09-27	West	Sorvi...	Pen	76	1.99	151.24

Figure 14-27. *SSMS view of all tables to verify that data exists*

Here are the SQL queries that have been executed in Figure 14-27:

```
SELECT [orderdate],
       [region],
       [rep],
       [item],
       [units],
       [unit cost],
       [total]
FROM   [dbo].[table1]

SELECT [orderdate],
       [region],
       [rep],
       [item],
       [units],
       [unit cost],
       [total]
FROM   [dbo].[table2]

SELECT [orderdate],
       [region],
       [rep],
       [item],
       [units],
       [unit cost],
       [total]
FROM   [dbo].[table3]
```

Summary

In this chapter, I have demonstrated how to dynamically load an Excel spreadsheet residing in ADLS Gen2 containing multiple sheets into a single Azure SQL table and also into multiple tables for every sheet using Azure Data Factory's Excel connector, along with parameterized pipelines.

## CHAPTER 15

Delta Lake

While working with Azure Data Lake Storage Gen2 and Apache Spark, users have learned about both the limitations of Apache Spark and the many data lake implementation challenges. The need for an ACID-compliant feature set is critical within the data lake landscape, and Delta Lake offers many solutions to the current limitations of the standard Azure Data Lake Storage Gen2 accounts.

Delta Lake is an open source storage layer that guarantees data atomicity, consistency, isolation, and durability in the lake. In short, a Delta Lake is ACID compliant. In addition to providing ACID transactions, scalable metadata handling, and more, Delta Lake runs on an existing data lake and is compatible with Apache Spark APIs. There are a few methods of getting started with Delta Lake. Databricks offers notebooks along with compatible Apache Spark APIs to create and manage Delta Lakes. Alternatively, Azure Data Factory's Mapping Data Flows, which uses scaled-out Apache Spark clusters, can be used to perform ACID-compliant CRUD operations through GUI-designed ETL pipelines. This chapter will demonstrate how to get started with Delta Lake using Azure Data Factory's Delta Lake connector through examples of how to create, insert, update, and delete in a Delta Lake.

Why an ACID Delta Lake

There are many advantages to introducing Delta Lake into a modern cloud data architecture. Traditionally, data lakes and Apache Spark are not ACID compliant. Delta Lake introduces this ACID compliance to solve many the following ACID compliance issues:

1. **Atomicity**: *Write either all data or nothing.* Apache Spark *save modes* do not utilize any locking and are not atomic. With this, a failed job may leave an incomplete file and may corrupt data. Additionally, a failing job may remove the old file and corrupt the new file.

© Ron C. L'Esteve 2021
R. C. L'Esteve, *The Definitive Guide to Azure Data Engineering*, https://doi.org/10.1007/978-1-4842-7182-7_15

While this seems concerning, Spark does have built-in data frame writer APIs that are not atomic but behave so for append operations. This however does come with performance overhead for use with cloud storage. The currently available Apache Spark save modes include ErrorIfExists, Append, Overwrite, and Ignore.

2. **Consistency**: *Data is always in a valid state*. If the Spark API writer deletes an old file and creates a new one and the operation is not transactional, then there will always be a period of time when the file does not exist between the deletion of the old file and creation of the new. In that scenario, if the overwrite operation fails, this will result in data loss of the old file. Additionally, the new file may not be created. This is a typical Spark overwrite operation issue related to consistency.

3. **Isolation**: *Multiple transactions occur independently without interference*. This means that when writing to a dataset, other concurrent reads or writes on the same dataset should not be impacted by the write operation. Typical transactional databases offer multiple isolation levels, such as read uncommitted, read committed, repeatable read, snapshot, and serializable. While Spark has task- and job-level commits, since it lacks atomicity, it does not have isolation types.

4. **Durability**: *Committed data is never lost.* When Spark does not correctly implement a commit, then it overwrites all the great durability features offered by cloud storage options and either corrupts and/or loses the data. This violates data durability.

Prerequisites

Now that you have an understanding of the current data lake and Spark challenges along with benefits of an ACID-compliant Delta Lake, let's get started with the exercise.

For this exercise, be sure to successfully create the following prerequisites:

1. **Create a Data Factory V2**: Data Factory will be used to perform the ELT orchestrations. Additionally, ADF's Mapping Data Flows Delta Lake connector will be used to create and manage the Delta Lake.

2. **Create a Data Lake Storage Gen2**: ADLS Gen2 will be the data lake storage, on top of which the Delta Lake will be created.

3. **Create Data Lake Storage Gen2 container and zones**: Once your Data Lake Storage Gen2 account is created, also create the appropriate containers and zones. Revisit Chapter 3 for more information on designing ADLS Gen2 zones where I discuss how to design an ADLS Gen2 storage account. This exercise will use the Raw zone to store a sample source parquet file. Additionally, the Staging zone will be used for Delta Updates, Inserts, and Deletes and additional transformations. Though the Curated zone will not be used in this exercise, it is important to mention that this zone may contain the final ETL, advanced analytics, or data science models that are further transformed and curated from the Staging zone. Once the various zones are created in your ADLS Gen2 account, they would look similar to the illustration in Figure 15-1.

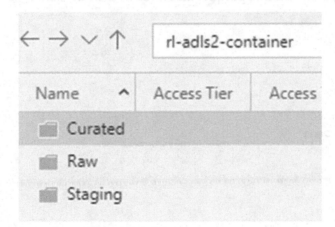

Figure 15-1. *ADLS Gen2 zones/folders*

4. **Upload data to the Raw zone**: Finally, you'll need some data for this exercise. By searching for "sample parquet files" online or within publicly available datasets, you'll obtain access to a number of free sample parquet files. For this exercise, you could download the sample parquet files within the following GitHub

repo (https://github.com/Teradata/kylo/tree/master/
samples/sample-data/parquet) and then upload it to your ADLS
Gen2 storage account, as shown in Figure 15-2.

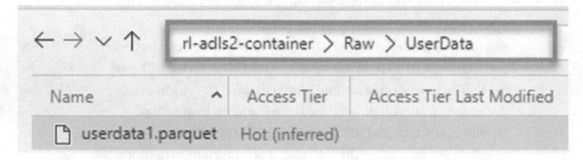

Figure 15-2. *Sample userdata1.parquet file*

5. **Create a Data Factory parquet dataset pointing to the Raw zone**:
   The final prerequisite would be to create a Parquet format dataset
   in the newly created instance of ADF V2, as shown in Figure 15-3,
   pointing to the sample parquet file stored in the Raw zone.

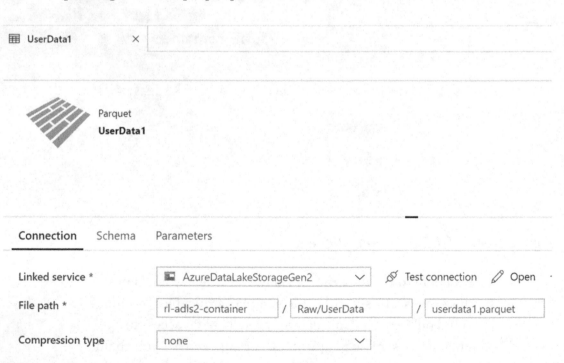

Figure 15-3. *userData1 connection setting in ADF*

324

Create and Insert into Delta Lake

Now that all prerequisites are in place, you are ready to create the initial delta tables and insert data from your Raw zone into the delta tables.

Begin the process by creating a new Data Factory pipeline and adding a new "Mapping Data Flow" to it. Also remember to name the pipeline and data flow sensible names, much like the sample shown in Figure 15-4.

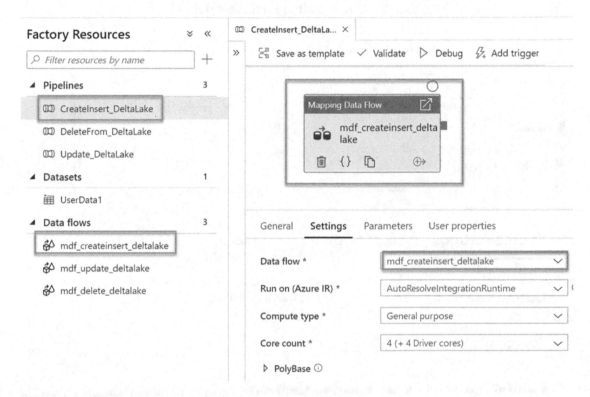

Figure 15-4. *Mapping Data Flow canvas for insert*

Within the data flow, add a source with the configuration settings shown in Figure 15-5. Also, check the option to "Allow schema drift." When metadata related to fields, columns, and types change frequently, this is referred to as schema drift. Without a proper process to handle schema drift, an ETL pipeline might fail. ADF supports flexible schemas that change often. ADF treats schema drift as late binding. Therefore, the drifted schemas will not be available for you to view in the data flow.

When schema drift is enabled, all of the incoming fields are read from your source during execution and passed through the entire flow to the sink. All newly detected columns arrive as a string data type by default. If you need to automatically infer data types of your drifted columns, then you'll need to enable Infer drifted column types in your source settings.

Allow auto-detection of drifted column types. Sampling offers a method to limit the number of rows from the source, mainly used for testing and debugging purposes.

Figure 15-5. *Mapping Data Flows ETL flow source settings for inserts*

Since Delta Lake leverages Spark's distributed processing power, it is capable of partitioning data appropriately. However, for purposes of exploring the capability of manually setting partitioning, configure 20 Hash partitions on the ID column, as shown in Figure 15-6.

Figure 15-6. Optimize settings for MDF partitioning

After adding the destination activity, ensure that the sink type, shown in Figure 15-7, is set to Delta format in Azure Data Factory (`https://docs.microsoft.com/en-us/azure/data-factory/format-delta`). Note that Delta Lake is available as both a source and sink in Mapping Data Flows. Also, you will be required to select the linked service once the sink type of Delta Lake is selected.

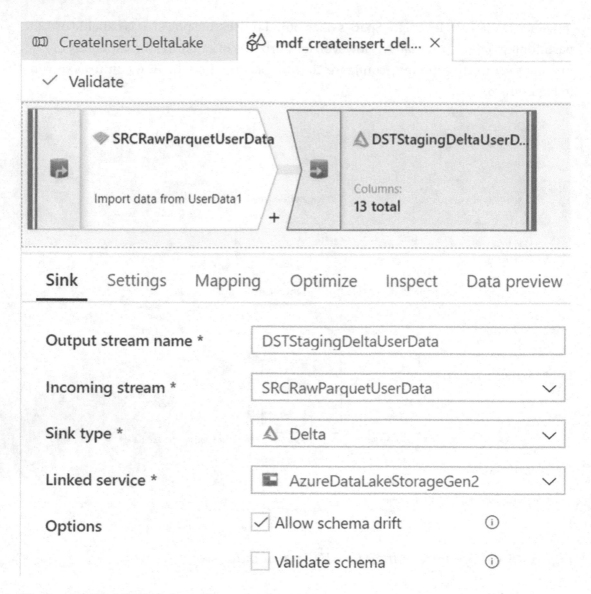

Figure 15-7. *MDF sink settings*

Under the Settings tab shown in Figure 15-8, ensure that the Staging folder is selected and select Allow insert for the update method. Also, select the Truncate table option if there is a need to truncate the delta table before loading it.

The process of vacuuming a delta table within the lake (https://docs.databricks. com/spark/latest/spark-sql/language-manual/delta-vacuum.html) will remove files that are no longer referenced by delta tables and are older than the retention threshold in hours. The default is 30 days if the value is left at 0 or empty.

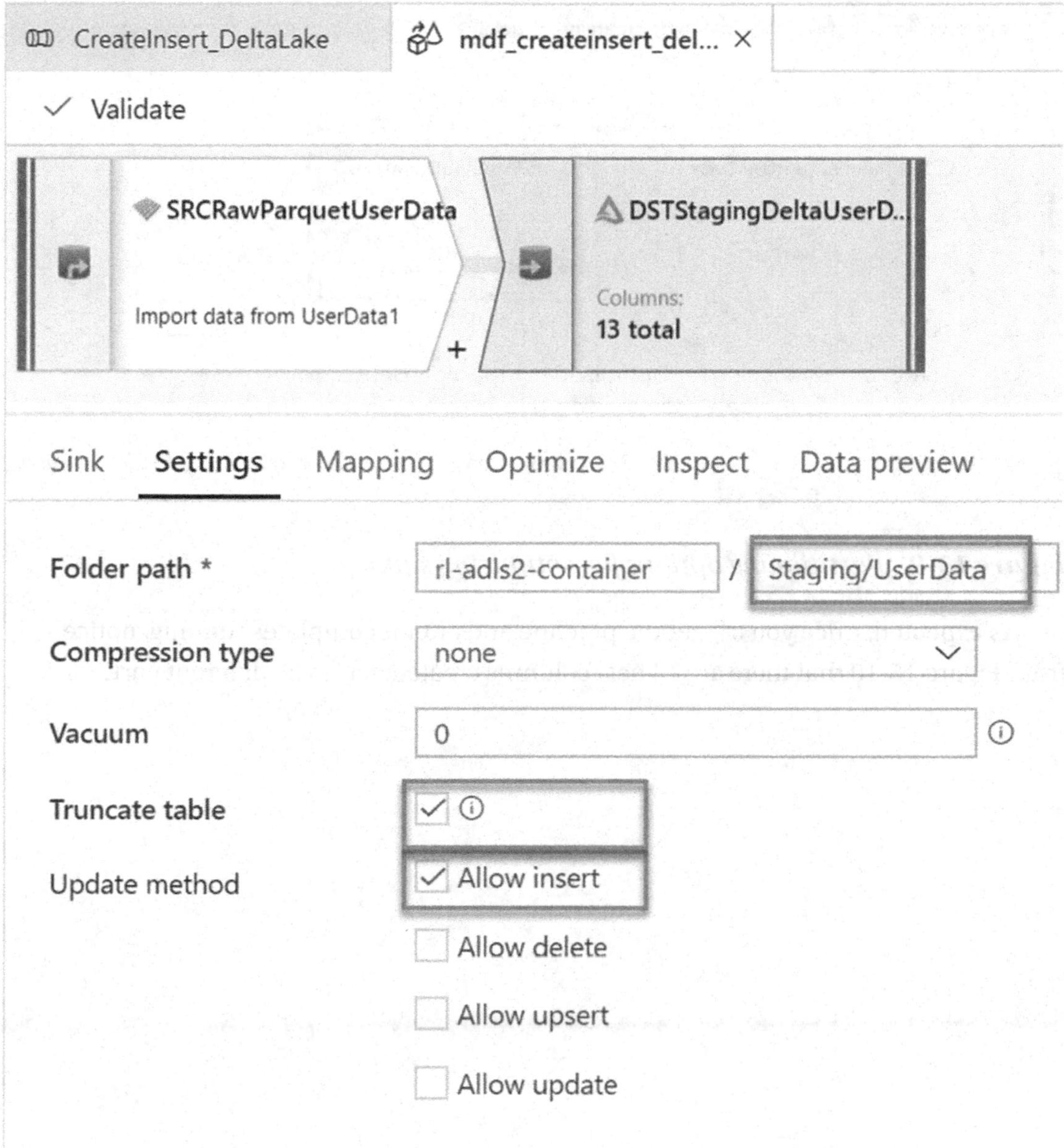

Figure 15-8. *Settings for insert sink in MDF*

Finally, within the Optimize tab shown in Figure 15-9, use the current partitioning since the source partitioning will flow downstream to the sink.

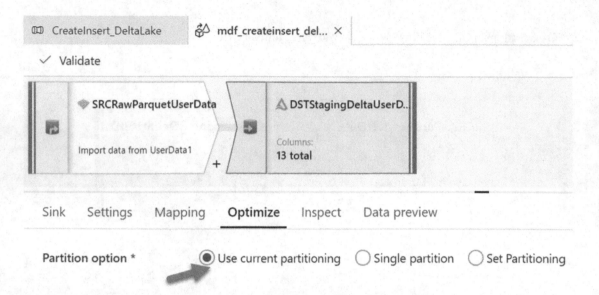

Figure 15-9. *Partitioning options for optimizing sink*

As expected, once you trigger the pipeline and after it completes running, notice from Figure 15-10 that there are 13 new columns created across 20 different partitions.

SRCRawParquetUserData
Source

Total columns	13
New columns	13
Updated columns	0
Dropped columns	0
Drifted columns	0

Stream information

Rows calculated	1,000
Total partition	20
Stage time	2s 160ms
Last update (CDT)	7/14/2020, 4:53:13 PM

Partition chart

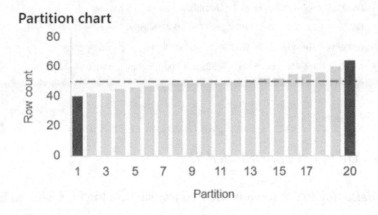

Skewness	0.4787
Kurtosis	2.8690

Figure 15-10. *ADF MDF pipeline run details*

While looking at the ADLS Gen2 Staging folder, notice from Figure 15-11 that a *delta_log* folder along with 20 snappy compressed parquet files has been created.

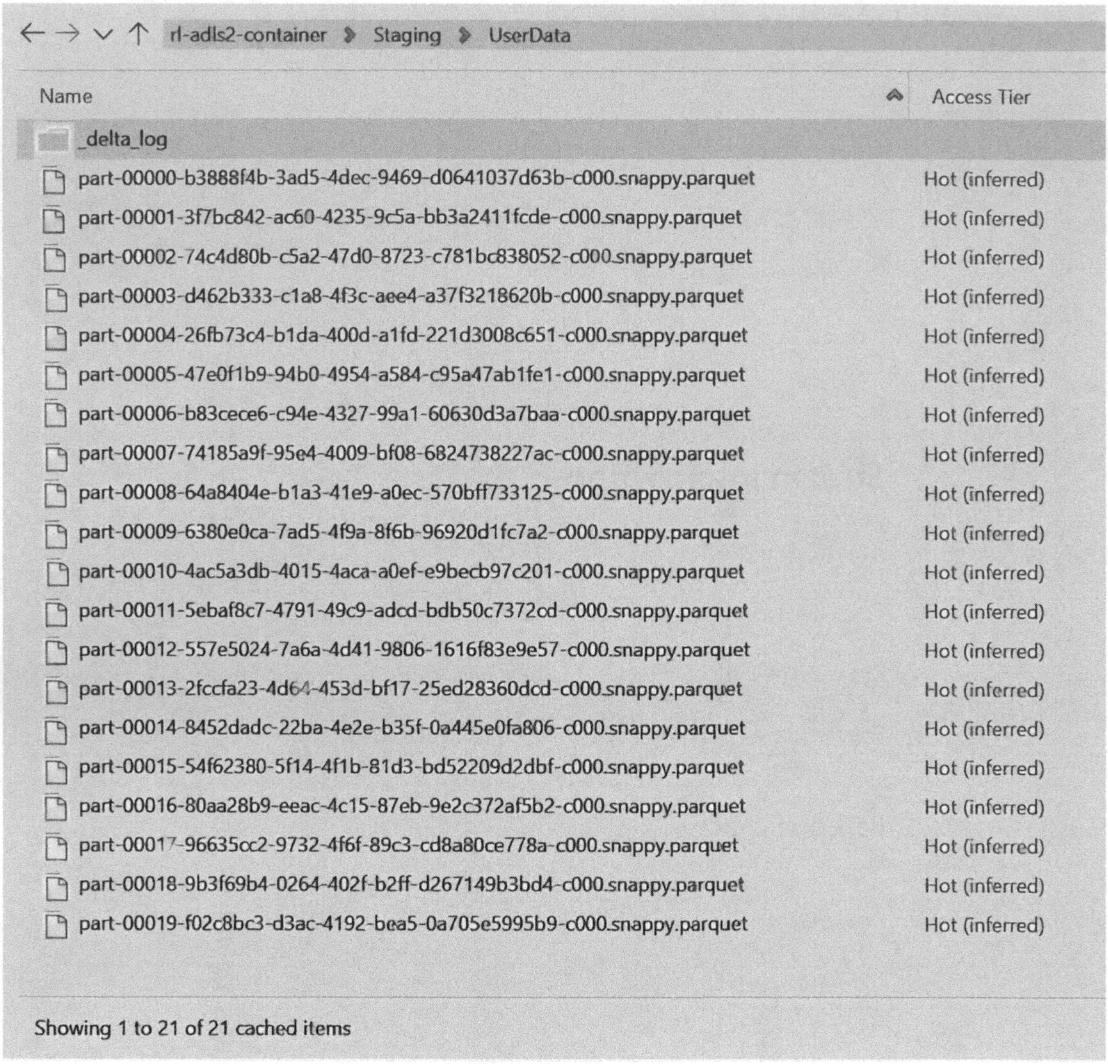

← → ∨ ↑ rl-adls2-container ❯ Staging ❯ UserData		
Name		**Access Tier**
_delta_log		
part-00000-b3888f4b-3ad5-4dec-9469-d0641037d63b-c000.snappy.parquet		Hot (inferred)
part-00001-3f7bc842-ac60-4235-9c5a-bb3a2411fcde-c000.snappy.parquet		Hot (inferred)
part-00002-74c4d80b-c5a2-47d0-8723-c781bc838052-c000.snappy.parquet		Hot (inferred)
part-00003-d462b333-c1a8-4f3c-aee4-a37f3218620b-c000.snappy.parquet		Hot (inferred)
part-00004-26fb73c4-b1da-400d-a1fd-221d3008c651-c000.snappy.parquet		Hot (inferred)
part-00005-47e0f1b9-94b0-4954-a584-c95a47ab1fe1-c000.snappy.parquet		Hot (inferred)
part-00006-b83cece6-c94e-4327-99a1-60630d3a7baa-c000.snappy.parquet		Hot (inferred)
part-00007-74185a9f-95e4-4009-bf08-6824738227ac-c000.snappy.parquet		Hot (inferred)
part-00008-64a8404e-b1a3-41e9-a0ec-570bff733125-c000.snappy.parquet		Hot (inferred)
part-00009-6380e0ca-7ad5-4f9a-8f6b-96920d1fc7a2-c000.snappy.parquet		Hot (inferred)
part-00010-4ac5a3db-4015-4aca-a0ef-e9becb97c201-c000.snappy.parquet		Hot (inferred)
part-00011-5ebaf8c7-4791-49c9-adcd-bdb50c7372cd-c000.snappy.parquet		Hot (inferred)
part-00012-557e5024-7a6a-4d41-9806-1616f83e9e57-c000.snappy.parquet		Hot (inferred)
part-00013-2fccfa23-4d64-453d-bf17-25ed28360dcd-c000.snappy.parquet		Hot (inferred)
part-00014-8452dadc-22ba-4e2e-b35f-0a445e0fa806-c000.snappy.parquet		Hot (inferred)
part-00015-54f62380-5f14-4f1b-81d3-bd52209d2dbf-c000.snappy.parquet		Hot (inferred)
part-00016-80aa28b9-eeac-4c15-87eb-9e2c372af5b2-c000.snappy.parquet		Hot (inferred)
part-00017-96635cc2-9732-4f6f-89c3-cd8a80ce778a-c000.snappy.parquet		Hot (inferred)
part-00018-9b3f69b4-0264-402f-b2ff-d267149b3bd4-c000.snappy.parquet		Hot (inferred)
part-00019-f02c8bc3-d3ac-4192-bea5-0a705e5995b9-c000.snappy.parquet		Hot (inferred)

Showing 1 to 21 of 21 cached items

Figure 15-11. *Delta Lake partitioned files*

Open the *delta_log* folder to view the two transaction log files, shown in Figure 15-12. The transaction log captures many important features, including ACID transactions, scalable metadata handling, time travel, and more (`https://databricks.com/blog/2019/08/21/diving-into-delta-lake-unpacking-the-transaction-log.html`).

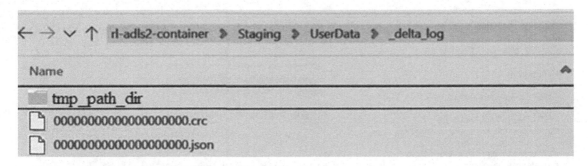

Figure 15-12. Delta log files

After checking the new data in the Staging Delta Lake, notice that there are new records inserted. To easily explore the Delta Lake from ADF with Mapping Data Flows, simply toggle the Data Preview tab to view the details of the data flow, as shown in Figure 15-13.

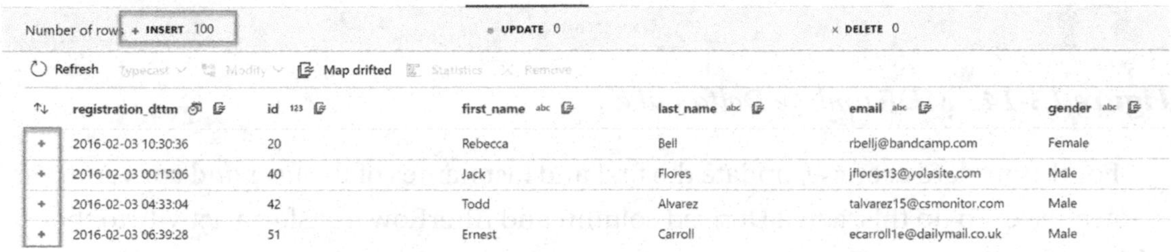

	registration_dttm	id	first_name	last_name	email	gender
+	2016-02-03 10:30:36	20	Rebecca	Bell	rbellj@bandcamp.com	Female
+	2016-02-03 00:15:06	40	Jack	Flores	jflores13@yolasite.com	Male
+	2016-02-03 04:33:04	42	Todd	Alvarez	talvarez15@csmonitor.com	Male
+	2016-02-03 06:39:28	51	Ernest	Carroll	ecarroll1e@dailymail.co.uk	Male

Number of rows + INSERT 100   ≡ UPDATE 0   × DELETE 0

Figure 15-13. Insert data results reading from parquet file

Update Delta Lake

So far, inserts into the Delta Lake have been covered in the previous section. Next, let's take a look at how Data Factory can handle updates to your delta tables. Similar to what you did in the previous section for inserts, create a new ADF pipeline with a Mapping Data Flow for updates, shown in Figure 15-14.

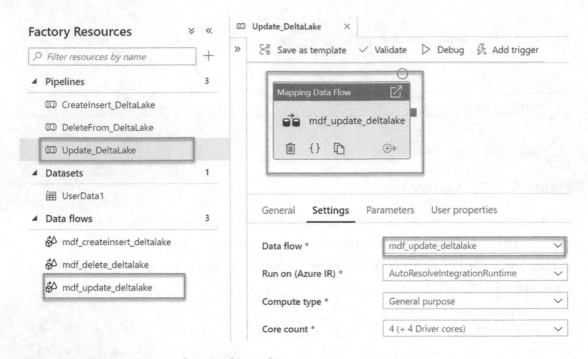

Figure 15-14. *MDF update Delta Lake*

For this update exercise, update the first and last names of the user and convert them to lowercase. To do this, add a Derived column and AlterRow transform activity to the update Mapping Data Flow canvas, as shown in Figure 15-15.

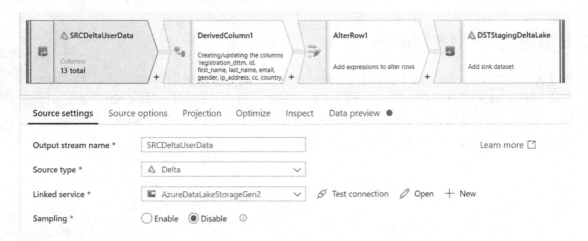

Figure 15-15. *Source settings for updating Mapping Data Flows parquet*

Within the Source options tab shown in Figure 15-16, the source data is still your Staging Delta Lake that was also configured for the inserts. The Delta Lake source connector within ADF also introduced delta time travel for large-scale data lakes to audit data changes, reproduce results, handle rollbacks, and more. Time travel allows you to query data by timestamp or version (`https://databricks.com/blog/2019/02/04/introducing-delta-time-travel-for-large-scale-data-lakes.html`).

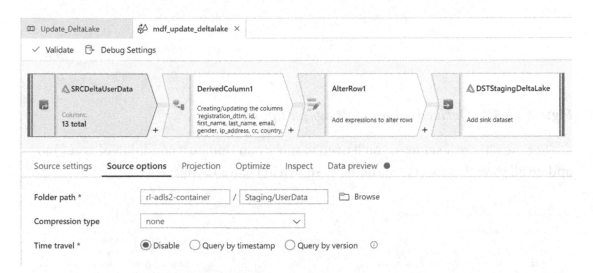

Figure 15-16. *Update source options*

The Derived column transform activity converts first and last names to lowercase using the expressions `lower(first_name)` and `lower(last_name)`, as shown in Figure 15-17. Mapping Data Flows is capable of handling extremely complex transformations in this stage.

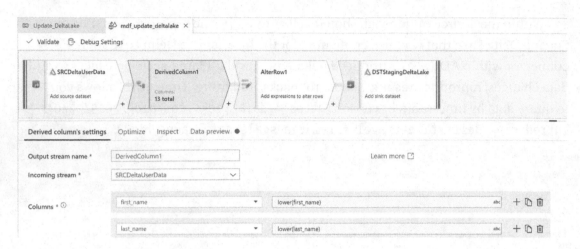

Figure 15-17. *Derived column settings for update*

For the alter row settings, you'll need to specify an Update if condition of `true()` to update all rows that meet the criteria, as shown in Figure 15-18.

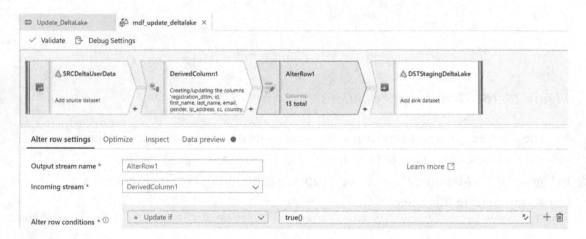

Figure 15-18. *Settings for the AlterRow updates*

Ensure that you verify the Sink tab's configuration settings, as shown in Figure 15-19.

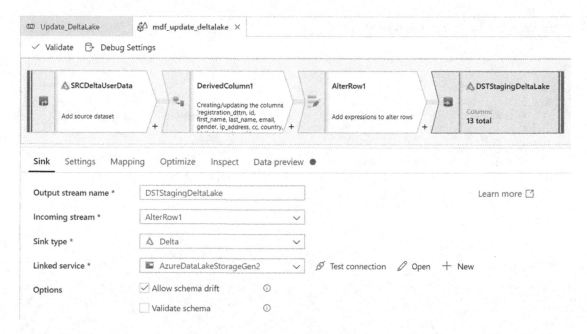

Figure 15-19. *Sink settings for update MDF*

Ensure that the sink is still pointing to the Staging Delta Lake data. Also, select Allow update as the update method. To show that multiple key columns can be simultaneously selected, there are three columns selected in Figure 15-20.

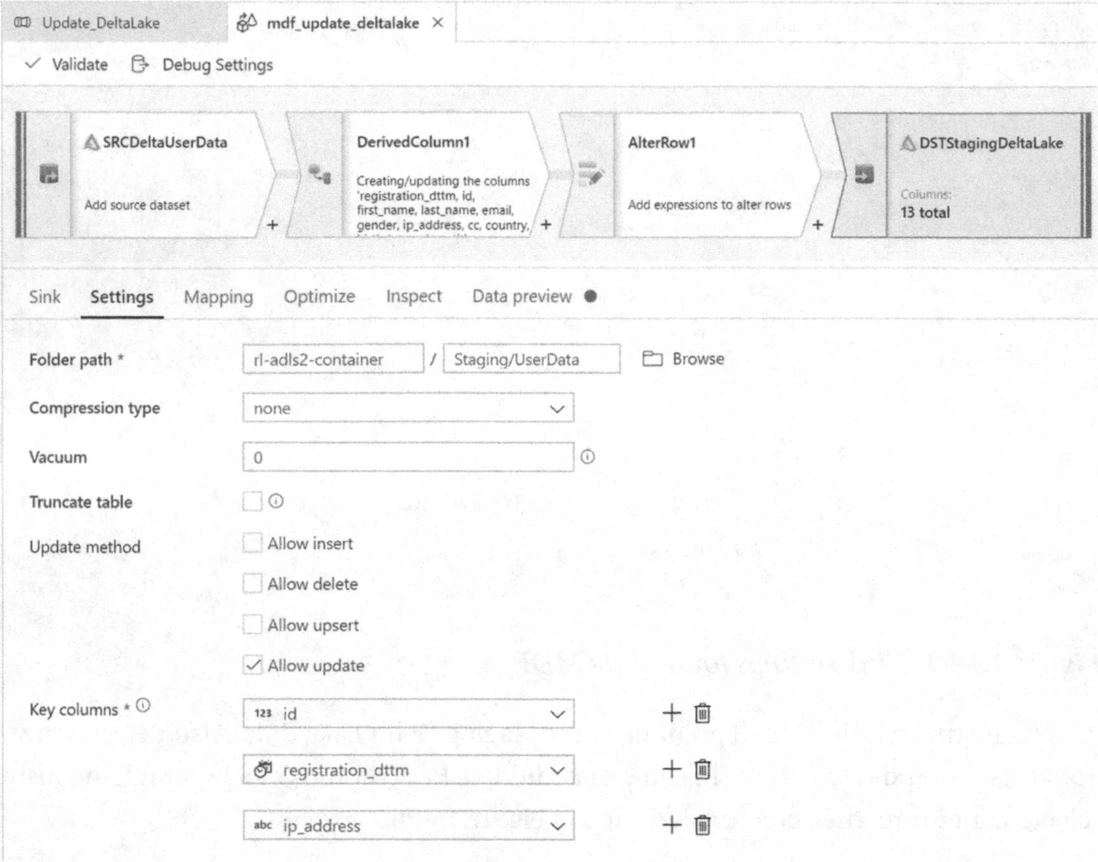

***Figure 15-20.** Update method settings for MDF sink*

After the pipeline is saved and triggered, notice the results from the ADF Data Preview tab shown in Figure 15-21. The first and last names have been updated to lowercase values.

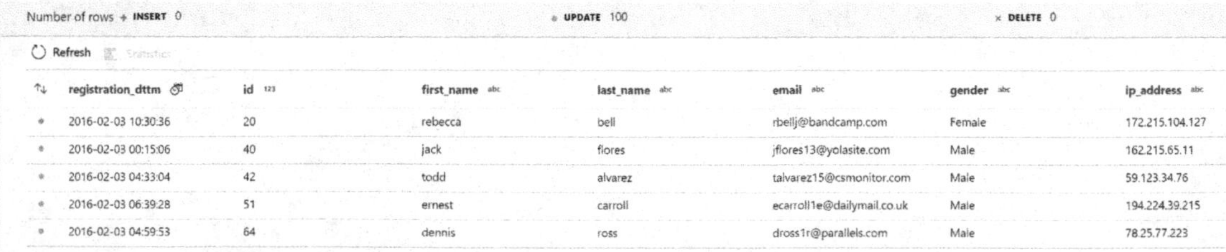
Number of rows + INSERT 0 · UPDATE 100 × DELETE 0

registration_dttm	id	first_name	last_name	email	gender	ip_address
2016-02-03 10:30:36	20	rebecca	bell	rbellj@bandcamp.com	Female	172.215.104.127
2016-02-03 00:15:06	40	jack	flores	jflores13@yolasite.com	Male	162.215.65.11
2016-02-03 04:33:04	42	todd	alvarez	talvarez15@csmonitor.com	Male	59.123.34.76
2016-02-03 06:39:28	51	ernest	carroll	ecarroll1e@dailymail.co.uk	Male	194.224.39.215
2016-02-03 04:59:53	64	dennis	ross	dross1r@parallels.com	Male	78.25.77.223

***Figure 15-21.** Data showing updates as expected*

Delete from Delta Lake

To recap, inserts and updates have been covered till now. Next, let's look at an example of how Mapping Data Flows handles deletes in Delta Lake. Similar to the process of setting up the insert and update MDF pipelines, create a new Data Factory Mapping Data Flow, as shown in Figure 15-22.

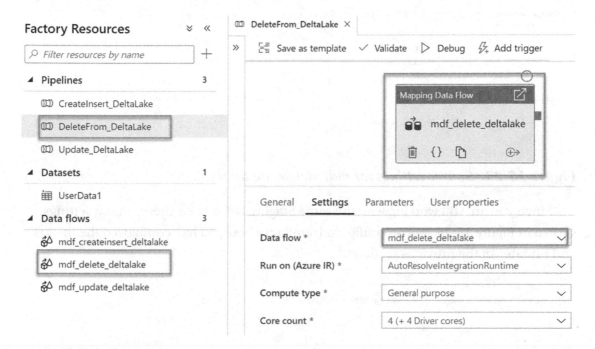

Figure 15-22. *MDF for deleting from Delta Lake*

Configure the Delta source settings as desired, shown in Figure 15-23.

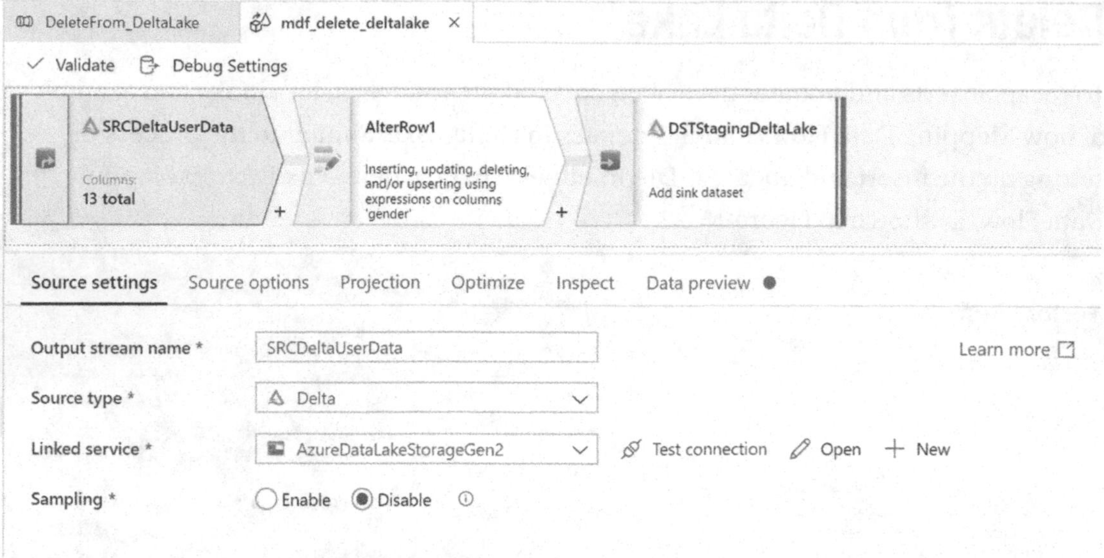

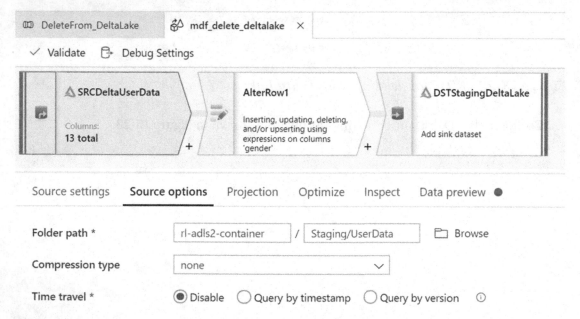

Figure 15-23. *Source settings for mdf_delete_deltalake*

Since you are still working with the same Staging Delta Lake, these source settings, shown in Figure 15-24, will be configured similar to how you had configured the inserts and updates in the previous sections.

Figure 15-24. *Source options for mdf_delete_deltalake*

For this example, delete all records where gender = male. To do this, you'll need to configure the alter row conditions to *Delete if gender == 'Male'*, as shown in Figure 15-25.

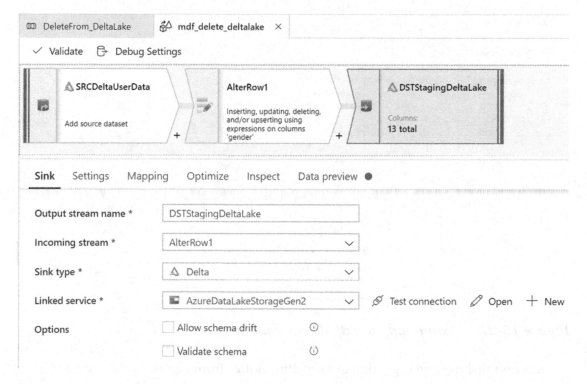

Figure 15-25. *Alter row settings for mdf_delete_deltalake delta*

Finally, Figure 15-26 shows the sink delta configuration settings.

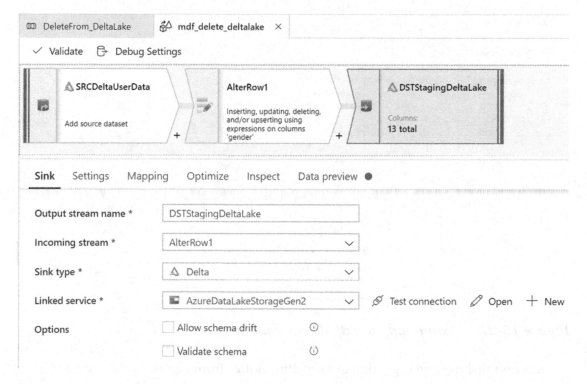

Figure 15-26. *Sink settings for mdf_delete_deltalake*

Select the Staging Delta Lake for the sink and select *"Allow delete"* along with the desired key columns of id, registration_dttm, and ip_address, as shown in Figure 15-27.

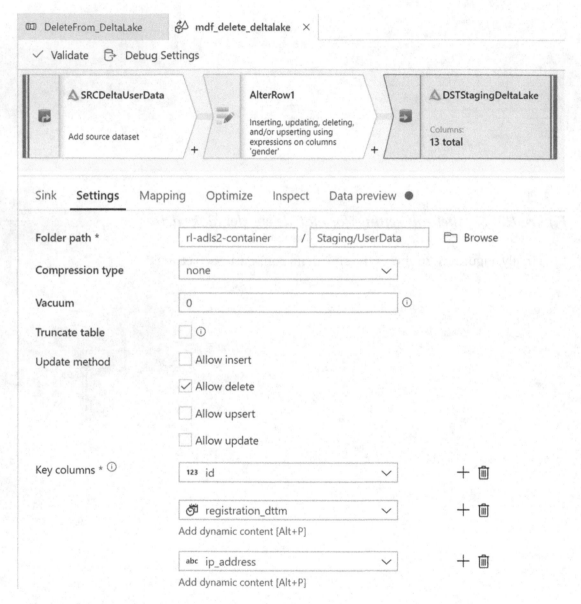

Figure 15-27. *Destination for mdf_delete_deltalake*

After publishing and triggering this pipeline, notice from Figure 15-28 how all records where gender = Male have been deleted.

first_name	last_name	email	gender
kimberly	burton	kburtonng@geocities.jp	Female
frances	ray	frayoc@npr.org	Female
carolyn	nelson	cnelson23@tiny.cc	Female
deborah	diaz	ddiaz3j@digg.com	Female
diane	andrews	dandrews7w@ovh.net	Female

Figure 15-28. *Delete data file as expected*

Explore Delta Logs

Lastly, let's take a look at the delta logs to briefly understand how the logs have been created and populated. The main commit info files are generated and stored in the Insert, Update, and Delete JSON commit files. Additionally, CRC files are created. CRC is a popular technique for checking data integrity as it has excellent error detection abilities, uses little resources, and is easily used. The delta logs that have been created from the ADF Delta insert, update, and delete MDF pipelines are stored in the *_delta_log* folder within your ADLS Gen2 account, as shown in Figure 15-29.

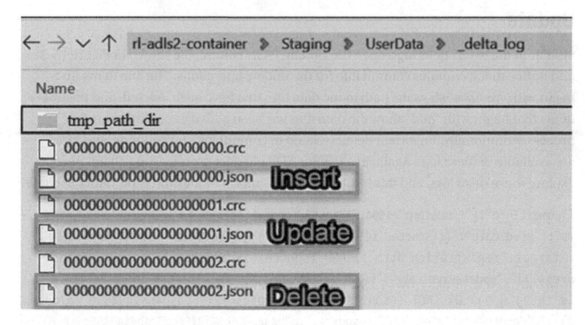

Figure 15-29. *Delta logs after insert, update, and delete*

Insert

Open the Insert JSON commit file shown in Figure 15-30, and notice that it contains commit info for the insert operations, referenced by line items that begin with the word *"add"*. You will usually not need to work with or open these files while building ETL pipelines and processes. However, the logs will always be persisted to this folder in JSON format, so you will always have the option to open the logs when needed. For this exercise, opening these logs will give you a deeper understanding as to how these logs capture information to help with better understanding the mechanics of this process.

```
{"commitInfo":{"timestamp":1594782281467,"operation":"WRITE","operationPara
meters":{"mode":"Append","partitionBy":"[]"},"isolationLevel":"WriteSeriali
zable","isBlindAppend":true}}
```

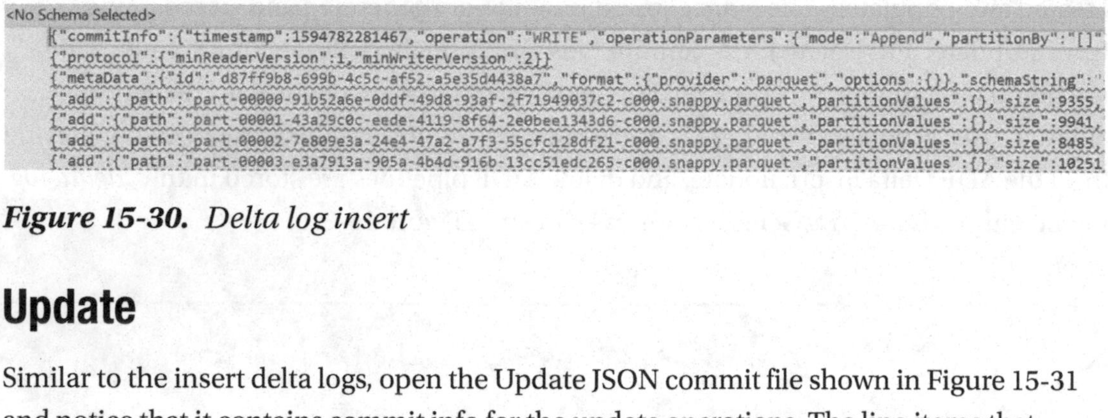

Figure 15-30. *Delta log insert*

Update

Similar to the insert delta logs, open the Update JSON commit file shown in Figure 15-31 and notice that it contains commit info for the update operations. The line items that begin with *"remove"* show the path to the data files that have been deleted, and the line items that begin with *"add"* show the data that has been added as part of the update process. Additionally, important details related to commit info, timestamps, and more are available in these logs. Again, in a regular ADF pipeline process, you will not need to explore these delta logs, and this is meant to be an informational and exploratory exercise.

```
{"commitInfo":{"timestamp":1594782711552,"operation":"MERGE","operationParamete
rs":{"predicate":"(((source.`id` = target.`id`) AND (source.`registration_dttm`
= target.`registration_dttm`)) AND (source.`ip_address` = target.`ip_ad
dress`))","updatePredicate":"((NOT ((source.`ra2b434a305b34f2f96cd5b4b414945
5e` & 2) = 0)) OR (NOT ((source.`ra2b434a305b34f2f96cd5b4b4149455e` & 8) =
0)))","deletePredicate":"(NOT ((source.`ra2b434a305b34f2f96cd5b4b4149455e` & 4) =
0))"},"readVersion":0,"isolationLevel":"WriteSerializable","isBlindAppend":false}}
```

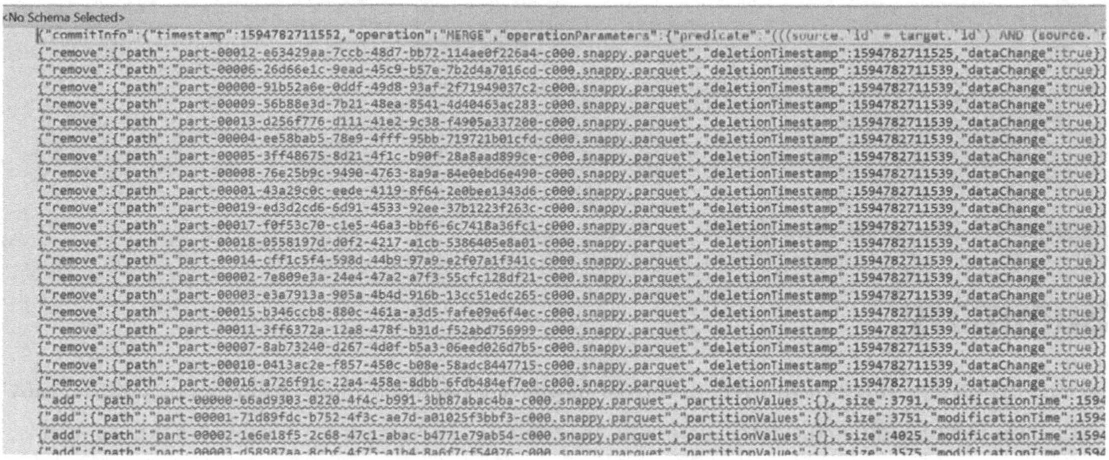

<No Schema Selected>
{"commitInfo":{"timestamp":1594782711552,"operation":"MERGE","operationParameters":{"predicate":"(((source.`id` = target.`id`) AND (source.`r
{"remove":{"path":"part-00012-e63429aa-7ccb-48d7-bb72-114ae0f226a4-c000.snappy.parquet","deletionTimestamp":1594782711525,"dataChange":true}}
{"remove":{"path":"part-00006-26d66e1c-9ead-45c9-b57e-7b2d4a7016cd-c000.snappy.parquet","deletionTimestamp":1594782711539,"dataChange":true}}
{"remove":{"path":"part-00000-91b52a6e-0ddf-49d8-93af-2f71949037c2-c000.snappy.parquet","deletionTimestamp":1594782711539,"dataChange":true}}
{"remove":{"path":"part-00009-56b88e3d-7b21-48ea-8541-4d40463ac283-c000.snappy.parquet","deletionTimestamp":1594782711539,"dataChange":true}}
{"remove":{"path":"part-00013-d256f776-d111-41e2-9c38-f4905a337200-c000.snappy.parquet","deletionTimestamp":1594782711539,"dataChange":true}}
{"remove":{"path":"part-00004-ee58bab5-78e9-4fff-95bb-719721b01cfd-c000.snappy.parquet","deletionTimestamp":1594782711539,"dataChange":true}}
{"remove":{"path":"part-00005-3ff48675-8d21-4f1c-b90f-28a8aad899ce-c000.snappy.parquet","deletionTimestamp":1594782711539,"dataChange":true}}
{"remove":{"path":"part-00008-76e25b9c-9490-4763-8a9a-84e0ebd6e490-c000.snappy.parquet","deletionTimestamp":1594782711539,"dataChange":true}}
{"remove":{"path":"part-00001-43a29c0c-eede-4119-8f64-2e0bee1343d6-c000.snappy.parquet","deletionTimestamp":1594782711539,"dataChange":true}}
{"remove":{"path":"part-00019-ed3d2cd6-6d91-4533-92ee-37b1223f263c-c000.snappy.parquet","deletionTimestamp":1594782711539,"dataChange":true}}
{"remove":{"path":"part-00017-f0f53c70-c1e5-46a3-bbf6-6c7418a36fc1-c000.snappy.parquet","deletionTimestamp":1594782711539,"dataChange":true}}
{"remove":{"path":"part-00018-0558197d-d0f2-4217-a1cb-5386405e8a01-c000.snappy.parquet","deletionTimestamp":1594782711539,"dataChange":true}}
{"remove":{"path":"part-00014-cff1c5f4-598d-44b9-97a9-e2f07a1f341c-c000.snappy.parquet","deletionTimestamp":1594782711539,"dataChange":true}}
{"remove":{"path":"part-00002-7e809e3a-24e4-47a2-a7f3-55cfc128df21-c000.snappy.parquet","deletionTimestamp":1594782711539,"dataChange":true}}
{"remove":{"path":"part-00003-e3a7913a-905a-4b4d-916b-13cc51edc265-c000.snappy.parquet","deletionTimestamp":1594782711539,"dataChange":true}}
{"remove":{"path":"part-00015-b346ccb8-880c-461a-a3d5-fafe09e6f4ec-c000.snappy.parquet","deletionTimestamp":1594782711539,"dataChange":true}}
{"remove":{"path":"part-00011-3ff6372a-12a8-478f-b31d-f52abd756999-c000.snappy.parquet","deletionTimestamp":1594782711539,"dataChange":true}}
{"remove":{"path":"part-00007-8ab73240-d267-4d0f-b5a3-06eed026d7b5-c000.snappy.parquet","deletionTimestamp":1594782711539,"dataChange":true}}
{"remove":{"path":"part-00010-0413ac2e-f857-450c-b08e-58adc8447715-c000.snappy.parquet","deletionTimestamp":1594782711539,"dataChange":true}}
{"remove":{"path":"part-00016-a726f91c-22a4-458a-8dbb-6fdb484ef7e0-c000.snappy.parquet","deletionTimestamp":1594782711539,"dataChange":true}}
{"add":{"path":"part-00000-66ad9303-0220-4f4c-b991-3bb87abac4ba-c000.snappy.parquet","partitionValues":{},"size":3791,"modificationTime":1594
{"add":{"path":"part-00001-71d89fdc-b752-4f3c-ae7d-a01025f3bbf3-c000.snappy.parquet","partitionValues":{},"size":3751,"modificationTime":1594
{"add":{"path":"part-00002-1e6e18f5-2c68-47c1-abac-b4771e79ab54-c000.snappy.parquet","partitionValues":{},"size":4025,"modificationTime":1594
{"add":{"path":"part-00003-d58987aa-8cbf-4f75-a1b4-8a6f7cf54076-c000.snappy.parquet","partitionValues":{},"size":3575,"modificationTime":1594

Figure 15-31. *Delta log update*

Delete

Lastly, open the Delete JSON commit file shown in Figure 15-32 and notice that it contains commit info for the delete operations. The files that have been deleted are captured by line items that begin with *"remove"*. Now that you have an understanding of these delta transaction logs, along with how to open and interpret them, you'll better understand the relevance of Delta Lake and how it is positioned to handle ACID-compliant transactions within a data lake.

"commitInfo":{"timestamp":1594783812366,"operation":"MERGE","oper
ationParameters":{"predicate":"(((source.`id` = target.`id`) AND
(source.`registration_dttm` = target.`registration_dttm`)) AND (source.`ip_
address` = target.`ip_address`))","updatePredicate":"((NOT ((source.`ra079d
97a688347b581710234d2cc4b63` & 2) = 0)) OR (NOT ((source.`ra079d97a688347b5
81710234d2cc4b63` & 8) = 0)))","deletePredicate":"(NOT ((source.`ra079d97a6
88347b581710234d2cc4b63` & 4) = 0))"},"readVersion":1,"isolationLevel":"Wri
teSerializable","isBlindAppend":false}}

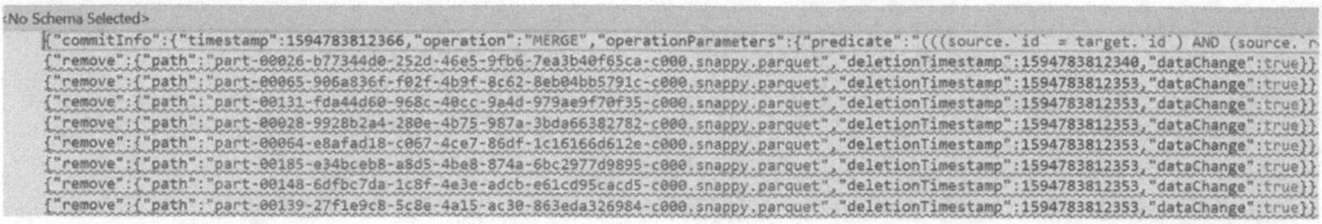

<No Schema Selected>

{"commitInfo":{"timestamp":1594783812366,"operation":"MERGE","operationParameters":{"predicate":"((((source.`id` = target.`id`) AND (source. r
{"remove":{"path":"part-00026-b77344d0-252d-46e5-9fb6-7ea3b40f65ca-c000.snappy.parquet","deletionTimestamp":1594783812340,"dataChange":true}}
{"remove":{"path":"part-00065-906a836f-f02f-4b9f-8c62-8eb04bb5791c-c000.snappy.parquet","deletionTimestamp":1594783812353,"dataChange":true}}
{"remove":{"path":"part-00131-fda44d60-968c-48cc-9a4d-979ae9f70f35-c000.snappy.parquet","deletionTimestamp":1594783812353,"dataChange":true}}
{"remove":{"path":"part-00028-9928b2a4-280e-4b75-987a-3bda66382782-c000.snappy.parquet","deletionTimestamp":1594783812353,"dataChange":true}}
{"remove":{"path":"part-00064-e8afad18-c067-4ce7-86df-1c16166d612e-c000.snappy.parquet","deletionTimestamp":1594783812353,"dataChange":true}}
{"remove":{"path":"part-00185-e34bceb8-a8d5-4be8-874a-6bc2977d9895-c000.snappy.parquet","deletionTimestamp":1594783812353,"dataChange":true}}
{"remove":{"path":"part-00148-6dfbc7da-1c8f-4e3e-adcb-e61cd95cacd5-c000.snappy.parquet","deletionTimestamp":1594783812353,"dataChange":true}}
{"remove":{"path":"part-00139-27f1e9c8-5c8e-4a15-ac30-863eda326984-c000.snappy.parquet","deletionTimestamp":1594783812353,"dataChange":true}}

Figure 15-32. *Delta log delete*

Summary

In this chapter, I have demonstrated how to get started with Delta Lake using Azure Data Factory's Delta Lake connector through examples of how to create, insert, update, and delete in a Delta Lake by using Azure Data Lake Storage Gen2 as the storage account. Since Delta Lake is an open source project that is meant to enable building a lakehouse architecture on top of existing storage systems, it can certainly be used with other storage systems such as Amazon S3, Google Cloud Storage, HDFS, and others. Additionally, you could just as easily work with Delta Lake by writing Spark, Python, and/or Scala code within Databricks notebooks.

With such flexibility, the lakehouse data management paradigm is gaining momentum with a vision of becoming an industry standard and the evolution of both the data lake and data warehouse. The low-cost storage of data lakes makes this option very attractive for organizations that embrace a cost-sensitive and growth mindset. Additional advantages of data lakehouse include reduced data redundancy, elimination of simple ETL jobs, decoupled compute from storage, real-time streaming support, ease of data governance, and the capability to connect directly to modern-day BI tools. While still in their infancy, data lakehouse and Delta Lake do have certain limitations that prevent them from being a full-fledged replacement of traditional data warehousing appliances such as SQL Server databases and warehouses. Nevertheless, this chapter demonstrates the capability of working directly with Delta Lake in a code-free manner and shows how to easily work with Delta Lake to begin exploring the building blocks of a data lakehouse.

PART III

Real-Time Analytics in Azure

CHAPTER 16

Stream Analytics Anomaly Detection

The need to process big data real-time streams is becoming an increasing need for many businesses. Customers in numerous industries are seeking to harness the power of real-time big data to unlock valuable insights. They are seeking an easy-to-use, flexible, reliable, and scalable solution to process and transform streams of real-time data for their IoT projects. Additionally, along with these big real-time data streams come anomalies in the data. Customers in a variety of industries are interested in the concept of real-time anomaly detection using machine learning algorithms and operators.

Azure Stream Analytics is an event-processing engine that allows examining high volumes of data streaming from devices, sensors, websites, social media feeds, applications, etc. It is easy to use and based on simple SQL. Additionally, it is a fully managed (PaaS) offering on Azure that can run large-scale analytics jobs that are optimized for cost since users only pay for streaming units that are consumed.

Azure Stream Analytics now offers built-in machine learning–based anomaly detection capabilities to monitor temporary and persistent anomalies. This anomaly detection capability coupled with Power BI's real-time streaming service makes for a powerful real-time anomaly detection service. In this chapter, I will demonstrate a practical example of how to create real-time anomaly detection using Azure Stream Analytics for processing the stream and Power BI for visualizing the data.

Prerequisites

To begin implementing the example solution for this chapter, you will need to create and run a few Azure resources. These are

© Ron C. L'Esteve 2021
R. C. L'Esteve, *The Definitive Guide to Azure Data Engineering*, https://doi.org/10.1007/978-1-4842-7182-7_16

- **Stream Analytics job**: Azure Stream Analytics is a real-time analytics and complex event-processing engine that is designed to analyze and process high volumes of fast-streaming data from multiple sources simultaneously.

- **IoT Hub**: IoT Hub is a managed service, hosted in the cloud, that acts as a central message hub for bidirectional communication between your IoT application and the devices it manages.

- **Power BI service**: For small and medium businesses, Power BI Pro is optimal to deliver full business intelligence capabilities to all users. Power BI Premium is best for large enterprise organizations who need a large number of people across the business to be able use Power BI to view dashboards and reports. Please revisit Chapter 1, where I discuss the pros and cons of Power BI Pro vs. Premium at the end of the chapter. For the purposes of this exercise, Power BI Premium will be used.

- **Device Simulator**: The Device Simulator app, which is a Visual Studio project that can be downloaded from the following GitHub location (`https://github.com/Azure/azure-stream-analytics/tree/master/Samples/DeviceSimulator`), is used to simulate anomalies being sent from a device to IoT Hub. Once you download and open this Device Simulator Visual Studio project, you will be able to run it to see the Device Simulator UI, which can be configured to pass events from the simulator to the IoT Hub that you'll create in Azure. The schema of the simulator's data uses temperature and a sensor ID. These events can then be consumed by an Azure Stream Analytics job that is configured to read from this IoT Hub.

The following sections walk you through creating these prerequisites.

Create an Azure Stream Analytics Job

Let's begin by creating a new Stream Analytics job in Azure Portal by simply searching for Stream Analytics, as shown in Figure 16-1.

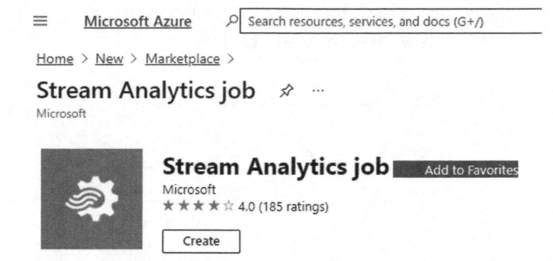

Figure 16-1. *New Stream Analytics job in Azure Portal*

Figure 16-2 illustrates that you will need to name the job as asa-001. Also, select the respective subscription, resource group, and location. I would recommend creating all of your resources for this project in a single resource group. That said, you could create a new resource group for this Stream Analytics job. Additionally, try to create all of your resources within the same location as much as possible to reduce data and network latency, especially since this is a real-time streaming solution. For this exercise, select "Cloud" for deploying the job to the Azure cloud and since this is the typical selection. Select "Edge" to deploy the job to an on-premises IoT Gateway Edge device. After the selections have been made, click Create.

Home > New > Marketplace > Stream Analytics job >

New Stream Analytics job   ⋯

Job name *

| asa-001 | ✓ |

Subscription *

| MSDN Platforms Subscription | ⌄ |

Resource group *

| rg-001 | ⌄ |
Create new

Location *

| Central US | ⌄ |

Hosting environment ⓘ

(Cloud) Edge

| Create |

Figure 16-2. New Stream Analytics job details

You have now created a Stream Analytics job. You will also need to create an IoT Hub, which will be used to bridge the gap between the Device Simulator and the Stream Analytics job.

Create an IoT Hub

IoT Hub is a managed service that is hosted in the cloud and acts as a central message hub for bidirectional communication between IoT applications and the devices they manage. An IoT Hub is essentially an Event Hub with additional features including per-device identity, cloud to device messaging, and several additional features. In Chapter 1, I have discussed the similarities and differences between IoT and Event Hubs in greater detail. Figure 16-3 shows how to create the IoT Hub from Azure Portal.

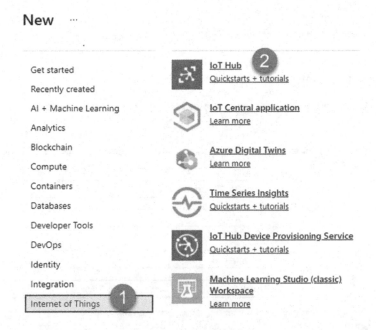

Figure 16-3. *Create an Azure Portal IoT Hub*

In the Basics tab shown in Figure 16-4, give your IoT Hub a name and ensure that the subscription, resource group, and region are configured correctly. Recall from the previous section that it would be best to have the IoT Hub reside in the same resource group and region as the Stream Analytics job.

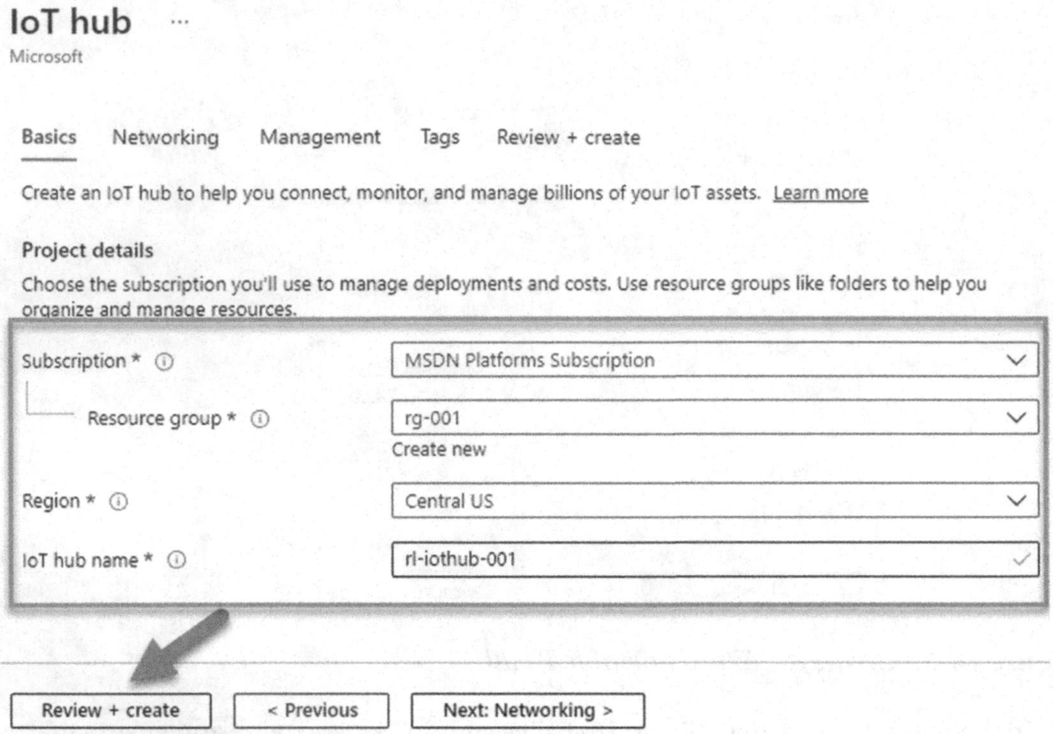

IoT hub ⋯

Microsoft

Basics   Networking   Management   Tags   Review + create

Create an IoT hub to help you connect, monitor, and manage billions of your IoT assets. Learn more

Project details

Choose the subscription you'll use to manage deployments and costs. Use resource groups like folders to help you organize and manage resources.

Subscription * ⓘ	MSDN Platforms Subscription ∨
Resource group * ⓘ	rg-001 ∨
	Create new
Region * ⓘ	Central US ∨
IoT hub name * ⓘ	rl-iothub-001 ✓

Review + create   < Previous   Next: Networking >

Figure 16-4. *IoT Hub Basics tab*

There is also an option to size and scale the IoT Hub, as shown in Figure 16-5. For this exercise, use the free tier since this is best suited for testing scenarios. The standard tier of IoT Hub enables all features and is required for any IoT solutions that want to make use of the bidirectional communication capabilities. The basic tier enables a subset of the features and is intended for IoT solutions that only need unidirectional communication from devices to the cloud. Both tiers offer the same security and authentication features. Read Microsoft's documentation on when and how to choose the right tier based on capability needs (`https://docs.microsoft.com/en-us/azure/iot-hub/iot-hub-scaling`).

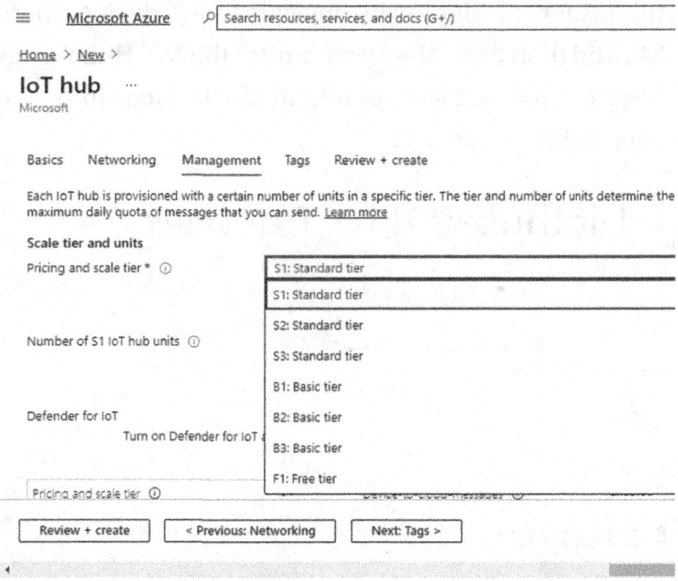

Figure 16-5. *IoT Hub Management tab*

After selecting the desired tier, review the selections shown in Figure 16-6 and then click Create to deploy the IoT resource.

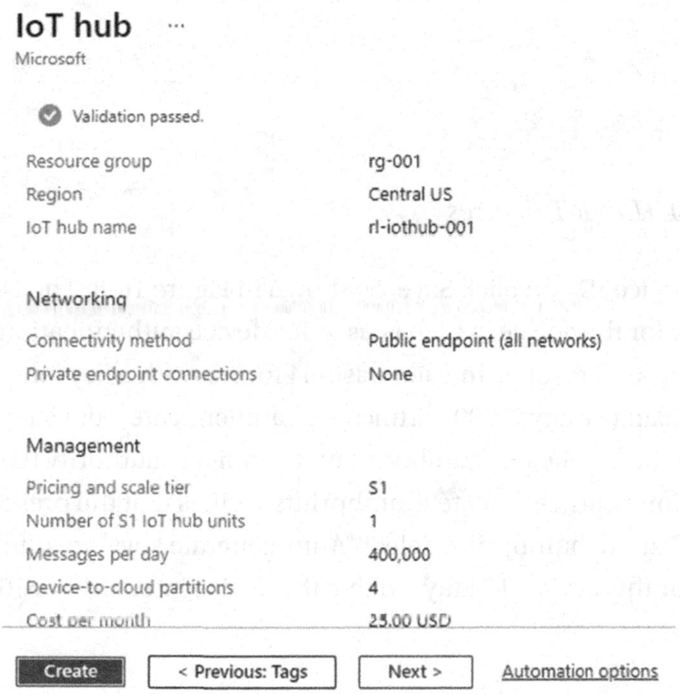

Figure 16-6. *IoT Hub Review + create*

Once the IoT Hub resource is deployed, navigate to IoT devices under Explorers shown in Figure 16-7, and then click New from under the IoT Hub navigation menu to add a new device that we can then use to configure the Device Simulator, which will send simulated events to this IoT device.

Figure 16-7. *IoT Hub IoT devices*

Then add a Device ID and click Save, as shown in Figure 16-8. The Device ID is simply an identity for the device, which is used for device authentication and access control. Symmetric keys must be in valid base-64 format with a key length between 16 and 64 bytes. You can use any X.509 certificate to authenticate a device with IoT Hub by uploading either a certificate thumbprint or a certificate authority (CA) to Azure IoT Hub. Authentication using certificate thumbprints verifies that the presented thumbprint matches the configured thumbprint. Select "Auto-generate keys" to auto-generate symmetric keys for this device. Finally, enable the device interaction with the IoT Hub.

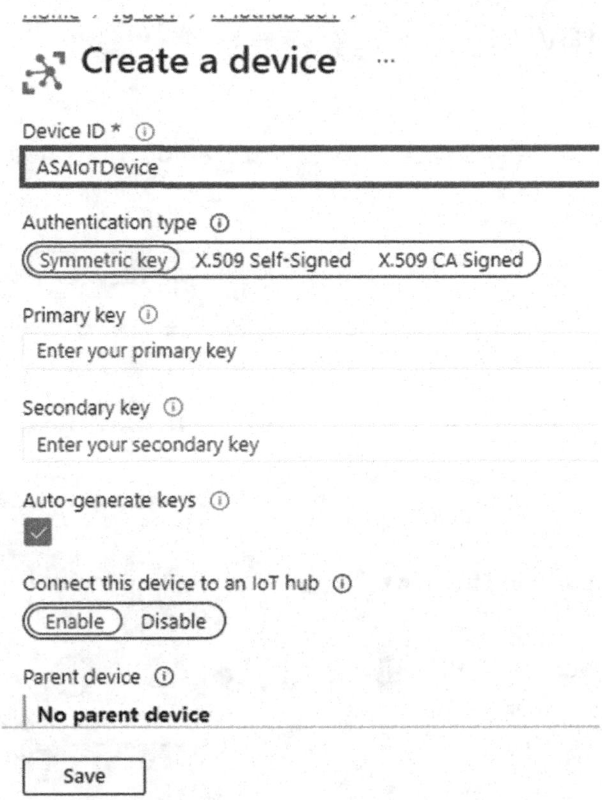

Create a device ···

Device ID * ⓘ

ASAIoTDevice

Authentication type ⓘ

(Symmetric key)   X.509 Self-Signed   X.509 CA Signed

Primary key ⓘ

Enter your primary key

Secondary key ⓘ

Enter your secondary key

Auto-generate keys ⓘ

☑

Connect this device to an IoT hub ⓘ

(Enable)   Disable

Parent device ⓘ

No parent device

Save

Figure 16-8. *Create an IoT Hub device*

Once the device is added, it will show a status of "Enabled," as shown in Figure 16-9. Click the device to open the device details including keys, identities, and additional configurations.

Device ID   Status

ASAIoTDevice   Enabled

Figure 16-9. *ASA device status and ID*

Next, copy the connection string for the primary key shown in Figure 16-10, which will be used as the connection for the IoT device.

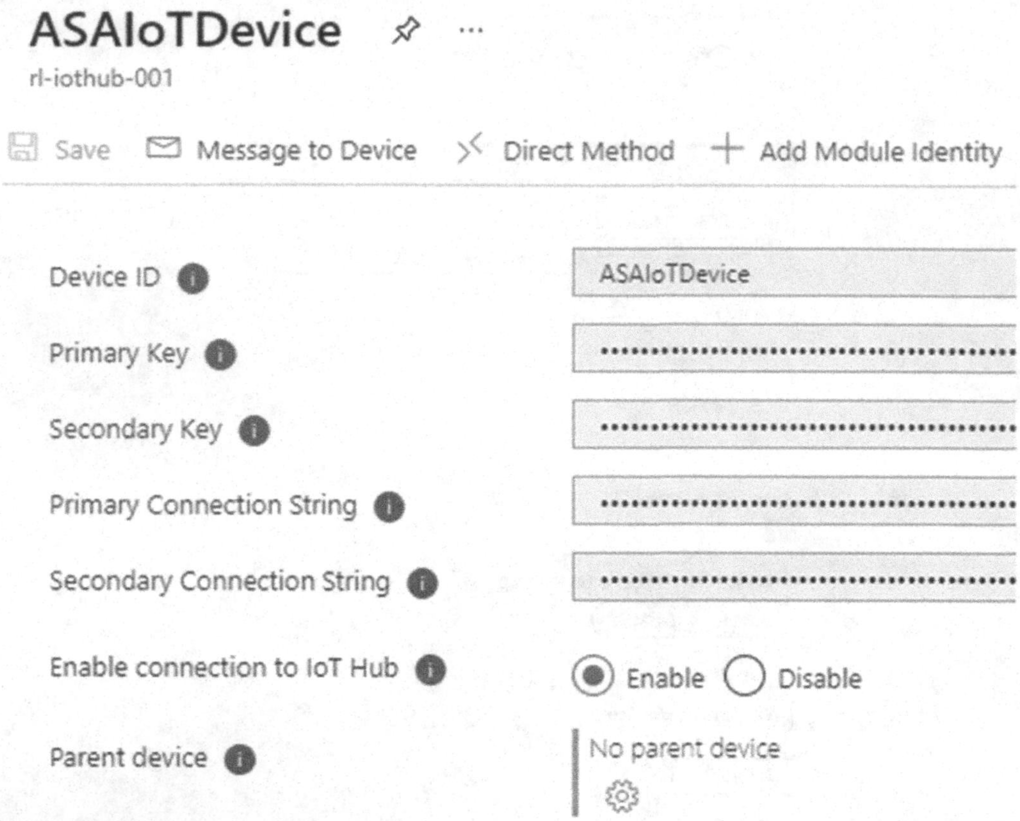

ASAIoTDevice 📌 ...

rl-iothub-001

🖫 Save   ✉ Message to Device   ✕ Direct Method   ＋ Add Module Identity

Device ID ⓘ	ASAIoTDevice
Primary Key ⓘ	••••••••••••••••••••••••••••••••••••••
Secondary Key ⓘ	••••••••••••••••••••••••••••••••••••••
Primary Connection String ⓘ	••••••••••••••••••••••••••••••••••••••
Secondary Connection String ⓘ	••••••••••••••••••••••••••••••••••••••
Enable connection to IoT Hub ⓘ	◉ Enable ○ Disable
Parent device ⓘ	No parent device ⚙

Figure 16-10. *Connection details for the ASA IoT device*

Create a Power BI Service

To get started with Power BI, download the free desktop version from the following URL:
https://powerbi.microsoft.com/en-us/downloads/. When considering a production-
ready Power BI service, explore the Pro and Premium options. In Chapter 1,
I have briefly compared both Pro and Premium. Pro has a $9.99/user license,
and Premium has a $20/user license. For the purposes of this exercise, Power BI
Premium has been used. Please review the various options and choose what is best
for you and your organization.

Download the Device Simulator

The Device Simulator is used to simulate anomalies being sent from a device to IoT Hub.
The schema uses temperature and a sensor ID. These events can then be consumed
by an Azure Stream Analytics job configured to read from this IoT Hub. Download the

Device Simulator from the following URL (`https://github.com/Azure/azure-stream-analytics/tree/master/Samples/DeviceSimulator`) and then open the corresponding Visual Studio solution file and run the simulator. The Device Simulator solution file should look similar to Figure 16-11 when it is opened.

Figure 16-11. *VS view of the Device Simulator project*

When the Device Simulator begins running, notice the various settings available. Be sure to review the readme.md GitHub file to understand the various configurations and settings that are available. For example, Mock Mode is a method of experimenting with the simulator and various anomaly patterns without sending data to a live IoT Hub.

Once you have the available IoT Hub Namespace (e.g., rl-iothub-001), Device ID (ASAIoTDevice), and Device Key (Primary Key), enter it in the IoT Hub config section, as shown in Figure 16-12, prior to running the simulator to ensure messages and data are sent to the IoT device.

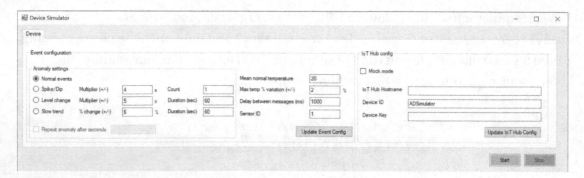

Figure 16-12. Device SimulatorDevice Simulator

Create a Stream Analytics Input and Output

A Stream Analytics job consists of an input, a query, and an output. It will need to be defined and run in order to take the IoT Hub device input, process it using the query, and output it to a Power BI real-time streaming dashboard.

Add Stream Input

Your first step is to capture the input. Begin by adding an IoT Hub stream input as shown in Figure 16-13. Notice that there are additional options to use Event Hub, Blob Storage, and ADLS Gen2 – they are stream inputs as well.

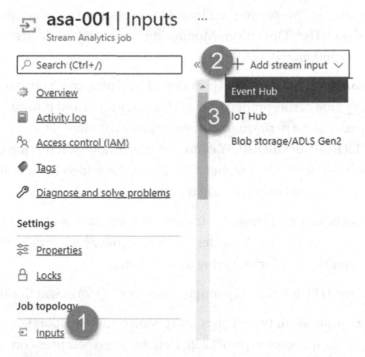

Figure 16-13. *Adding an IoT Hub stream input*

Next, Figure 16-14 shows you how to configure the input stream details. Here are some additional details for the advanced configuration options:

- **Consumer group**: IoT Hubs limit the number of readers within one consumer group to five. Microsoft recommends using a separate group for each job. Leaving this field empty will use the '$Default' consumer group.

- **Shared access policy name**: When you create an IoT Hub, you can also create shared access policies in the IoT Hub settings. Each shared access policy will have a name, permissions that you set, and access keys.

- **Shared access policy key**: When you create an IoT Hub, you can also create shared access policies in the IoT Hub settings. Each shared access policy will have a name, permissions that you set, and access keys.

- **Endpoint**: Use "Messaging" endpoint for messages from devices to the cloud. Use "Operations Monitoring" endpoint for device telemetry and metadata.

- **Partition key**: If your input is partitioned by a property, you can add the name of this property here. This is optional and is used for improving the performance of your query if it includes a PARTITION BY or GROUP BY clause on this property. If this job uses compatibility level 1.2 or higher, "PartitionId" is added by default, so you do not need to explicitly add it here.

- **Event serialization format**: To make sure your queries work the way you expect, Stream Analytics needs to know which serialization format you're using for incoming data streams.

- **Encoding**: UTF-8 is the only supported encoding format at this time.

- **Event compression type**: The compression option enables you to specify a compression type of Gzip, Deflate, or no compression.

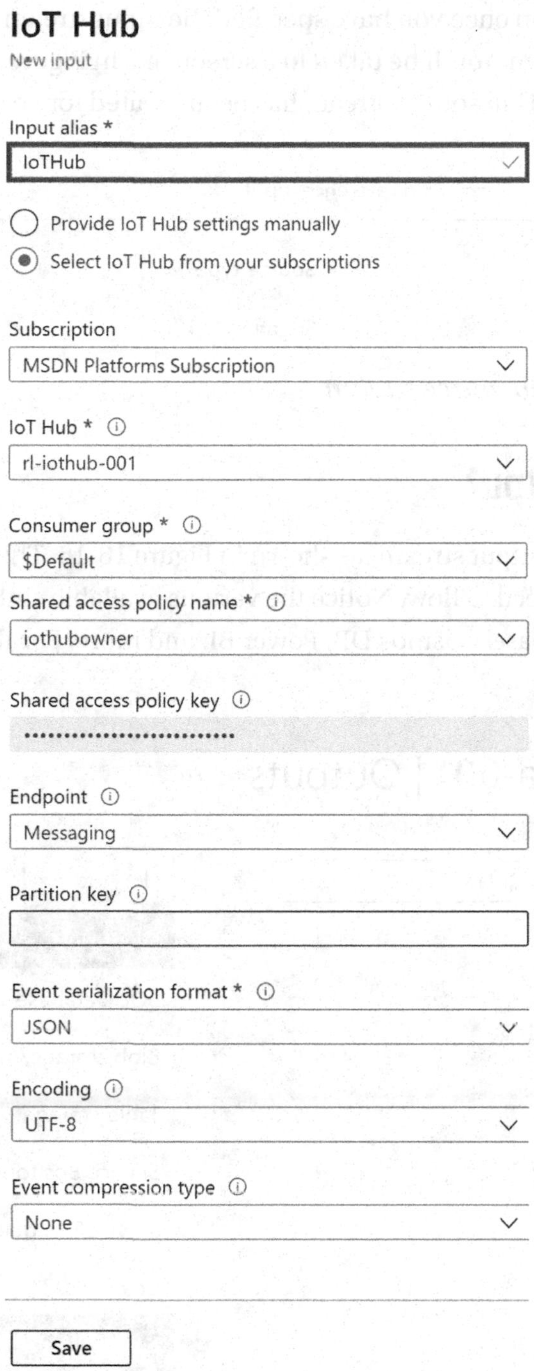

IoT Hub
New input

Input alias *

| IoTHub | ✓ |

○ Provide IoT Hub settings manually
◉ Select IoT Hub from your subscriptions

Subscription

| MSDN Platforms Subscription | ∨ |

IoT Hub * ⓘ

| rl-iothub-001 | ∨ |

Consumer group * ⓘ

| $Default | ∨ |

Shared access policy name * ⓘ

| iothubowner | ∨ |

Shared access policy key ⓘ

| •••••••••••••••••••••••••• |

Endpoint ⓘ

| Messaging | ∨ |

Partition key ⓘ

| |

Event serialization format * ⓘ

| JSON | ∨ |

Encoding ⓘ

| UTF-8 | ∨ |

Event compression type ⓘ

| None | ∨ |

| Save |

Figure 16-14. *Configure the input stream details*

Click the Save button once you have specified the input stream details and have them how you want them. You'll be taken to a screen like in Figure 16-15 where you'll be able to see that the IoT Hub source stream has been created for you.

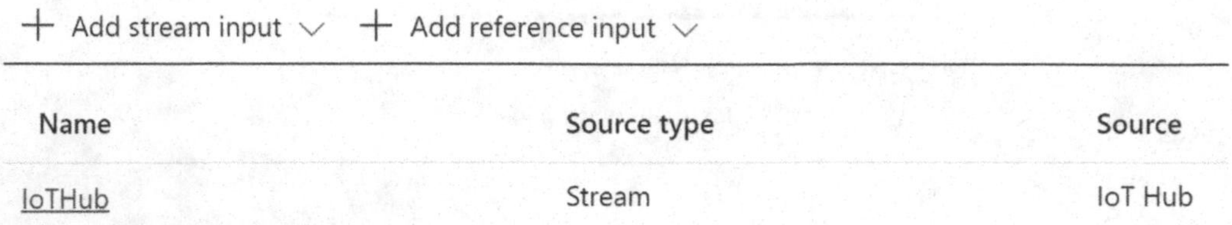

+ Add stream input ∨   + Add reference input ∨

Name	Source type	Source
IoTHub	Stream	IoT Hub

Figure 16-15. *IoT Hub source stream*

Add Stream Output

Similarly, also add an output stream, as shown in Figure 16-16. This will define the sink where the events will need to flow. Notice the various available sink options including ADLS Gen2, SQL Database, Cosmos DB, Power BI, and more. For this exercise, choose Power BI.

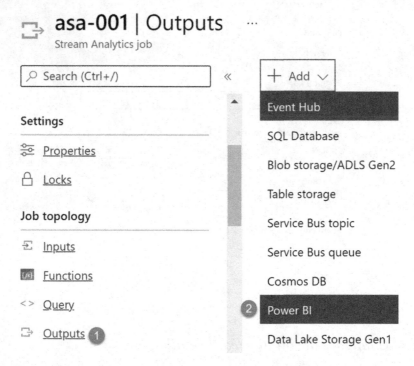

Figure 16-16. *IoT output stream to PBI*

Next, authorize Power BI to access the Stream Analytics job, as shown in Figure 16-17.

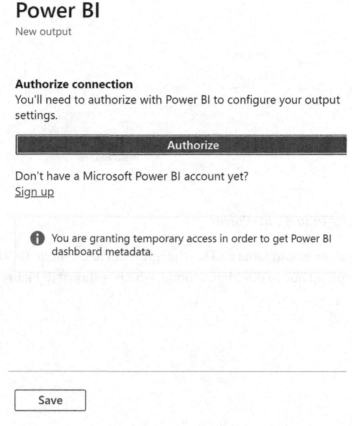

Figure 16-17. Authorize Power BI to access ASA

This will display a prompt to enter your Azure Portal credential and then click next, shown in Figure 16-18. Consider using a service account for production authorization and authentication rather than simply using a personal account for this purpose.

Figure 16-18. *Log in to Azure Portal*

Configure the dataset and table and set the authentication mode to "User token" since this will be run ad hoc in dev/demo mode, which is shown in Figure 16-19.

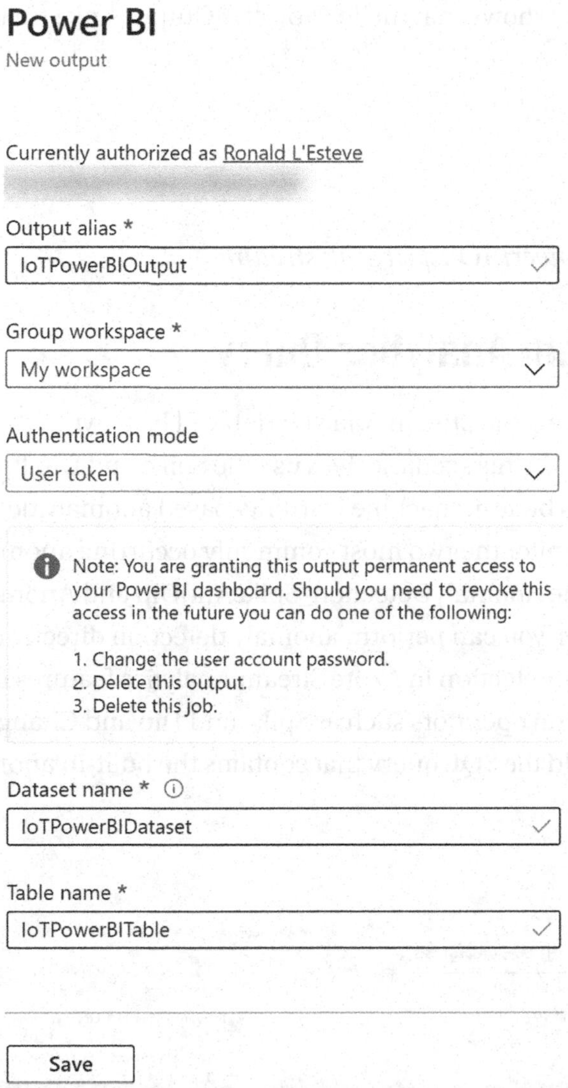

Power BI
New output

Currently authorized as <u>Ronald L'Esteve</u>

Output alias *

    IoTPowerBIOutput   ✓

Group workspace *

    My workspace   ⌄

Authentication mode

    User token   ⌄

> ⓘ Note: You are granting this output permanent access to
> your Power BI dashboard. Should you need to revoke this
> access in the future you can do one of the following:
>
> 1. Change the user account password.
> 2. Delete this output.
> 3. Delete this job.

Dataset name * ⓘ

    IoTPowerBIDataset   ✓

Table name *

    IoTPowerBITable   ✓

    ┌──────────┐
    │   Save   │
    └──────────┘

Figure 16-19. *Configure the dataset and table and set the authentication mode*

You could also use Managed Identity to authenticate your Azure Stream Analytics job to Power BI. Managed Identity authentication for output to Power BI gives Stream Analytics jobs direct access to a workspace within your Power BI account. This feature allows for deployments of Stream Analytics jobs to be fully automated, since it is no longer required for a user to interactively log in to Power BI via Azure Portal. Additionally, long-running jobs that write to Power BI are now better supported, since you will not need to periodically reauthorize the job (`https://docs.microsoft.com/en-us/azure/stream-analytics/powerbi-output-managed-identity`).

Lastly, Figure 16-20 shows that the IoTPowerBIOutput sink stream has been created.

Name	Sink
IoTPowerBIOutput	Power BI

Figure 16-20. *IoTPowerBIOutput sink stream*

Write the Stream Analytics Query

The final piece to creating the Stream Analytics job will be to write the SQL query for the anomaly detection. In this scenario, let's use the Spike and Dip function. Azure Stream Analytics offers built-in machine learning–based anomaly detection capabilities that can be used to monitor the two most commonly occurring anomalies: temporary and persistent. With the AnomalyDetection_SpikeAndDip and AnomalyDetection_ ChangePoint functions, you can perform anomaly detection directly in your Stream Analytics job. Anomaly detection in Azure Stream Analytics features machine learning–based anomaly detection operators such as Spike and Dip and Change Point. Figure 16-21 shows you where to add the SQL query that contains the built-in anomaly detection Spike and Dip function.

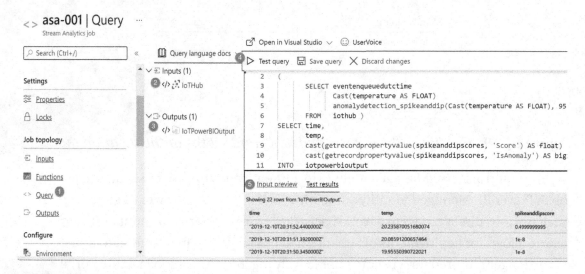

Figure 16-21. *ASA SQL query for the anomaly detection*

Here is the code that is being used as the source Stream Analytics query shown in Figure 16-21. This source query will take the incoming streaming events and apply a SQL query to them, which contains the addition of aliases, casting, and the usage of the Spike and Dip anomaly detection function that is a part of the Stream Analytics SQL query functions:

```
WITH anomalydetectionstep AS
(
        SELECT eventenqueuedutctime
AS time,
            Cast(temperature AS FLOAT)
AS temp,
            anomalydetection_spikeanddip(Cast(temperature AS FLOAT), 95,
            120, 'spikesanddips') OVER(limit duration(second, 120)) AS
            spikeanddipscores
        FROM    iothub )
SELECT time,
        temp,
        cast(getrecordpropertyvalue(spikeanddipscores, 'Score') AS float)
        AS spikeanddipscore,
        cast(getrecordpropertyvalue(spikeanddipscores, 'IsAnomaly') AS
        bigint) AS isspikeanddipanomaly
INTO    iotpowerbioutput
FROM    anomalydetectionstep
```

Start the Stream Analytics Job

The final step to creating a successful Stream Analytics job is to start the job. This will ensure that events are being received from the IoT Hub device and passed to the Power BI service in real time. After all the necessary components of the job have been configured, start the Stream Analytics job, as shown in Figure 16-22. Notice that there is one IoT Hub input and one Power BI output.

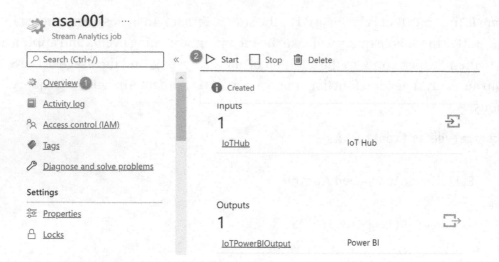

Figure 16-22. *Start the ASA job from Azure Portal*

When starting the job, you will be prompted to set the job output start time, as shown in Figure 16-23. This job will start with a specific number of streaming units. You can change streaming units under the Scale section. The job will default to 3 streaming units. You can configure to run this job in either the standard multi-tenant environment that Stream Analytics provides or a dedicated Stream Analytics cluster that you own. The job will default to the standard environment. The job might need to read input data ahead of time to ensure results are accurate. To resume a stopped job without losing data, choose Last stopped. Note that this option isn't available if you're running a job for the first time. For the purposes of this exercise, ensure that the job output start time is set to "Now" and click Start.

Start job

asa-001

Streaming units 🔘
3

Environment ⓘ
Standard

Job output start time ⓘ
① (Now) Custom

② Start

Figure 16-23. *ASA start job details*

Notice the status of "Running" once the job has started, as shown in Figure 16-24.

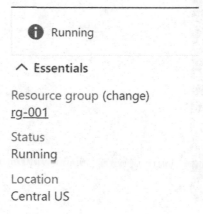

ⓘ Running

⌃ Essentials

Resource group (change)
rg-001

Status
Running

Location
Central US

Figure 16-24. *ASA job in Running status*

Create a Real-Time Power BI Dashboard

Once the job is running, head over to Power BI to begin building the real-time dashboard. It is on this dashboard that you'll be able to see the results of your monitoring query. Keeping an eye on the dashboard is how you will know when an anomaly is present.

Create a Dataset

Start by navigating to the workspace containing the IoTPowerBIDataSet in the Datasets section shown in Figure 16-25. Note that the Stream Analytics job must be running and must have processed at least one event for the dataset to be created.

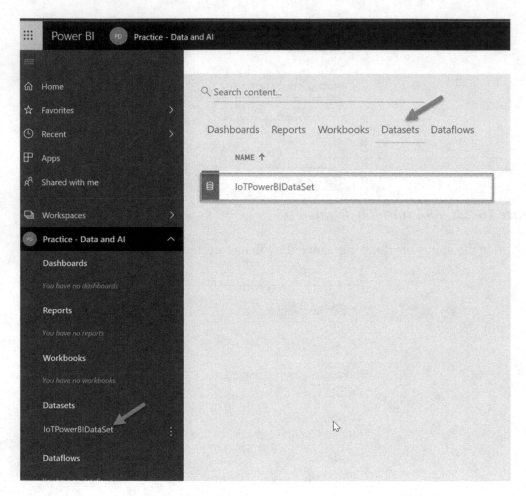

Figure 16-25. *PBI create datasets*

Create a Dashboard

After verifying that the dataset was created, also create a new dashboard, as shown in Figure 16-26.

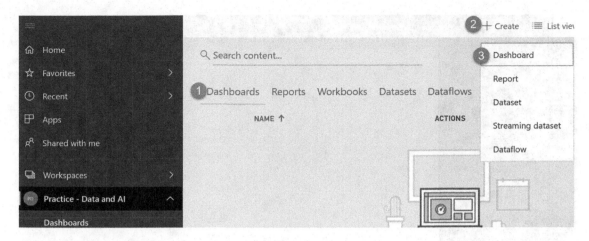

Figure 16-26. *PBI create a dashboard*

Give the dashboard a name. In this exercise, it is called IoTPowerBIDashboard, as shown in Figure 16-27.

Figure 16-27. *PBI name the dashboard*

Add a Tile

A tile is a snapshot of your data, pinned to the dashboard. A tile can be created from a report, dataset, dashboard, and more. Dashboards and dashboard tiles are a feature of Power BI service, not Power BI Desktop, so you'll need to ensure you have the Power BI service up and running. Add a tile for the real-time custom streaming data, as shown in Figure 16-28.

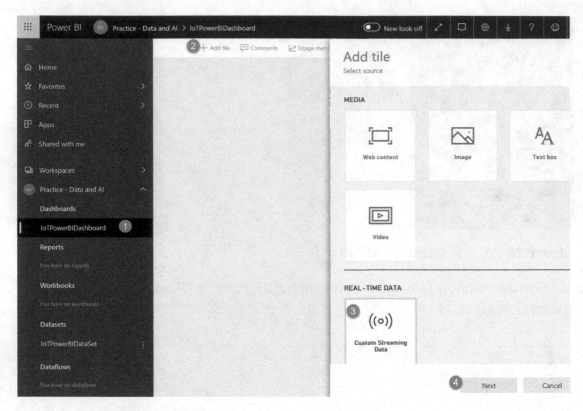

Figure 16-28. PBI add a tile

Select the dataset, as shown in Figure 16-29. Note that a dataset is a collection of data that you import or connect to. Power BI lets you connect to and import all sorts of datasets and bring all of them together in one place. Datasets are associated with workspaces, and a single dataset can be part of many workspaces.

Figure 16-29. *PBI add a custom streaming data tile*

A card visualization is a solitary number within your Power BI dashboard and mostly the most critical thing you need to track in your dashboard or report. In this exercise, anomalies are the most critical metric that will need to be tracked. Add a card visualization to track the count of Spike and Dip anomalies, as shown in Figure 16-30.

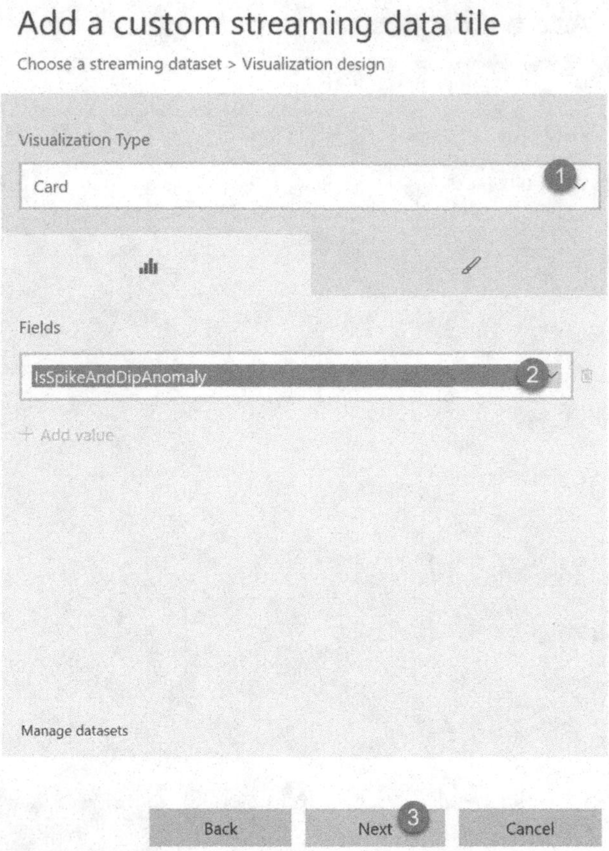

Add a custom streaming data tile

Choose a streaming dataset > Visualization design

Visualization Type

Card   ①

⏺   ✎

Fields

IsSpikeAndDipAnomaly   ②

+ Add value

Manage datasets

Back   Next ③   Cancel

Figure 16-30. *PBI add a card visualization*

A line chart is a series of data points that are represented by dots and connected by straight lines. A line chart may have one or many lines. Line charts have an X and a Y axis. Also add a line chart visualization to track the Spike and Dip scores, as shown in Figure 16-31.

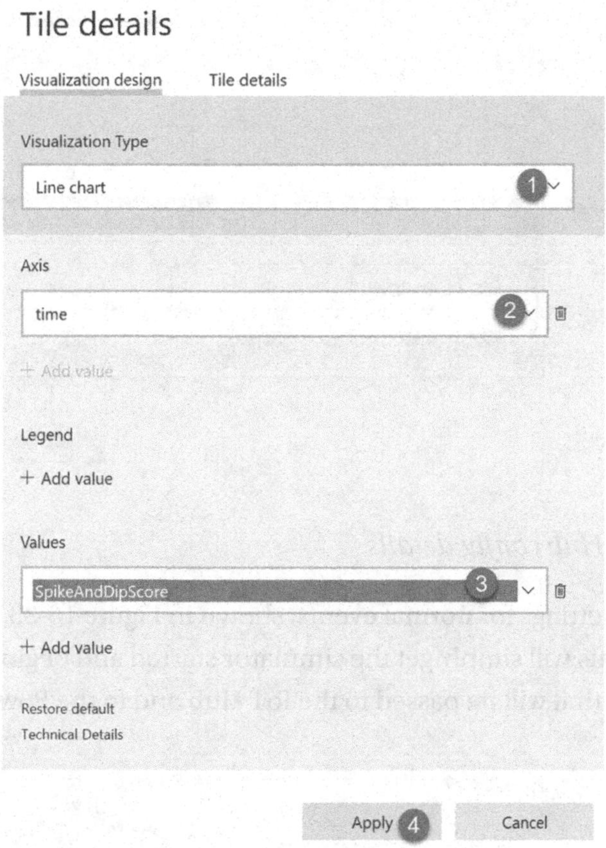

Tile details

Visualization design   Tile details

Visualization Type

Line chart   1 ⌄

Axis

time   2 ⌄   🗑

+ Add value

Legend

+ Add value

Values

SpikeAndDipScore   3 ⌄   🗑

+ Add value

Restore default
Technical Details

Apply 4   Cancel

Figure 16-31. *PBI Tile details*

Run the Device Simulator

Now head back to the Device Simulator and configure the IoT Hub details that you've accumulated from the previous sections of this chapter. Once the details are configured, click "Update IoT Hub Config" in the Device Simulator, as shown in Figure 16-32.

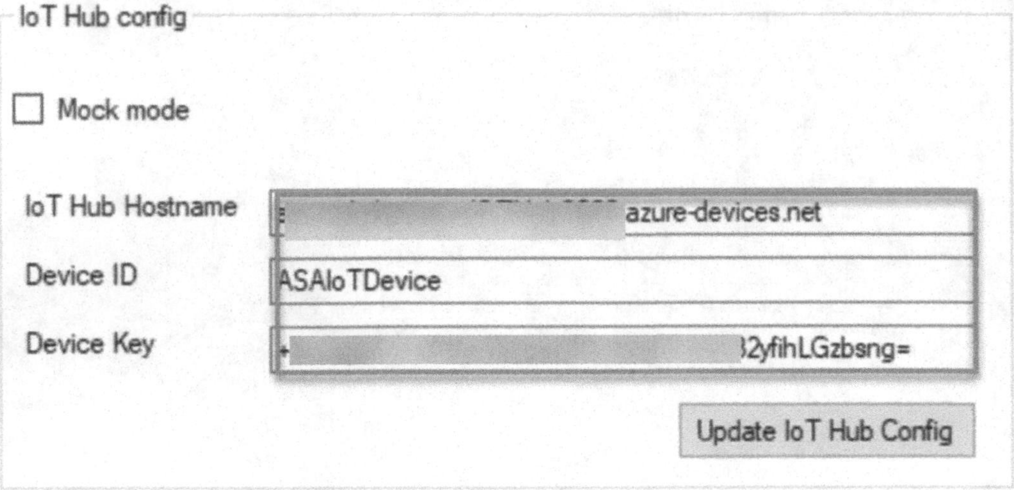

IoT Hub config

☐ Mock mode

IoT Hub Hostname _____azure-devices.net

Device ID   ASAIoTDevice

Device Key   _____)2yfihLGzbsng=

Update IoT Hub Config

Figure 16-32. *IoT Hub config details*

Select anomaly settings for normal events, shown in Figure 16-33, and start the Device Simulator. This will simply get the simulator started and begin generating a normal set of events that will be passed to the IoT Hub and to the Power BI dashboard.

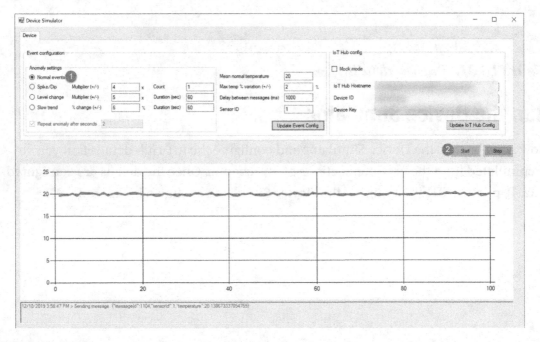

Figure 16-33. *Device Simulator anomaly settings for normal events*

Monitor Real-Time Power BI Streaming

Once the device has been started, head back to your Power BI dashboard to begin monitoring the stream. Notice from Figure 16-34 that the streaming details are now available on the Power BI dashboard. Note also how the SpikeandDip Anomalies card is still 0 since we are currently only streaming normal events.

Figure 16-34. *PBI real-time streaming dashboard with normal events*

Then head back to the simulator and change the anomaly settings to Spike/Dip and have it repeat every 2 seconds. Notice from Figure 16-35 that the anomalies have begun in the simulator.

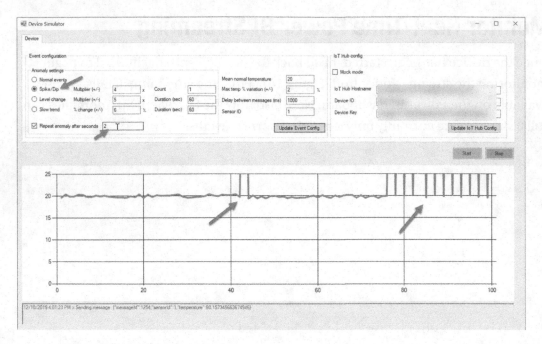

Figure 16-35. *Device Simulator creating anomaly events*

After heading back to the real-time Power BI dashboard, notice that the Spike and Dip anomalies have begun flowing into the dashboard and the SpikeandDip Anomalies card is beginning to display the incoming anomalies, shown in Figure 16-36.

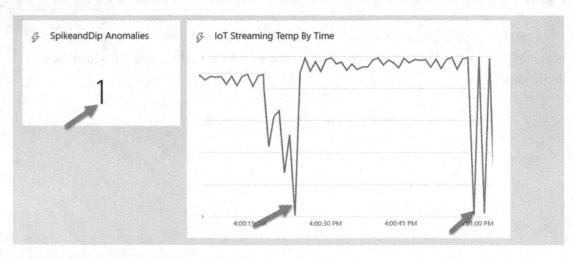

Figure 16-36. *PBI Spike and Dip anomalies*

There are a few additional anomaly settings in the Device Simulator that could be explored as a next step, such as Level change and Slow trend, as shown in Figure 16-37.

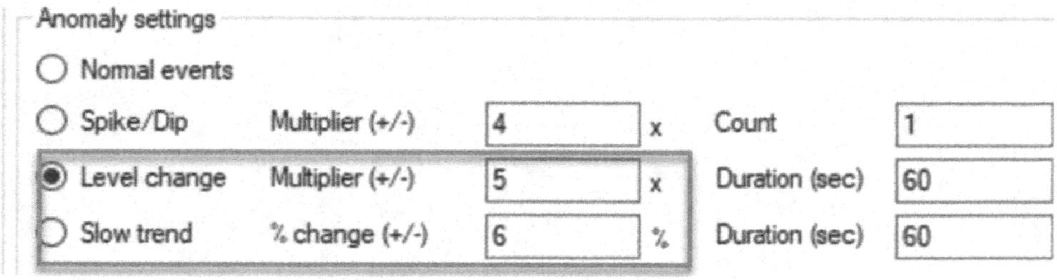

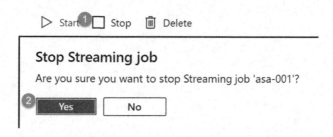

Figure 16-37. *PBI anomaly settings in Device Simulator*

Once the tests are complete, ensure that you head back to Azure Portal and stop the Stream Analytics job and Device Simulator as well, as shown in Figure 16-38.

▷ Start ① □ Stop   🗑 Delete

Stop Streaming job
Are you sure you want to stop Streaming job 'asa-001'?

② [Yes]   [No]

Figure 16-38. *Stop the streaming job*

Summary

In this chapter, I have demonstrated a practical end-to-end example of how to create real-time events using a Device Simulator and send those events to an IoT Hub, which will collect those events and process for downstream anomaly detection using Azure Stream Analytics' built-in Spike and Dip anomaly detection function. Finally, you learned how to use Power BI for visualizing the real-time streaming data into a dashboard. You also learned about some of the capabilities of real-time anomaly detection and reporting in Azure.

## CHAPTER 17

Real-Time IoT Analytics Using Apache Spark

Real-time IoT analytics, advanced analytics, and real-time machine learning insights are all areas of interest that many organizations are eager to pursue to advance their business and goals. Apache Spark's advancing API offerings have opened many opportunities for advanced and streaming analytics for big data workloads. One such API offering from Apache Spark is centered around Structured Streaming, which supports big data and real-time advanced analytics capabilities.

Apache Spark's Structured Streaming, which fits into the overall Databricks Unified Data Analytics Platform, as shown in Figure 17-1, is a stream processing framework built on the Spark SQL engine. Once a computation along with the source and destination is specified, the Structured Streaming engine will run the query incrementally and continuously as new data is available. Structured Streaming treats a stream of data as a table and continuously appends data. In this chapter, I will walk you through an end-to-end exercise to implement a Structured Streaming solution using a Device Simulator that will generate random device data that will be fed into an Azure IoT Hub and processed by Apache Spark through a Databricks notebook and into a Delta Lake to persist the data. Additionally, I will show you how to customize Structured Streaming output modes such as append vs. update vs. complete, along with introducing you to triggers in the code.

© Ron C. L'Esteve 2021
R. C. L'Esteve, *The Definitive Guide to Azure Data Engineering*, https://doi.org/10.1007/978-1-4842-7182-7_17

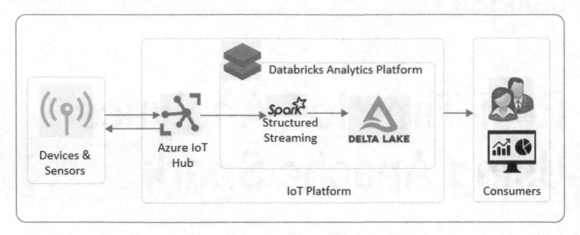

Figure 17-1. *Apache Spark's Structured Streaming and Delta Lake framework*

Prerequisites

As a basis for this exercise, ensure that you have read and understood Chapter 16, which discusses how to complete the following steps:

1. Install and run the IoT Device Simulator Visual Studio solution file. This Device Simulator will create a stream of random device data that will be fed into the IoT Hub device and used by Spark Structured Streaming.

2. Create and configure an IoT Hub device. This service will bridge the divide between the Device Simulator and the Spark Structured Streaming service.

3. Additionally, the Databricks service will need to be created in Azure Portal. Databricks' Spark compute clusters will be used for the Structured Streaming process. Alternatively, Synapse Analytics could also be used for this process.

Create an IoT Hub

Once an IoT Hub has been created along with an IoT device being added to the hub, add a new consumer group to the Built-in endpoints section of the IoT Hub, as shown in Figure 17-2. Consumer groups are used by applications to pull data from the IoT Hub. Hence, having a recognizable alias will be useful when you begin writing the Structured Streaming code.

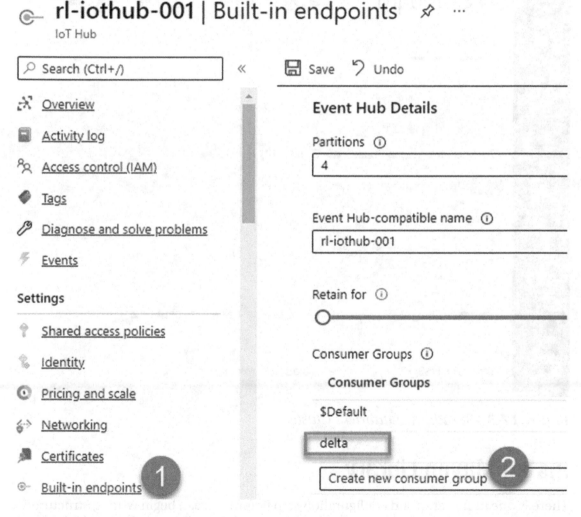

Figure 17-2. *Built-in endpoints Event Hub Details*

Create a Databricks Cluster

Next, a Databricks cluster will need to be created. For the purposes of this exercise, create a Standard Cluster with the configurations shown in Figure 17-3.

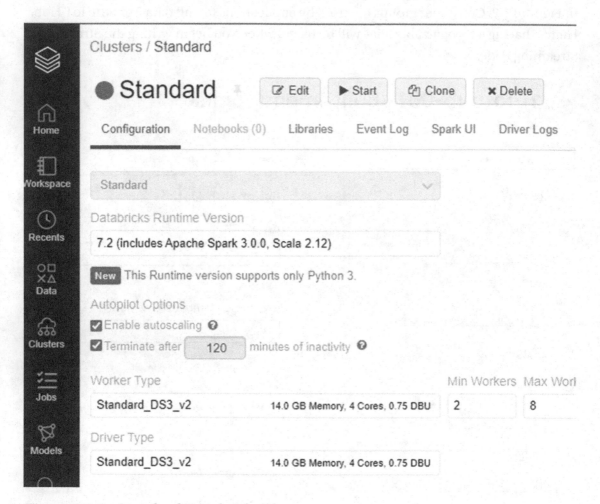

Figure 17-3. *Standard Databricks Cluster*

Install Maven Library

There is one more setup and configuration step before you can begin writing Structured Streaming code in the Databricks notebook. Install a Maven library with the coordinates listed in the following. These coordinates have been taken from MvnRepository, which can be found through the following URL: `https://mvnrepository.com/artifact/com.microsoft.azure/azure-eventhubs-spark`.

Enter the coordinates `com.microsoft.azure:azure-eventhubs-spark_2.12:2.3.15` in the Maven library within the cluster library configuration UI, as shown in Figure 17-4. Notice that there are various library source options to choose from and install. For this scenario, you will be using Maven. Optionally, you could also specify the source repository and dependencies to exclude.

Install Library   ✕

Library Source

| Upload | DBFS | PyPI | Maven | CRAN | Workspace |

Coordinates

com.microsoft.azure:azure-eventhubs-spark_2.12:2.3.15| Search Packages

Repository ❓

Optional

Exclusions

Dependencies to exclude (log4j:log4j,junit:junit)

Cancel   Install

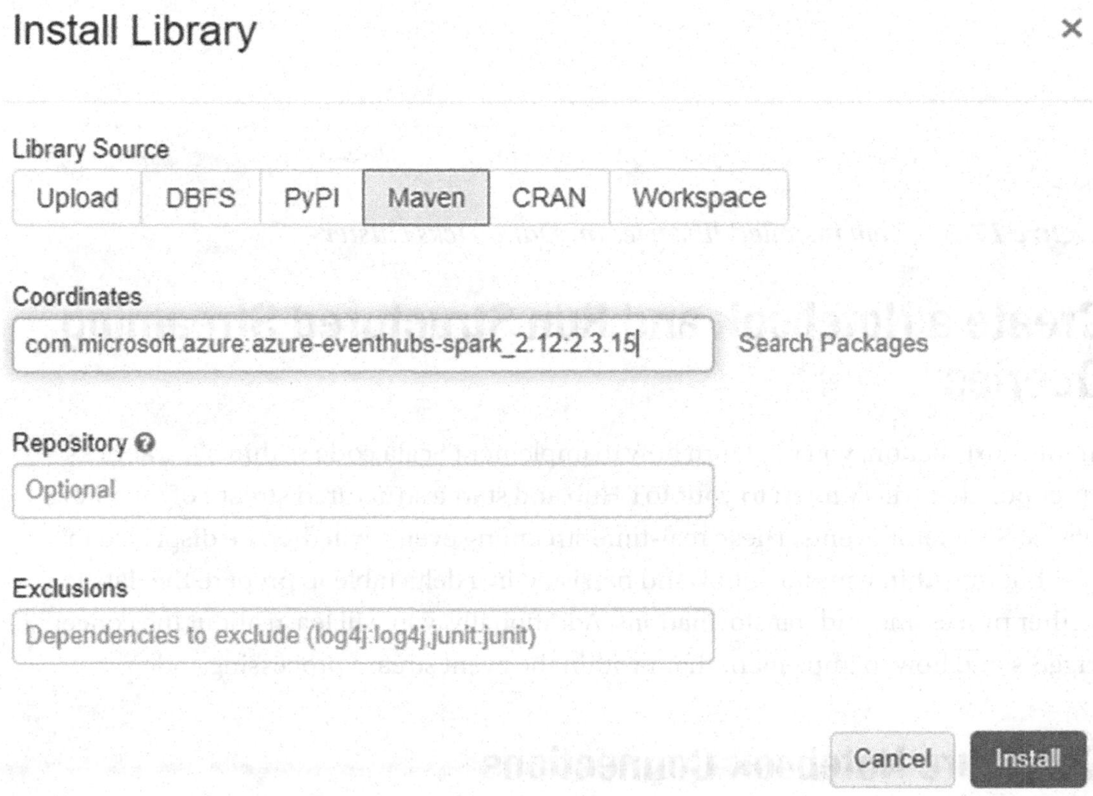

Figure 17-4. *Install Databricks Maven library*

Once the selected Maven library is installed on the cluster, it will display a status of "Installed," which is shown in Figure 17-5. Restart the cluster for the library to be properly installed on the cluster.

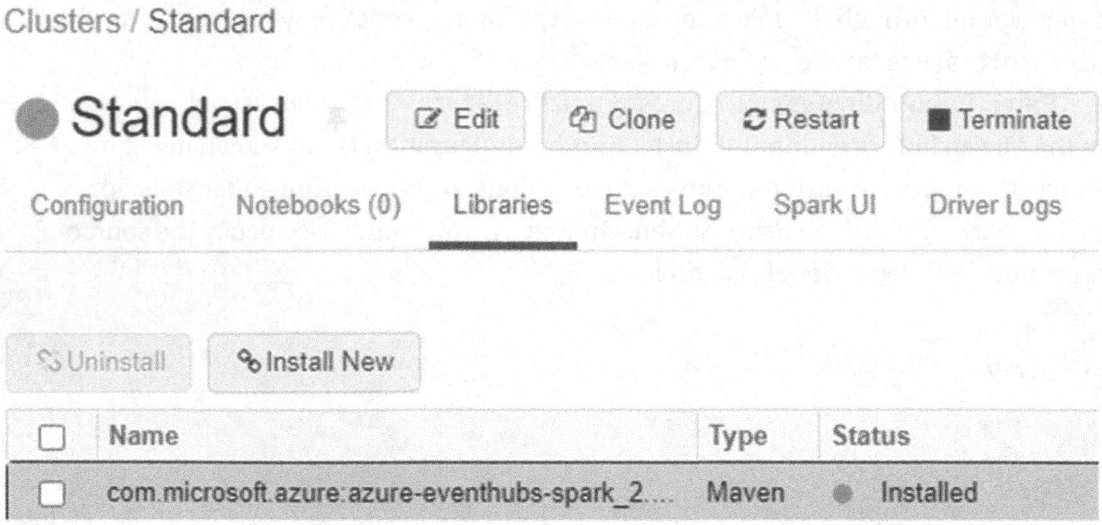

Clusters / Standard

● Standard   [☑ Edit]   [⊘ Clone]   [⟳ Restart]   [■ Terminate]

Configuration   Notebooks (0)   **Libraries**   Event Log   Spark UI   Driver Logs

---

[⟲ Uninstall]   [⚲ Install New]

☐	Name	Type	Status
☐	com.microsoft.azure:azure-eventhubs-spark_2....	Maven	● Installed

Figure 17-5. *View installed libraries on Databricks cluster*

Create a Notebook and Run Structured Streaming Queries

In this next section, you will learn how to implement Scala code within a Databricks notebook that will connect to your IoT Hub and start a structured stream of incoming Device Simulator events. These real-time streaming events will then be displayed in a dashboard within your notebook and persisted in a delta table to prepare the data for further processing and transformations. Additionally, you will learn about the concept of triggers and how to implement them within the event stream processing code.

Configure Notebook Connections

Now you are ready to create a new Databricks notebook, as shown in Figure 17-6, and attach the Standard Cluster with the Maven library installed on it. Additionally, use Scala as the language for the code that will be implemented in this Databricks notebook.

Figure 17-6. *Create a new Databricks notebook*

The following code will build a connection string using the IoT Hub connection details and start the structured stream. Throughout this section, you will need to refine this generic code, which currently includes placeholders to the connection configurations. Replace the IoT Hub connections in the following example before running the code. Also, remember to verify the consumer group in the code based on what was defined in the IoT Hub in Azure Portal:

```scala
import org.apache.spark.eventhubs._
import org.apache.spark.eventhubs.{ ConnectionStringBuilder, EventHubsConf,
EventPosition }
import org.apache.spark.sql.functions.{ explode, split }

// To connect to an Event Hub, EntityPath is required as part of the
connection string.
// Here, we assume that the connection string from the Azure portal does
not have the EntityPath part.
val connectionString = ConnectionStringBuilder("--Event Hub Compatible
Endpoint--")
  .setEventHubName("--Event Hub Compatible Name--")
  .build
val eventHubsConf = EventHubsConf(connectionString)
  .setStartingPosition(EventPosition.fromEndOfStream)
  .setConsumerGroup("delta")

val eventhubs = spark.readStream
  .format("eventhubs")
  .options(eventHubsConf.toMap)
  .load()
```

Within the Built-in endpoints section of the IoT Hub from Azure Portal, copy the Event Hub–compatible name shown in Figure 17-7 and replace it in the –Event Hub Compatible Name-- section of the code block provided.

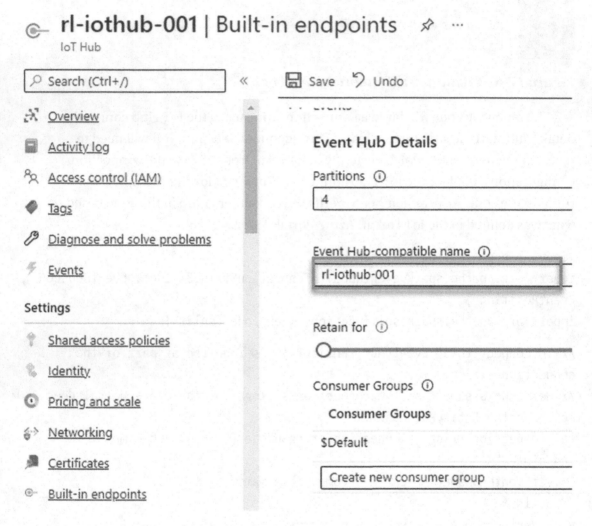

Figure 17-7. *Event Hub–compatible name within Built-in endpoints*

Next, copy the Event Hub–compatible endpoint shown in Figure 17-8, and replace it in the –Event Hub Compatible Endpoint-- section of the code block.

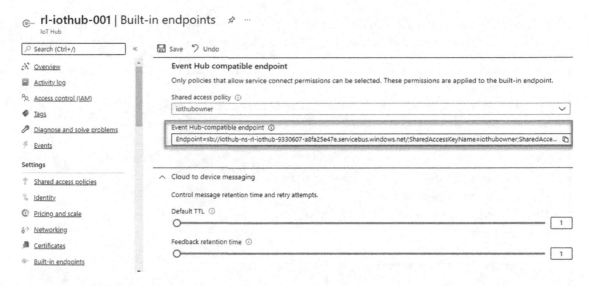

Figure 17-8. *Event Hub–compatible endpoint within Built-in endpoints*

Start the Structured Stream

The second section of this Databricks notebook will add the code provided in the following, which is intended to take the connections that were defined in the previous section, read the stream, and then persist the data to a delta table. The results in Figure 17-9 indicate that the stream has been read successfully.

IoTtoDbricks_StructuredStreaming (Scala)

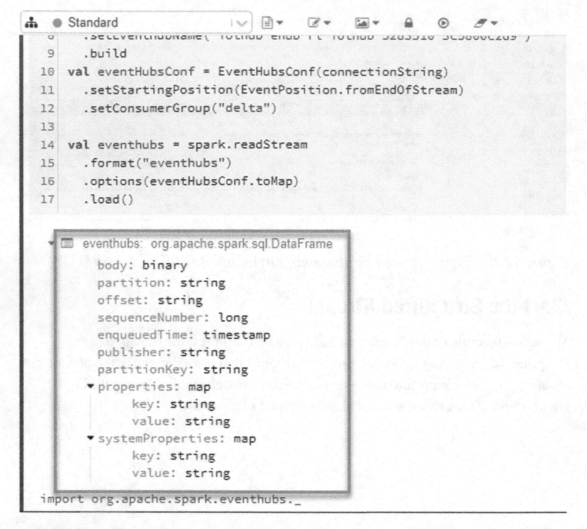

Figure 17-9. *Notebook code to read and verify the stream details*

Next, run the following code to display stream details:

```
display(eventhubs)
```

Figure 17-10 shows that the stream is initializing.

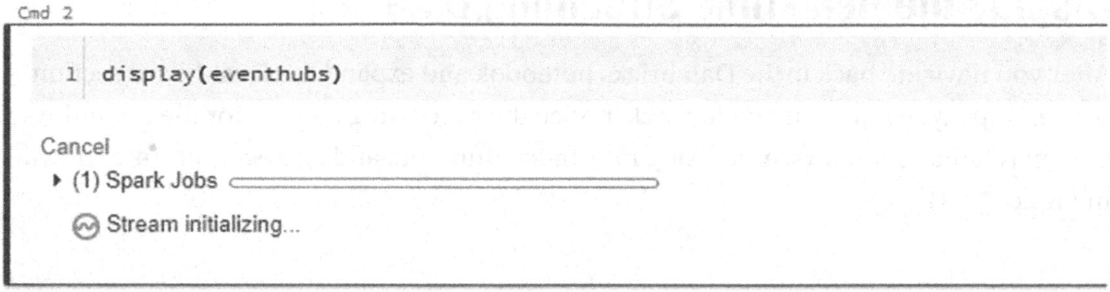

```
Cmd 2

1  display(eventhubs)
```

Cancel

▶ (1) Spark Jobs

◯ Stream initializing...

Figure 17-10. *Initializing the stream*

Start the IoT Device Simulator

After completing the previous step, head over to the Device Simulator and enter the IoT Hub device details related to Hub Namespace, Device ID, and Device Key and run the Device Simulator. Once the device is generating events, these events will begin to appear in the line chart toward the bottom of the Device Simulator shown in Figure 17-11. Notice that normal events are being generated, so the event data will be fairly consistent throughout the process.

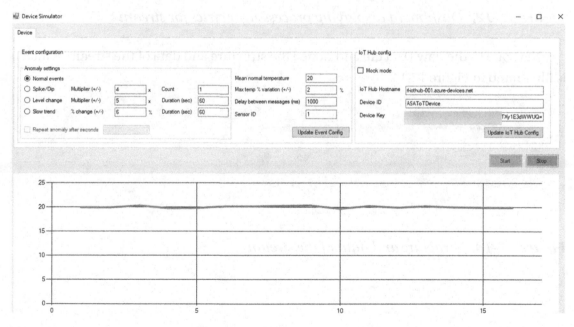

Figure 17-11. *Device Simulator details*

Display the Real-Time Streaming Data

After you navigate back to the Databricks notebook and expand the Dashboard section of the `display(eventhubs)` code block, notice the processing metrics for the incoming stream related to input vs. processing rate, batch duration, and aggregation state shown in Figure 17-12.

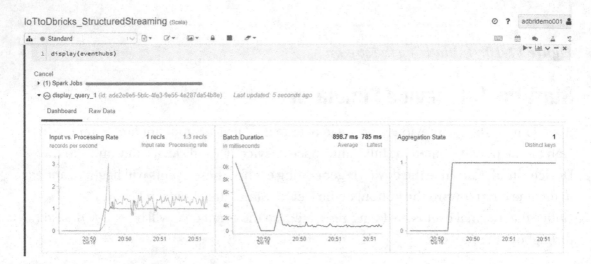

Figure 17-12. *Dashboard displaying processing metrics for stream*

Navigate to the Raw Data tab and notice the structure and data of the stream, which is illustrated in Figure 17-13.

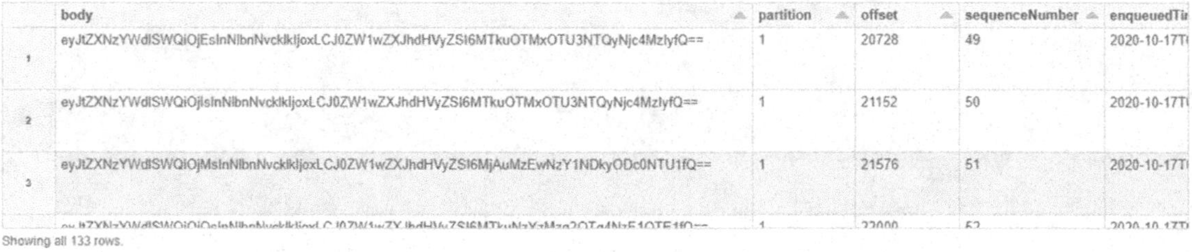

body	partition	offset	sequenceNumber	enqueuedTi
1	eyJtZXNzYWdlSWQiOjEsInNlbnNvcklkljoxLCJ0ZW1wZXJhdHVyZSI6MTkuOTMxOTU3NTQyNjc4MzIyfQ==	1	20728	49	2020-10-17T
2	eyJtZXNzYWdlSWQiOjIsInNlbnNvcklkljoxLCJ0ZW1wZXJhdHVyZSI6MTkuOTMxOTU3NTQyNjc4MzIyfQ==	1	21152	50	2020-10-17T
3	eyJtZXNzYWdlSWQiOjMsInNlbnNvcklkljoxLCJ0ZW1wZXJhdHVyZSI6MjAuMzEwMzEwNzY1NDEyODk0NTU1fQ==	1	21576	51	2020-10-17T
	ey JtZXNzYWdlSWQiOjQsInNlbnNvcklkljoxLCJ0ZW1wZXJhdHVyZSI6MTkuNzYxMzg2OTTodNzE1OTE1fQ==	1	22000	52	2020-10-17T

Showing all 133 rows.

Figure 17-13. *Structure and data of the stream*

Create a Spark SQL Table

Since data is now streaming, create a Spark SQL table by running the following code in a new code block within the same Databricks notebook. Remember to define the columns based on your IoT device data. For this exercise, use columns body and sequenceNumber from the Device Simulator:

```
import org.apache.spark.sql.types._
import org.apache.spark.sql.functions._
val schema = (new StructType)
    .add("body", DoubleType)
    .add("sequence_number", DoubleType)
val df = eventhubs.select(($"enqueuedTime").as("Enqueued_
Time"),($"systemProperties.iothub-connection-device-id")
                .as("Device_ID"),(from_json($"body".cast("string"),
                schema)
                .as("telemetry_json"))).select("Enqueued_Time",
                "Device_ID", "telemetry_json.*")
```

The expected output will display the results of the spark.sqlDataFrame, as shown in Figure 17-14.

```
Cmd 3
1  import org.apache.spark.sql.types._
2  import org.apache.spark.sql.functions._
3  val schema = (new StructType)
4      .add("body", DoubleType)
5      .add("sequence_number", DoubleType)
6  val df = eventhubs.select(($"enqueuedTime").as("Enqueued_Time"),($"systemProperties.iothub-connection-device-id")
7                  .as("Device_ID"),(from_json($"body".cast("string"), schema)
8                  .as("telemetry_json"))).select("Enqueued_Time","Device_ID", "telemetry_json.*")

▼ ▣ df: org.apache.spark.sql.DataFrame
      Enqueued_Time: timestamp
      Device_ID: string
      body: double
      sequence_number: double
```

Figure 17-14. *Results of the spark.sqlDataFrame*

Run the following code to create the Spark SQL table to store the device telemetry data:

```
df.createOrReplaceTempView("device_telemetry_data")
```

Figure 17-15 shows the execution results of this code within the Databricks notebook.

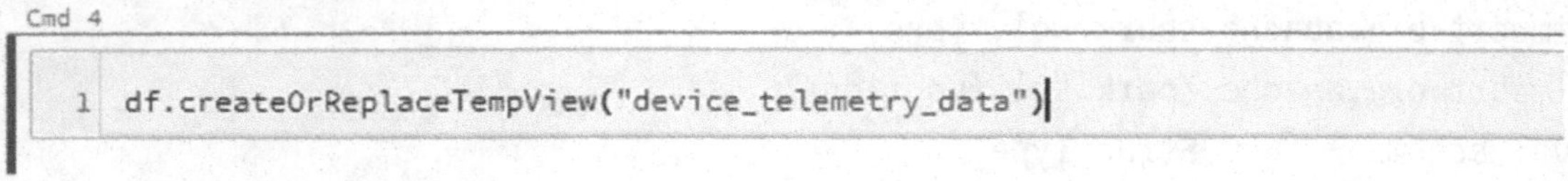

Figure 17-15.   *Create a Spark SQL table*

Write the Stream to a Delta Table

Write the stream to a delta table and start by running the following code to define the final data frame:

```
val finalDF = spark.sql("Select Date(Enqueued_Time) Date_Enqueued,
Hour(Enqueued_Time) Hour_Enqueued, Enqueued_Time, Device_ID, body AS
Body,sequence_number as Sequence_Number from device_telemetry_data")
```

Figure 17-16 shows the execution results of the code within the Databricks notebook.

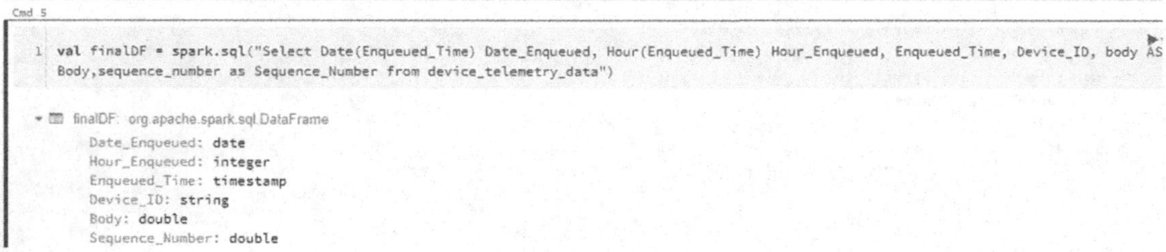

Figure 17-16.   *Code to define the final data frame*

Copy and paste this next block of code shown in Figure 17-17 into a new cell within the same Databricks notebook. This code will write the stream to the delta table. Notice that you can define the partitions, format, checkpoint location, and output mode.

A default checkpoint location is being used, which is defined and managed by Databricks, but you could just as easily define this location yourself and persist the data to a different folder:

```
finalDF.writeStream
  .outputMode("append")
  .option("checkpointLocation", "/delta/events/_checkpoints/device_delta")
  .format("delta")
  .partitionBy("Date_Enqueued", "Hour_Enqueued")
  .table("delta_telemetry_data")
```

Figure 17-17 shows the execution results of the code within the Databricks notebook.

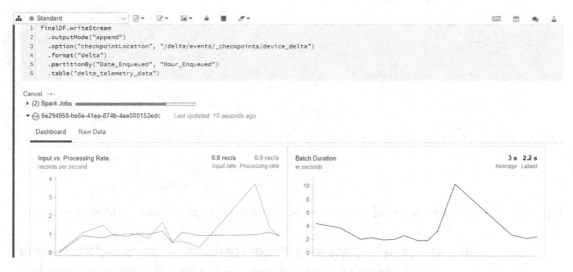

Figure 17-17. *Code to write the stream to the delta table*

Note that the output mode is set to append in Figure 17-17. The following output modes are supported:

- Append (only add new records to the output sink)

- Update (update changed records in place)

- Complete (rewrite the full output)

Triggers can also be added to the write stream to define the processing times of streaming data, along with whether the query is going to be executed as a micro-batch query with a fixed batch interval or as a continuous processing query.

Here are a few examples of triggers. Apache Spark's documentation contains more details about triggers and can be found within the following URL: https://spark. apache.org/docs/2.3.0/structured-streaming-programming-guide.html#triggers

```
.trigger(Trigger.ProcessingTime("2 seconds"))
.trigger(Trigger.Once())
.trigger(Trigger.Continuous("1 second"))
```

Notice from the Raw Data tab shown in Figure 17-18 that the device data is continuously streaming in, and you are now able to view the sample streaming data along with its structure.

Figure 17-18. *Raw Data tab showing device data is continuously streaming*

Finally, write and execute the following SQL query shown in Figure 17-19 to retrieve the delta table that the structured streaming data is writing into. This table can then be used to perform additional advanced analytics and/or build machine learning models to gain further valuable real-time insights into the IoT device data:

```
%sql
SELECT *
FROM   delta_telemetry_data
```

Figure 17-19 shows the execution results of the code within the Databricks notebook.

Cmd 7

```
1  %sql
2  SELECT * FROM delta_telemetry_data
```

▸ (4) Spark Jobs

	Date_Enqueued	Hour_Enqueued	Enqueued_Time	Device_ID	Body	Sequence_Number
1	2020-10-17	2	2020-10-17T02:04:50.505+0000	"ASAToTDevice"	null	null
2	2020-10-17	2	2020-10-17T02:03:36.918+0000	"ASAToTDevice"	null	null
3	2020-10-17	2	2020-10-17T02:03:43.202+0000	"ASAToTDevice"	null	null
4	2020-10-17	2	2020-10-17T02:03:44.249+0000	"ASAToTDevice"	null	null
5	2020-10-17	2	2020-10-17T02:03:45.296+0000	"ASAToTDevice"	null	null
6	2020-10-17	2	2020-10-17T02:03:46.358+0000	"ASAToTDevice"	null	null
7	2020-10-17	2	2020-10-17T02:03:47.405+0000	"ASAToTDevice"	null	null
8	2020-10-17	2	2020-10-17T02:03:48.452+0000	"ASAToTDevice"	null	null

Figure 17-19. *Query delta table that the structured streaming data is writing into*

Summary

In this chapter, I have demonstrated how to implement a Structured Streaming solution using a Device Simulator that generates random device data into an Azure IoT Hub that is processed by Apache Spark through a Databricks notebook and streamed into a Delta Lake to persist the data. I also showed you how to customize Structured Streaming output modes including append, update, and complete. Finally, I introduced the concept of triggers and how to implement them within your Databricks notebook code.

Real-time analytics using Apache Spark in Databricks is an alternative to Stream Analytics and best suited for big data scenarios or situations where the data is fairly unstructured and requires some of the advanced capabilities that are available within Databricks such as schema evolution and more. It is also great for integrating your real-time streams with advanced analytics and machine learning workflows. The potential is limitless, and this chapter merely scratches the surface of some of these capabilities.

## CHAPTER 18

Azure Synapse Link for Cosmos DB

The desire to get near-real-time insights on data stored in a transactional system such as Cosmos DB has been a long-standing goal and need for many organizations. Azure Synapse Link for Azure Cosmos DB is a cloud-native hybrid transactional and analytical processing (HTAP) capability that allows users to run near-real-time analytics over operational data in Azure Cosmos DB. Data engineers, business analysts, and data scientists now have the ability to use Spark or SQL pools to get near-real-time insights into their data without impacting the performance of their transactional workloads in Cosmos DB.

There are numerous advantages to Azure Synapse Link for Azure Cosmos DB including reduced complexity since a near-real-time analytical store either reduces or eliminates the need for complex ETL or change feed job processes. Additionally, there will be little to no impact on operational workloads since the analytical workloads are rendered independently of the transactional workloads and do not consume the provisioned operational throughput. Additionally, it is optimized for large-scale analytics workloads by leveraging the power of Spark and SQL on-demand pools, which makes it cost effective due to the highly elastic Azure Synapse Analytics compute engines. With a column-oriented analytical store for workloads on operational data including aggregations and more, along with decoupled performance for analytical workloads, Azure Synapse Link for Azure Cosmos DB enables and empowers self-service, near-real-time insights on transactional data, as shown in Figure 18-1.

R. C. L'Esteve, *The Definitive Guide to Azure Data Engineering*, https://doi.org/10.1007/978-1-4842-7182-7_18

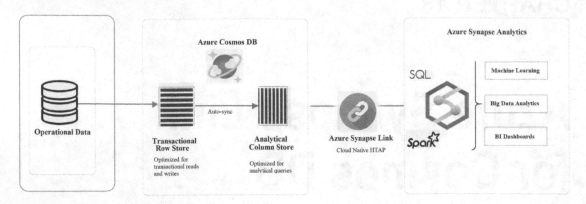

Figure 18-1. *Architecture diagram of Azure Synapse Link*

In this chapter, you will learn how to achieve the following through a practical end-to-end exercise:

1. Create a basic Azure Cosmos DB account enabled for analytical storage.

2. Create a Cosmos DB linked service in Azure Synapse Analytics.

3. Aggregate and query Cosmos DB data with Spark from a Synapse Workspace notebook.

Create an Azure Cosmos DB Account

Azure Cosmos DB is a fully managed NoSQL database service for modern app development. You will need to create an Azure Cosmos DB account from Azure Portal, as shown in Figure 18-2.

Azure Cosmos DB 📌 ···

Microsoft

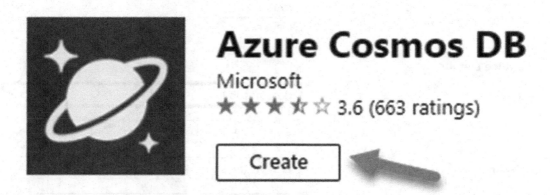

Figure 18-2. *Creating a Cosmos DB*

Ensure that the account details are configured as desired, shown in Figure 18-3, and create the Azure Cosmos DB account.

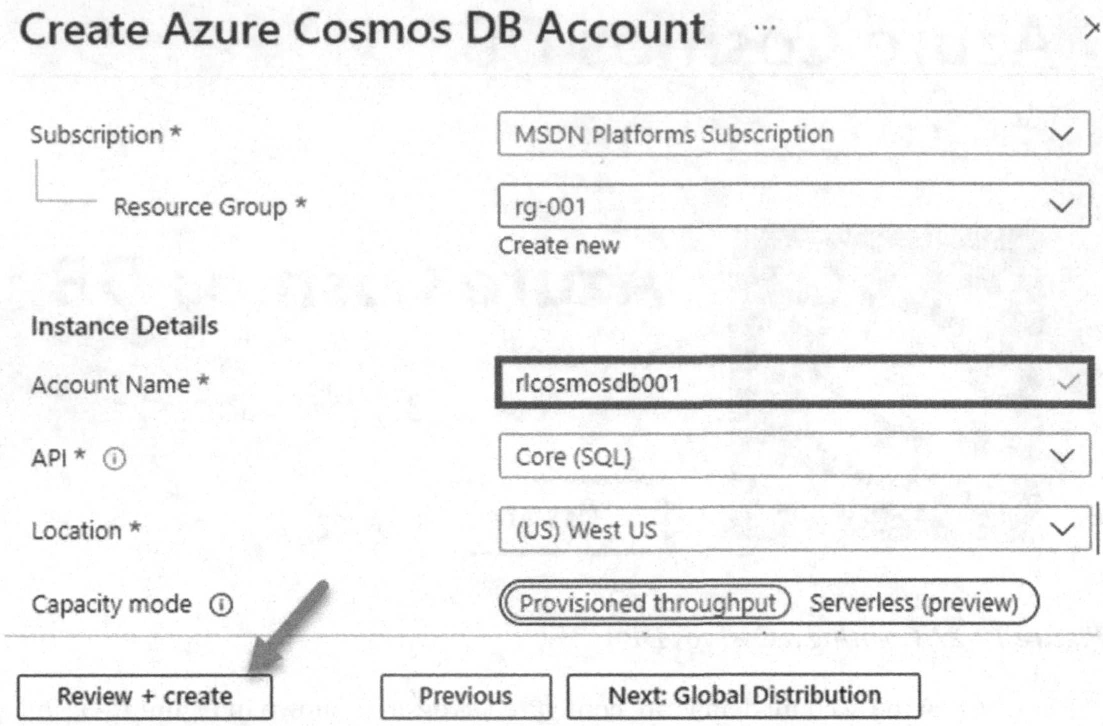

Figure 18-3. *Configure the Cosmos DB account details*

To recap, in this section, you created an Azure Cosmos DB account using the Core (SQL) API. There are many APIs to choose from including the native Core (SQL) API, API for MongoDB, Cassandra API, Gremlin API, and Table API. Also, choose Provisioned throughput as the capacity mode as it is best suited for workloads with sustained traffic requiring predictable performance, while Serverless is best suited for workloads with intermittent or unpredictable traffic and low average-to-peak traffic ratio. For more details on capacity modes, please see `https://docs.microsoft.com/en-us/azure/cosmos-db/throughput-serverless`.

Enable Azure Synapse Link

Once the Cosmos DB account is created, you will need to enable the Azure Synapse Link, which by default is set to "Off." This can be achieved by first clicking the Azure Synapse Link feature shown in Figure 18-4.

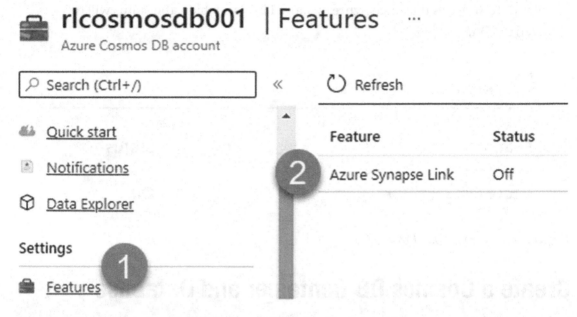

Figure 18-4. *Steps to enable Synapse Link for Cosmos DB*

Then click "Enable," as shown in Figure 18-5.

Azure Synapse Link   ✕

Azure Synapse Link for Cosmos DB creates a tight integration between Azure Cosmos DB and Azure Synapse Analytics enabling customers to run near real-time analytics over their operational data with no-ETL and full performance isolation from their transactional workloads.

By combining the distributed scale of Cosmos DB's transactional processing with build-in analytical store and the computing power of Azure Synapse Analytics, Azure Synapse Link enables Hybrid Transactional/Analytical Processing (HTAP) architectures for optimizing your business processes. This integration eliminates ETL processes, enabling business analysts, data engineers & data scientists to self-serve and run near real-time BI, analytics and ML pipelines over operational data.

Learn More

| Enable | Close |

Figure 18-5. *Enable Synapse Link*

Figure 18-6 shows that once Azure Synapse Link is enabled, the status will be changed to "On."

Figure 18-6. *Synapse Link is On*

Create a Cosmos DB Container and Database

Now that the Azure Cosmos DB account is enabled for Azure Synapse Link, create a database and container. To get started, the Quick start section within Cosmos DB allows for an easier experience of choosing a platform, creating a database and container, and then getting started with a Cosmos DB notebook to run the code to import data.

For the purposes of this exercise, use the Python platform. However, notice the additional wide range of platform options to choose from, which is shown in Figure 18-7.

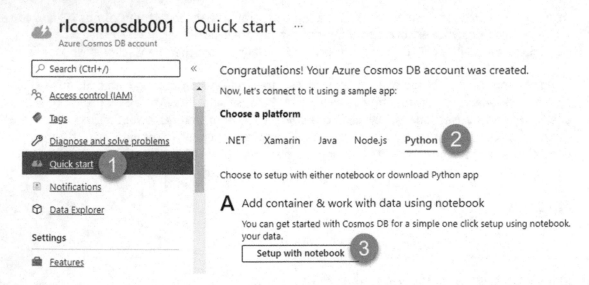

Figure 18-7. *Create a new notebook in Python*

Figure 18-8 shows how to begin by creating a database and container through the UI. There are also options available to start with a new notebook or a sample notebook with predefined code.

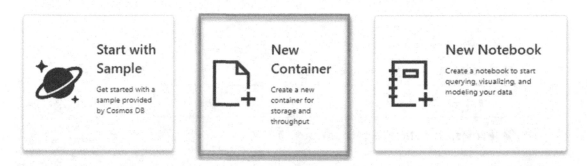

Figure 18-8. *Create a new Cosmos DB container*

You will need to configure the database name, throughput, container name, and partition key, as shown in Figure 18-9. You could always start with the free tier containing 400 RU/s and 5 GB of storage and then scale up accordingly.

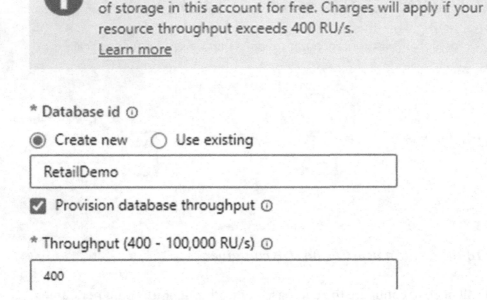

Figure 18-9. *Configure the container Database id and Throughput details*

Finally, it is important to remember to set your partition key and switch on the analytical store, as shown in Figure 18-10. The partition key is used to automatically partition data across multiple servers for scalability. It is the analytical store that will allow you to perform real-time analytics on your data.

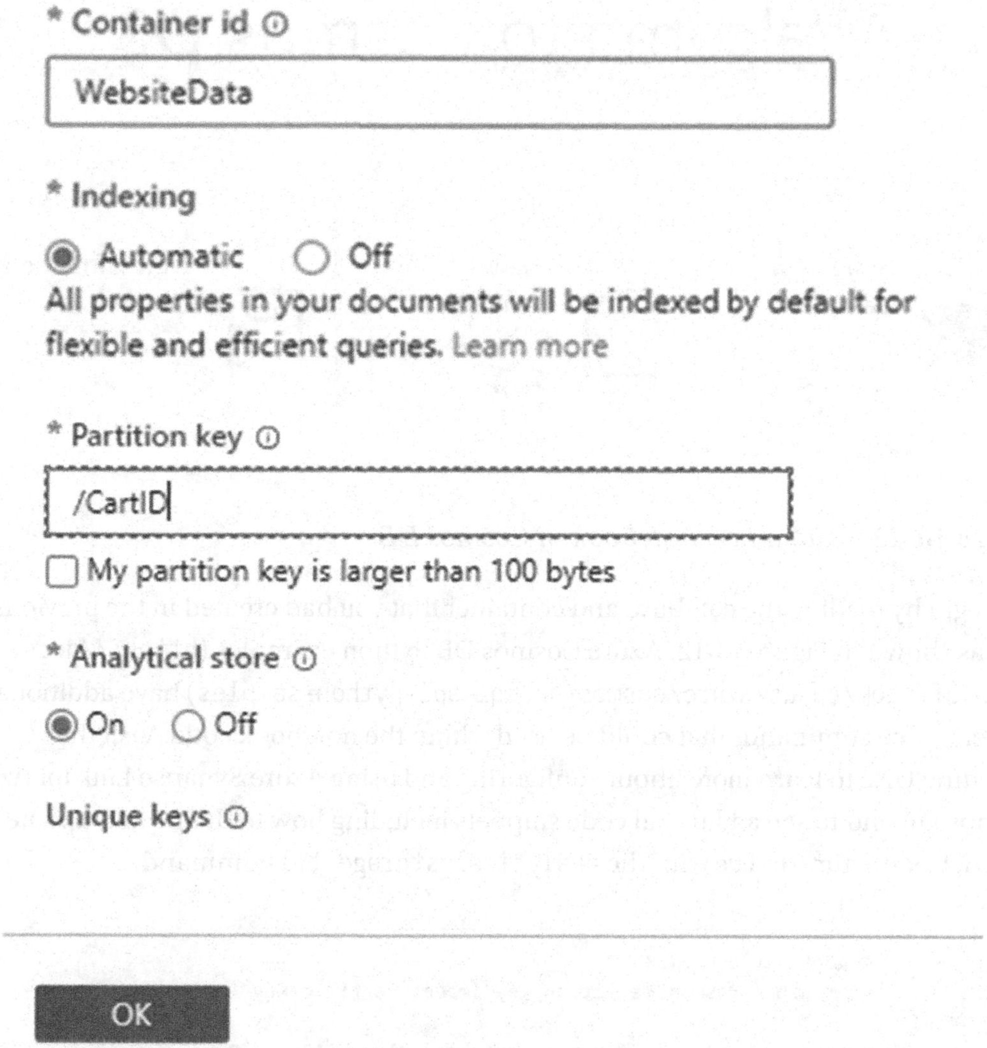

* Container id ⓘ

WebsiteData

* Indexing

● Automatic   ○ Off
All properties in your documents will be indexed by default for
flexible and efficient queries. Learn more

* Partition key ⓘ

/CartID

☐ My partition key is larger than 100 bytes

* Analytical store ⓘ

● On   ○ Off

Unique keys ⓘ

OK

Figure 18-10. Configure the remaining container details

Import Data into Azure Cosmos DB

Now that you have created and configured the details for the database and container,
create a new notebook to import data into the Cosmos database container, as shown in
Figure 18-11.

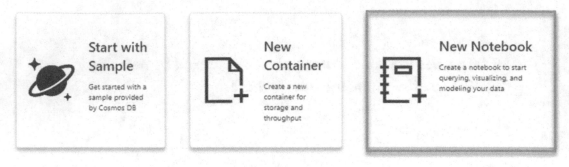

Welcome to Cosmos DB

Globally distributed, multi-model database service for any scale

Start with Sample

Get started with a sample provided by Cosmos DB

New Container

Create a new container for storage and throughput

New Notebook

Create a notebook to start querying, visualizing, and modeling your data

Figure 18-11. *Add a new notebook in Cosmos DB*

Begin by reading the database and container that you had created in the previous step, as shown in Figure 18-12. Azure Cosmos DB Python examples (`https://docs.microsoft.com/en-us/azure/cosmos-db/sql-api-python-samples`) have additional API reference commands that could be used within the notebook code. Visit the preceding URL to learn more about configuring and using Azure Synapse Link for Azure Cosmos DB and to see additional code snippets including how to define and update the analytical store time to live with the `analytical_storage_ttl` command.

```
[14]    1    import azure.cosmos
        2    from azure.cosmos.partition_key import PartitionKey
        3
        4    database = cosmos_client.get_database_client('RetailDemo')
        5    print('Database RetailDemo Read')
        6
        7    container = database.get_container_client('WebsiteData')
        8    print('Container WebsiteData Read')
        9
       10

Database RetailDemo Read
Container WebsiteData Read
```

Figure 18-12. *Create or read the new DB and container*

Here is the code that is shown in Figure 18-12 to read the database and container name that you created in the previous section:

```
import azure.cosmos
from azure.cosmos.partition_key import PartitionKey

database = cosmos_client.get_database_client('RetailDemo')
print('Database RetailDemo Read')

container = database.get_container_client('WebsiteData')
print('Container WebsiteData Read')
```

Once the data is read, you can update the throughput of the container to allow for a quicker upload by using the code shown in Figure 18-13.

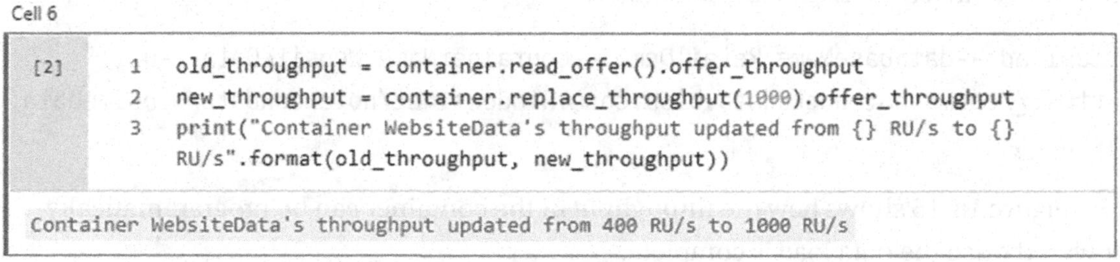

```
Cell 6

[2]    1   old_throughput = container.read_offer().offer_throughput
       2   new_throughput = container.replace_throughput(1000).offer_throughput
       3   print("Container WebsiteData's throughput updated from {} RU/s to {}
           RU/s".format(old_throughput, new_throughput))

Container WebsiteData's throughput updated from 400 RU/s to 1000 RU/s
```

Figure 18-13. *Scale up throughput before load*

Here is the code that is shown in Figure 18-13 to update the throughput of the container to allow for a quicker upload:

```
old_throughput = container.read_offer().offer_throughput
new_throughput = container.replace_throughput(1000).offer_throughput
print("Container WebsiteData's throughput updated from {} RU/s to {} RU/s".
format(old_throughput, new_throughput))
```

Next, use the %%upload magic function shown in Figure 18-14 to insert items into the container.

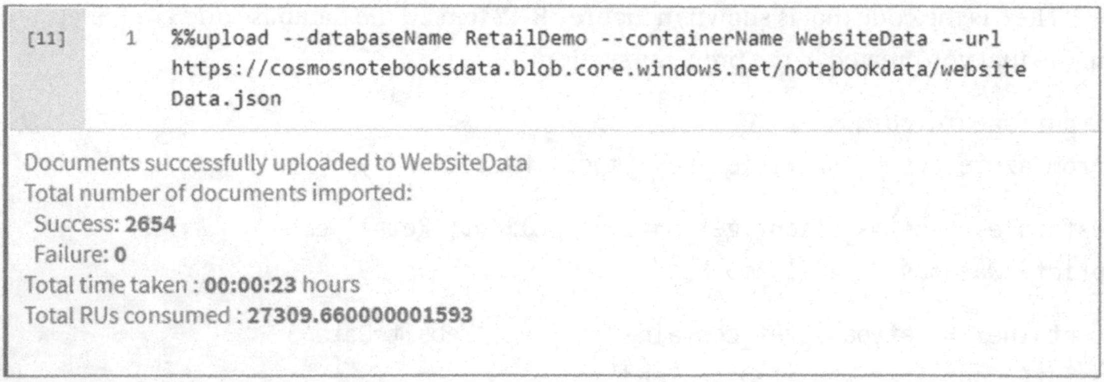

```
[11]    1    %%upload --databaseName RetailDemo --containerName WebsiteData --url
             https://cosmosnotebooksdata.blob.core.windows.net/notebookdata/website
             Data.json
```

Documents successfully uploaded to WebsiteData
Total number of documents imported:
  Success: **2654**
  Failure: **0**
Total time taken : **00:00:23** hours
Total RUs consumed : **27309.660000001593**

Figure 18-14. *Load data into Cosmos DB*

Here is the code that is shown in Figure 18-14 that uses the %%upload magic function to insert items into the container:

```
%%upload --databaseName RetailDemo --containerName WebsiteData --url
https://cosmosnotebooksdata.blob.core.windows.net/notebookdata/websiteData.
json
```

Figure 18-15 shows how the throughput of the container can be programmatically lowered once the data load is complete.

```
[3]    1    lowered_throughput =
            container.replace_throughput(400).offer_throughput
       2    print("Container WebsiteData's throughput lowered from {} RU/s to {}
            RU/s".format(new_throughput, lowered_throughput))
```

Container WebsiteData's throughput lowered from 1000 RU/s to 400 RU/s

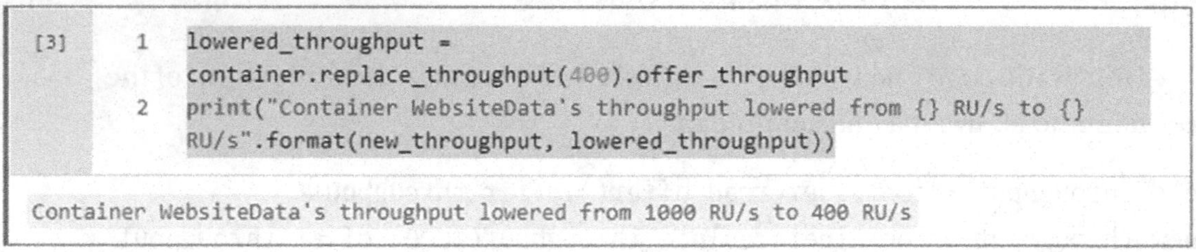

Figure 18-15. *Scale down the throughput once load is complete*

Here is the code that is shown in Figure 18-15 to scale down the throughput of the container once the data load is complete:

```
lowered_throughput =
container.replace_throughput(400).offer_throughput
print("Container WebsiteData's throughput lowered from {} RU/s to {} RU/s".
format(new_throughput, lowered_throughput))
```

Create a Cosmos DB Linked Service in Azure Synapse Analytics

After data is available in the Cosmos DB database container, create a linked service within the Azure Synapse Analytics workspace by following the steps illustrated in Figure 18-16.

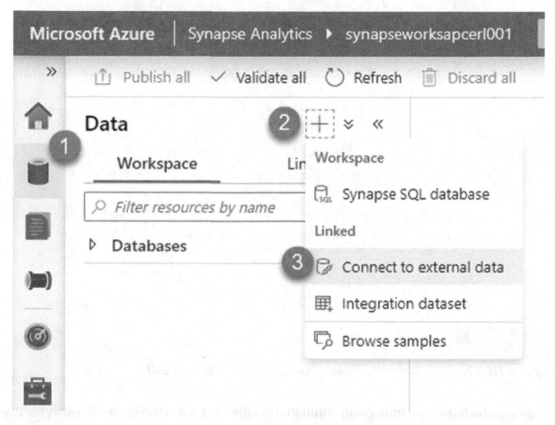

Figure 18-16. *Create Cosmos DB link in Synapse Analytics*

Remember to select Azure Cosmos DB (SQL API) shown in Figure 18-17 since that is the configuration of the Azure Cosmos DB API. Note that there is also an option for a MongoDB API.

Connect to external data

Once a connection is created, the underlying data of that connection will be available for analysis in the Data hub or for pipeline activities in the Orchestrate hub.

Continue   Cancel

Figure 18-17. *Step to create linked connection to Cosmos DB*

Fill in the required connection configuration details shown in Figure 18-18 and create the new linked service.

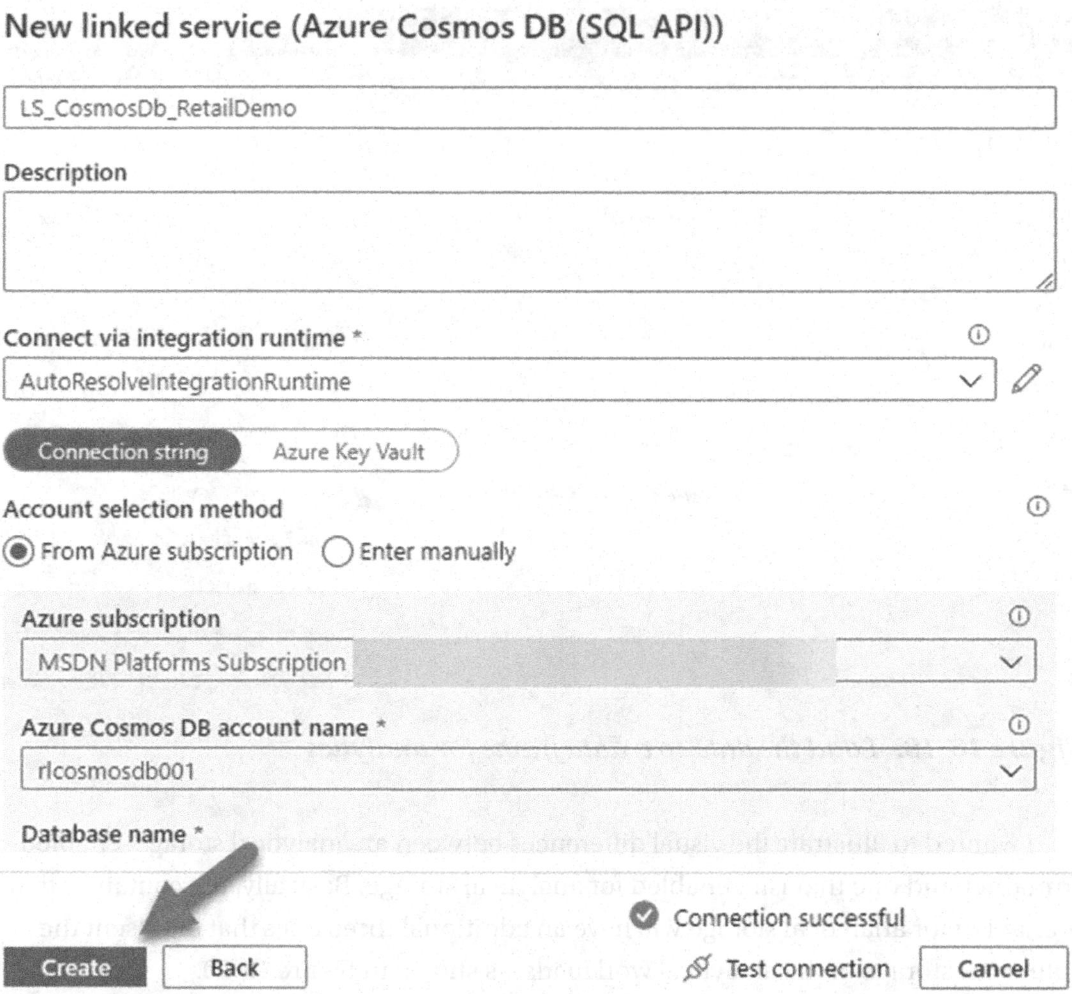

New linked service (Azure Cosmos DB (SQL API))

LS_CosmosDb_RetailDemo

Description

Connect via integration runtime *   ⓘ

AutoResolveIntegrationRuntime   ⌄   ✎

Connection string   Azure Key Vault

Account selection method   ⓘ

⦿ From Azure subscription   ◯ Enter manually

Azure subscription   ⓘ

MSDN Platforms Subscription   ⌄

Azure Cosmos DB account name *   ⓘ

rlcosmosdb001   ⌄

Database name *

✓ Connection successful

| Create | Back | 🖉 Test connection | Cancel |

Figure 18-18. *Create the new linked service*

Load and Query the Data Using Synapse Spark

After creating a new linked service to Cosmos DB from Azure Synapse Analytics, follow
the steps illustrated in Figure 18-19 to create a new notebook and load the data to a data
frame.

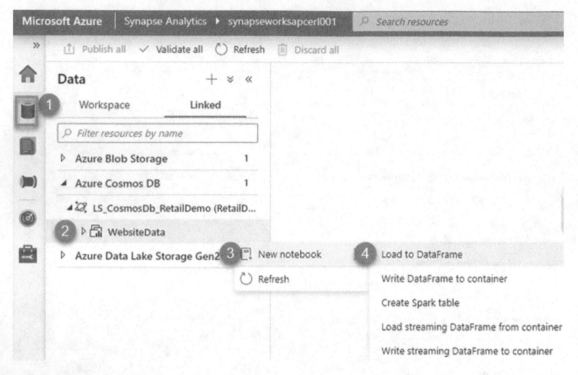

***Figure 18-19.** Load the data to a data frame for analytics*

I wanted to illustrate the visual differences between an analytical storage–enabled container and one that isn't enabled for analytical storage. Basically, the container that is enabled for analytical storage will have an additional three lines that represent the columnar storage of the analytical workloads, as shown in Figure 18-20.

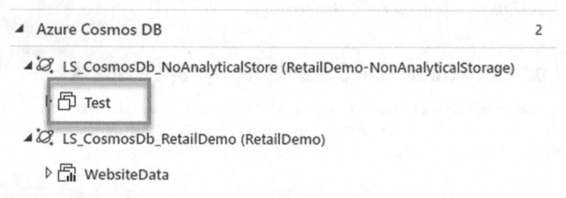

***Figure 18-20.** Non–analytical storage enabled*

After the linked service has been created, run the following code that will be auto-created when the data is loaded to a data frame from the previous step. Prior to running the code, remember to create a Synapse Spark pool and attach it to the notebook:

```
# Read from Cosmos DB analytical store into a Spark DataFrame and display
10 rows from the DataFrame
# To select a preferred list of regions in a multi-region
Cosmos DB account, add .option("spark.cosmos.preferredRegions",
"<Region1>,<Region2>")

df = spark.read\
    .format("cosmos.olap")\
    .option("spark.synapse.linkedService", "LS_CosmosDb_RetailDemo")\
    .option("spark.cosmos.container", "WebsiteData")\
    .load()

display(df.limit(10))
```

Notice from the illustration in Figure 18-21 that the code ran successfully and used two executors and eight cores for the job. Note that this can be customized to fit your desired workloads. Additionally, the job details can be viewed in the Spark UI.

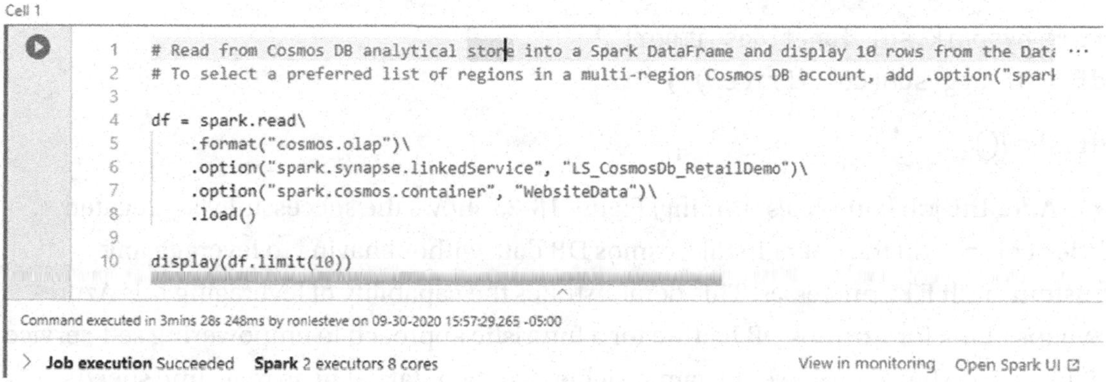

Figure 18-21. *Code to load to data frame*

Figure 18-22 shows a preview of the top ten records from the data frame, which confirms that the real-time querying capability is active.

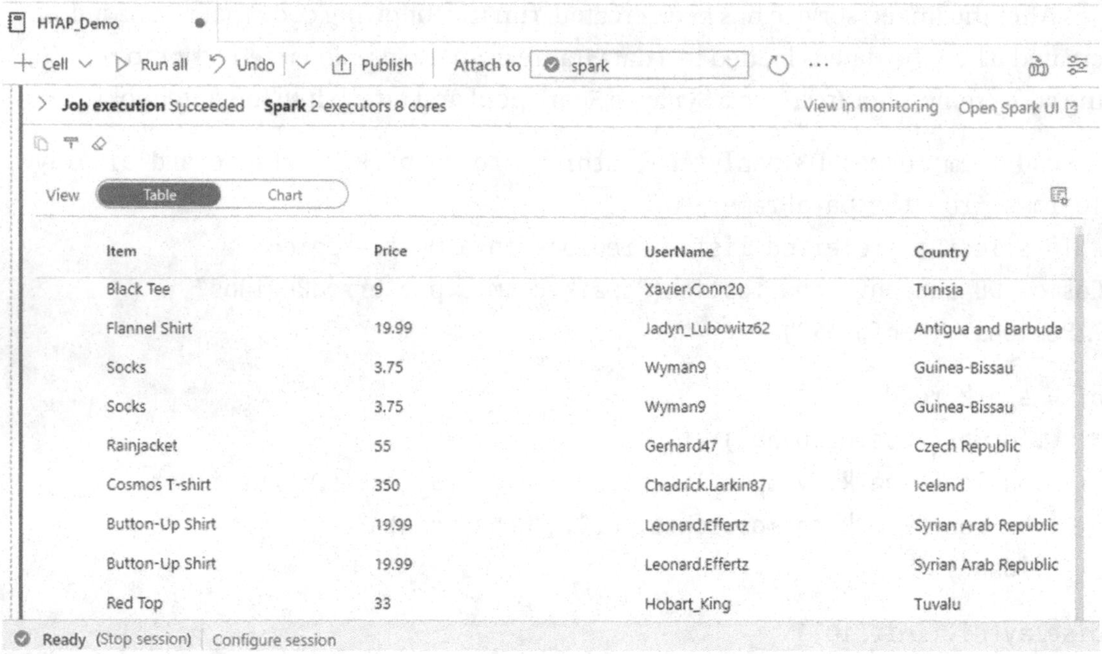

Figure 18-22. Image of sample data returned

Next, use the following code to aggregate the dataset to get a sum of the Price column and then display the data frame:

```
from pyspark.sql.functions import *
df = df.agg(sum(col('Price')))

df.show()
```

After the job completes running, Figure 18-23 shows the successfully aggregated Price column on the operational Cosmos DB data without having to leverage any custom-built ETL processes. This demonstrates the capability of leveraging this Azure Synapse Link for Cosmos DB feature for a futuristic approach to empowering self-service data users to have the power to gain insights into their data at near-real-time speeds.

Figure 18-23. *Aggregate the price data*

Summary

In this chapter, I have shown you how to create a basic Azure Cosmos DB account enabled for analytical storage, create a Cosmos DB linked service in Azure Synapse Analytics, and aggregate and query Cosmos DB data with Spark from a Synapse Workspace notebook.

By setting up this process, your organization has the power to gain valuable insights from real-time analytics on your transactional data, which allows business analysts, data scientists, data engineers, and other citizen developers to analyze large datasets without impacting the performance of transactional workloads. Among the various other benefits, this process reduces the need to build and manage complex ETL jobs.

PART IV

DevOps for Continuous Integration and Deployment

## CHAPTER 19

Deploy Data Factory Changes

The critical need to deploy Azure Data Factory from one environment to another using the best practices of the Azure DevOps continuous integration and deployment process presents a number of complexities to completing the deployment process successfully. In this chapter, I will cover how to utilize PowerShell scripts along with the Azure Resource Group Deployment task to start/stop ADF triggers and change ADF environment connection configuration properties through an end-to-end deployment of Azure Data Factory CI/CD changes with Azure DevOps.

Prerequisites

To continue the exercise, you'll need a few prerequisite Azure resources. This section shows what you must create and run prior to building the end-to-end CI/CD process.

Figure 19-1 shows an image of a DEV resource group containing a few Azure resources. Similarly, Figure 19-2 shows an image of a QA resource group containing identical Azure resources as the DEV resource group. For the purpose of this exercise, you'll be focusing on deploying ADF changes from one environment to another through an automated DevOps CI/CD pipeline. You would need to manually create the QA Data Factory for this exercise. You will also need a DEV and QA Key Vault, which will store pertinent credentials and secrets that will be used in this chapter. Note also that a SQL Server, SQL database, and ADLS Gen2 account have also been created, mainly to show consistency of resources from a lower environment (DEV) when compared to an upper environment (QA) and to use these resources as linked service connections within the DEV resource group's ADF, which will be deployed to the QA ADF.

© Ron C. L'Esteve 2021

R. C. L'Esteve, *The Definitive Guide to Azure Data Engineering*, https://doi.org/10.1007/978-1-4842-7182-7_19

Create the DEV resource group shown in Figure 19-1. This resource group will contain the lower environment (DEV) Azure Data Factory, ADLS Gen2, and SQL database whose changes will be continuously deployed to the upper environment (QA) as part of the DevOps CI/CD pipeline.

Figure 19-1.   *Azure Portal DEV resource group*

Then create the QA resource group shown in Figure 19-2. This resource group will contain the upper environment (QA) Azure Data Factory, ADLS Gen2, and SQL database.

Figure 19-2.   *Azure Portal QA resource group*

Create the following Key Vault account secrets in both the DEV resource group, shown in Figure 19-3, and the QA resource group, shown in Figure 19-4. Please note that the secret names in both the DEV and QA accounts must be the same for deployment purposes. However, the actual secrets are unique between DEV and QA:

- **ls-adls2-akv**: This will contain the access key for both the DEV and QA ADLS Gen2 accounts, which can be found in the Access Keys section of the respective DEV and QA Data Lake Storage accounts.

- **ls-sql-akv**: This will contain the admin password of the DEV and QA Azure SQL Servers and databases for authentication purposes.

Figure 19-3 shows the secrets that you have created and enabled in the DEV resource group.

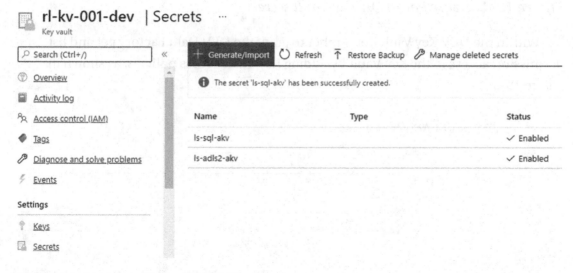

Figure 19-3. *Azure Portal DEV Key Vault secrets*

Similarly, Figure 19-4 shows the secrets you have created and enabled in the QA resource group.

Figure 19-4. *Azure Portal QA Key Vault secrets*

Within the DEV Key Vault, remember to grant the DEV Data Factory get and list permission access to the Key Vault by adding Key Vault access policies, as shown in Figure 19-5.

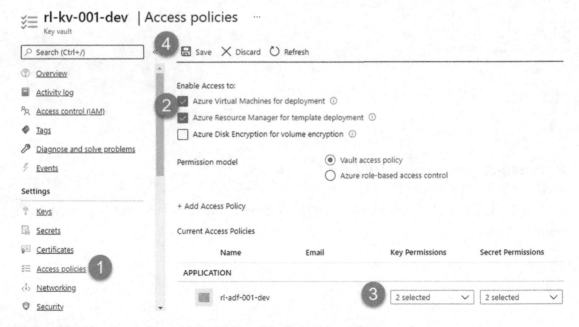

Figure 19-5. *Azure Portal DEV Key Vault access policies*

Similarly, within the QA Key Vault, remember to grant the QA Data Factory get and list permission access to the Key Vault by adding Key Vault access policies, as shown in Figure 19-6.

Figure 19-6. *Azure Portal QA Key Vault access policies*

Create the following linked services in the DEV Data Factory for testing purposes, as shown in Figure 19-7.

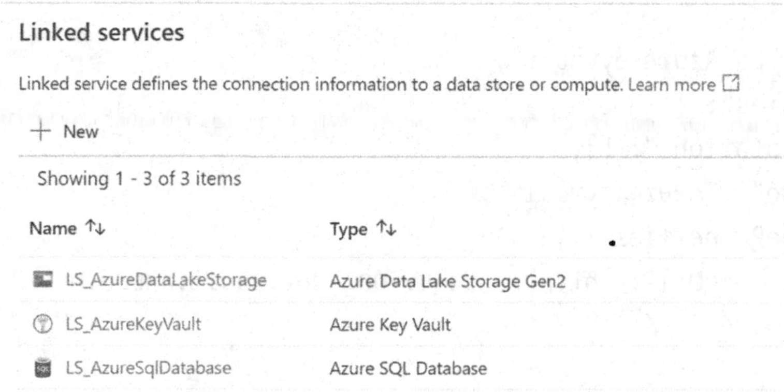

Figure 19-7. *Linked services in the DEV Data Factory*

The following is the JSON script for the Azure Data Lake Storage linked service connection:

```
{
    "name": "LS_AzureDataLakeStorage",
    "properties": {
        "annotations": [],
        "type": "AzureBlobFS",
        "typeProperties": {
            "url": "https://adls2dev.dfs.core.windows.net",
            "accountKey": {
                "type": "AzureKeyVaultSecret",
                "store": {
                    "referenceName": "LS_AzureKeyVault",
                    "type": "LinkedServiceReference"
                },
                "secretName": "ls-adls2-akv",
                "secretVersion": ""
            }
        }
    }
}
```

The following is the JSON script for the Azure Key Vault linked service connection:

```
{
    "name": "LS_AzureKeyVault",
    "properties": {
        "annotations": [],
        "type": "AzureKeyVault",
        "typeProperties": {
            "baseUrl": "https://rl-kv-001-dev.vault.azure.net/"
        }
    }
}
```

The following is the JSON script for the Azure SQL Database linked service connection:

```json
{
    "name": "LS_AzureSqlDatabase",
    "type": "Microsoft.DataFactory/factories/linkedservices",
    "properties": {
        "annotations": [],
        "type": "AzureSqlDatabase",
        "typeProperties": {
            "connectionString": "Integrated Security=False;Encrypt=True;
Connection Timeout=30;Data Source=rl-sqlserver-001-dev.database.windows.
net;Initial Catalog=rl-sql-001-dev;User ID=devadmin",
            "password": {
                "type": "AzureKeyVaultSecret",
                "store": {
                    "referenceName": "LS_AzureKeyVault",
                    "type": "LinkedServiceReference"
                },
                "secretName": "ls-sql-akv"
            }
        }
    }
}
```

Figure 19-8 shows the trigger that you will need to start in the DEV Data Factory for testing purposes and to demonstrate how triggers impact the deployment of ADF changes from one environment to another, such that it would need to be stopped and restarted by the Azure DevOps CI/CD pipeline. Figure 19-8 will help you with understanding where this trigger will exist.

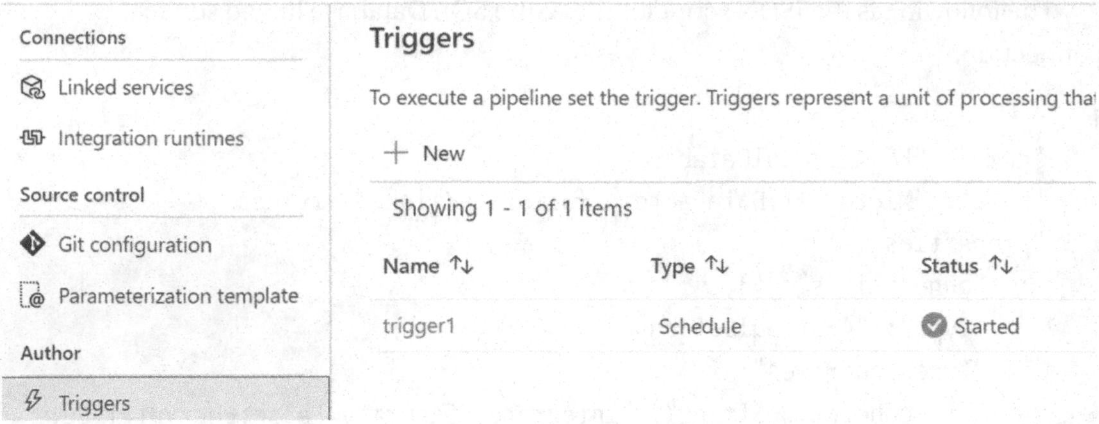

Connections	Triggers

Connections

- 🗟 Linked services
- 🕮 Integration runtimes

Source control

- ◈ Git configuration
- 🗟 Parameterization template

Author

- 𝄍 Triggers

Triggers

To execute a pipeline set the trigger. Triggers represent a unit of processing that

+ New

Showing 1 - 1 of 1 items

Name ↑↓	Type ↑↓	Status ↑↓
trigger1	Schedule	✅ Started

Figure 19-8. *Trigger in the DEV Data Factory*

Once you have added your linked service connections and triggers, publish the
DEV Data Factory to commit the changes to the *adf_publish* branch and to prepare for
the DevOps CI/CD process. Please note that this method will be using the *adf_publish*
branch rather than the master branch for the CI/CD deployment process. Please verify
that the Data Factory additions and changes have been published to the *adf_publish*
branch in the GitHub repo shown in Figure 19-9.

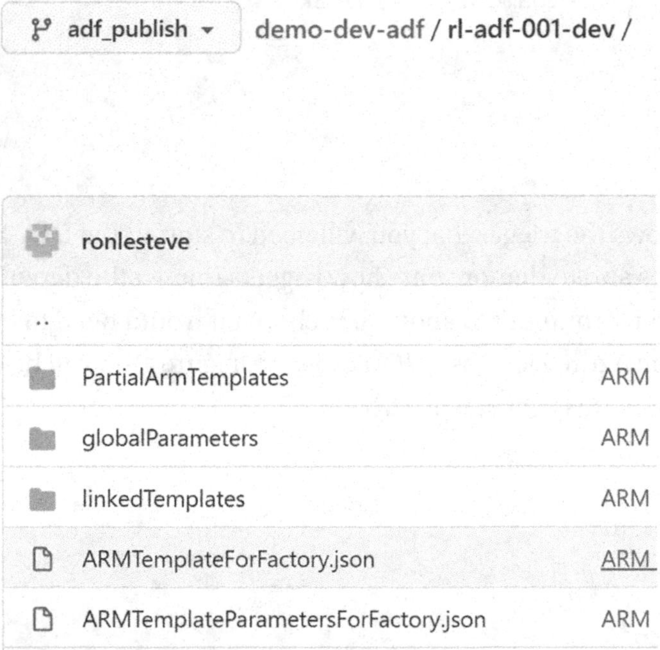

⑁ adf_publish ▾   demo-dev-adf / rl-adf-001-dev /

🖳 ronlesteve	
..	
📁 PartialArmTemplates	ARM
📁 globalParameters	ARM
📁 linkedTemplates	ARM
🗋 ARMTemplateForFactory.json	<u>ARM</u>
🗋 ARMTemplateParametersForFactory.json	ARM

Figure 19-9. *Changes published to adf_publish branch*

Finally, add the sample pre- and post-deployment script, which is included toward the bottom of Microsoft's article on continuous integration and delivery in ADF (`https://docs.microsoft.com/en-us/azure/data-factory/continuous-integration-deployment#script`). Also ensure that you add the file within the same folder as the DEV ADF resources shown in Figure 19-10 and name it cicd.ps1. The details of this file can also be found in my GitHub repo (`https://github.com/ronlesteve/demo-dev-adf/blob/adf_publish/rl-adf-001-dev/cicd.ps1`).

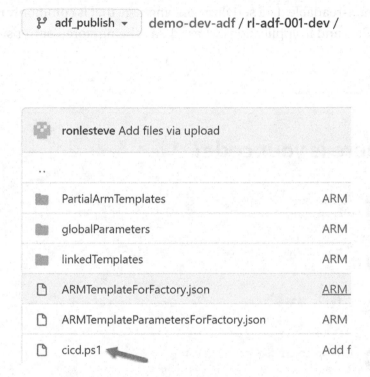

Figure 19-10. *Sample pre- and post-deployment script*

Create the DevOps Continuous Integration Build Pipeline

After creating all of the necessary prerequisites, you can begin creating the Azure DevOps continuous integration build pipeline by clicking "New pipeline" in Azure DevOps pipelines, as shown in Figure 19-11.

Figure 19-11. *New build pipeline*

Select the "Use the classic editor" link toward the bottom of the UI shown in Figure 19-12 to prevent having to create the pipeline with YAML code. For reference, YAML is a human-readable data serialization language that is commonly used for configuration files and in applications where data is being stored or transmitted.

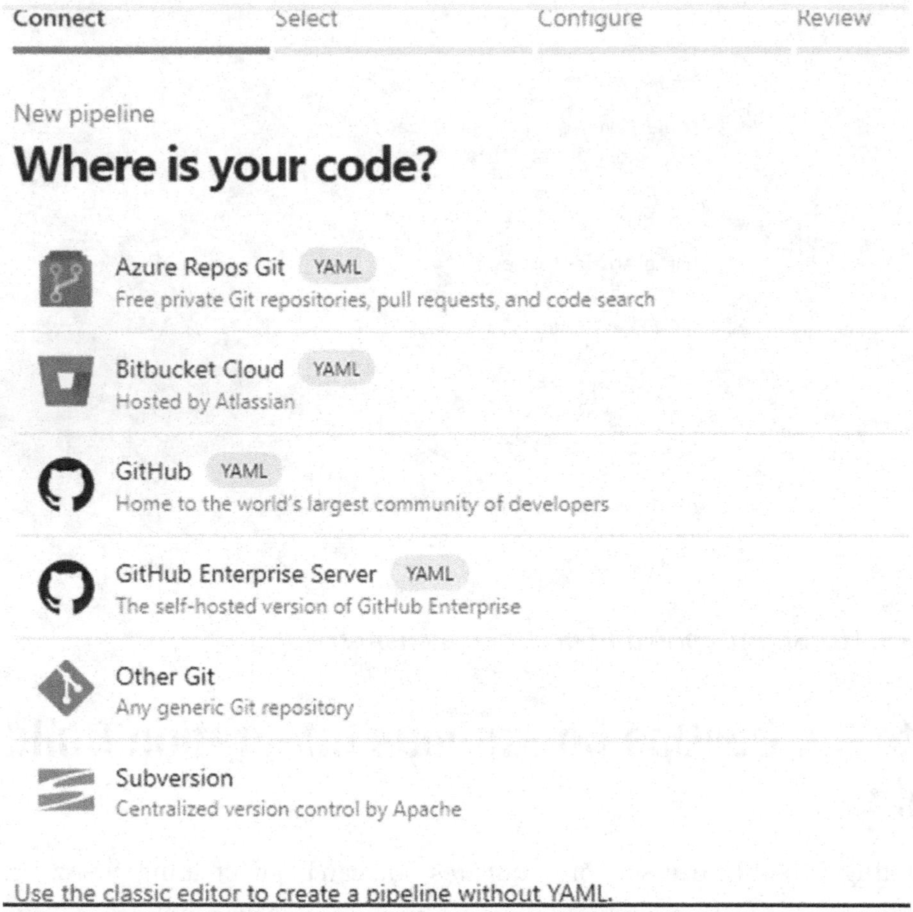

Figure 19-12. *Use the classic editor*

Select GitHub since that is where the code repo is saved. Also select the repo name along with the *adf_publish* branch, which is shown in Figure 19-13.

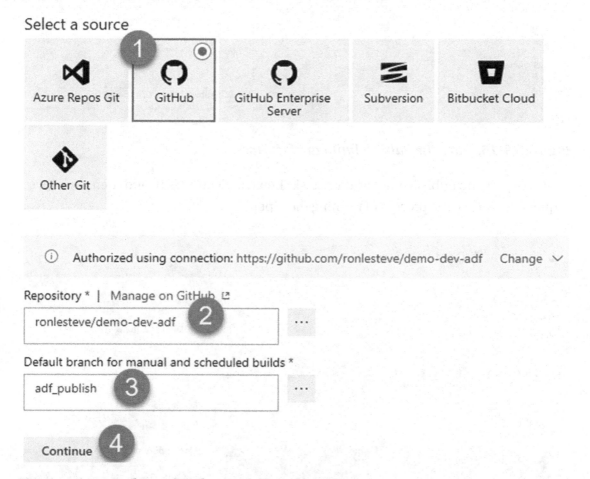

Figure 19-13. *Select GitHub source and repo details*

Start with an Empty job template, as shown in Figure 19-14.

Figure 19-14. *Start with an Empty job template*

Click the + icon on the Agent job to add a new task. Also add the Publish build artifacts task, which can be seen in Figure 19-15.

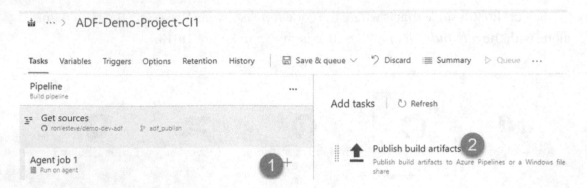

Figure 19-15. *Add the Publish build artifacts task*

Configure the Publish build artifacts task shown in Figure 19-16 and then click Save & queue to initiate the process of running the pipeline.

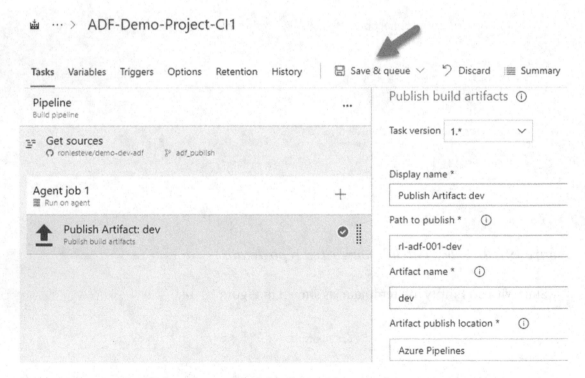

Figure 19-16. *Configure the Publish build artifacts task*

Verify the run pipeline details in Figure 19-17 and click Run to begin manually running the pipeline.

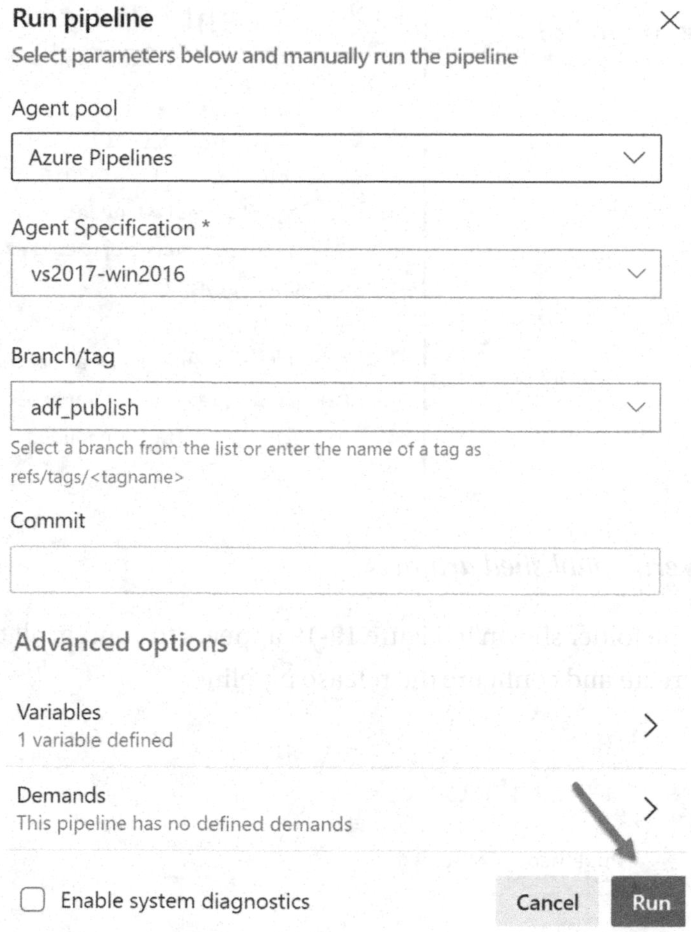

Run pipeline   ✕

Select parameters below and manually run the pipeline

Agent pool

Azure Pipelines	∨

Agent Specification *

vs2017-win2016	∨

Branch/tag

adf_publish	∨

Select a branch from the list or enter the name of a tag as
refs/tags/<tagname>

Commit

Advanced options

Variables   ❯
1 variable defined

Demands   ❯
This pipeline has no defined demands

☐ Enable system diagnostics   Cancel   Run

Figure 19-17. *Run the pipeline*

Once the job completes running successfully, verify that the artifacts have been
published, as shown in Figure 19-18.

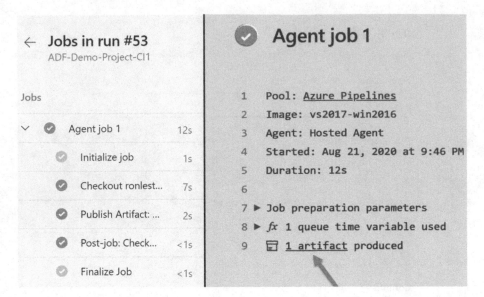

Figure 19-18. *Verify published artifacts*

As expected, the folder shown in Figure 19-19 appears to contain all the artifacts that you will need to create and configure the release pipeline.

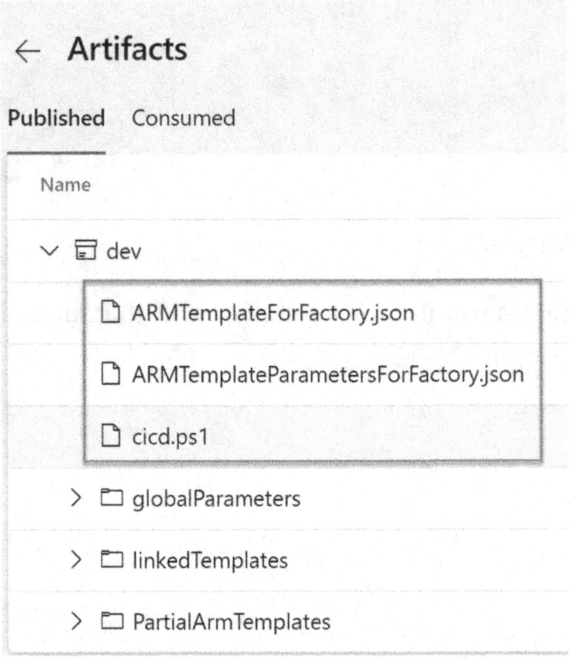

Figure 19-19. *Published artifacts in Azure DevOps*

Create the DevOps Continuous Deployment Release Pipeline

Now that the continuous integration (CI) build pipeline has been successfully completed and verified, it is time to create a new continuous deployment (CD) release pipeline. To begin this process, click New release pipeline as shown in Figure 19-20.

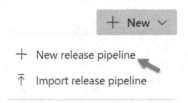

Figure 19-20. *New release pipeline*

Click Add an artifact, as shown in Figure 19-21. Notice within the diagram that you could also set an automated run schedule when necessary. Artifacts may consist of project source code, dependencies, binaries, or resources. For this scenario, the artifact will be coming directly from the output of what you created in the build stage.

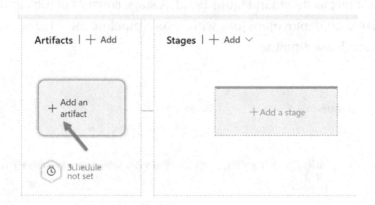

Figure 19-21. *Add an artifact*

Select Build as the source type. Also select the build pipeline that was created in the previous section and populate the additional required details, as shown in Figure 19-22. Finally, click Add to complete this step.

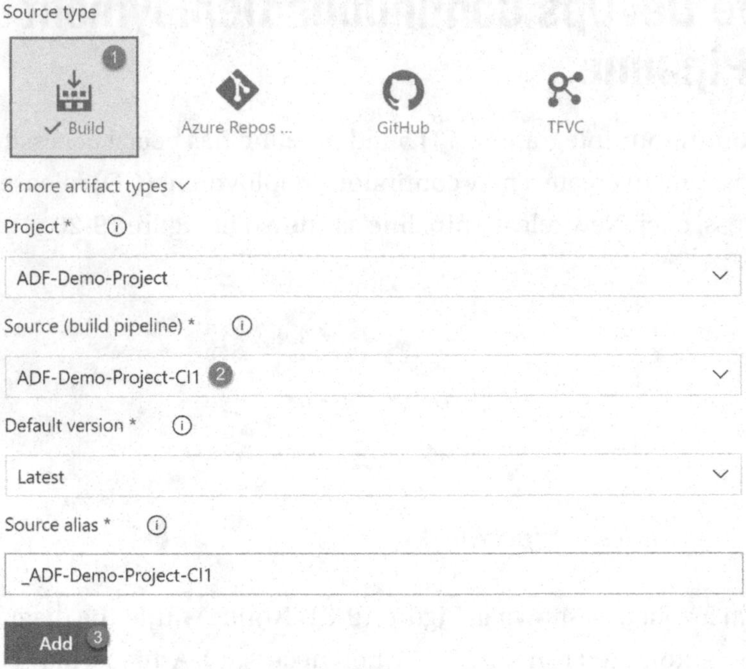

Source type
1
Build
Azure Repos ...
GitHub
TFVC
6 more artifact types
Project *
ADF-Demo-Project
Source (build pipeline) *
ADF-Demo-Project-CI1
2
Default version *
Latest
Source alias *
_ADF-Demo-Project-CI1
Add
3

***Figure 19-22.** Add build artifact details*

Next, add a stage, as shown in Figure 19-23. A stage consists of jobs and tasks, and you can organize your deployment jobs within your pipeline into stages, which are major divisions in your release pipeline.

Artifacts | + Add
Stages | + Add
_ADF-Demo-Project-CI1
+ Add a stage
Schedule not set

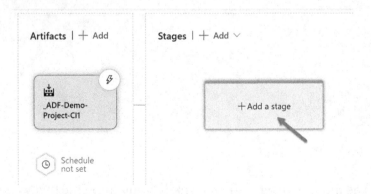

***Figure 19-23.** Add a stage*

Start with an Empty job template shown in Figure 19-24. A job is a series of steps that run sequentially as a unit, and every pipeline must have at least one job.

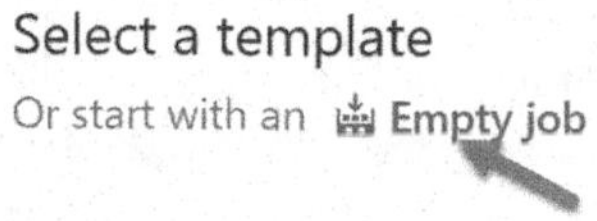

Select a template

Or start with an 🏭 Empty job

Figure 19-24. *Start with an Empty job template*

Next, click the link to add a task, as shown in Figure 19-25. A task is the building block for defining automation in a pipeline. A task is simply a packaged script or procedure that has been abstracted with a set of inputs.

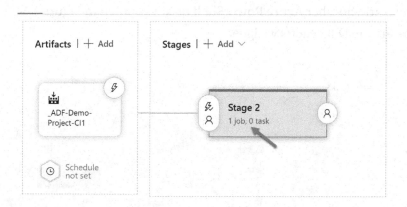

Figure 19-25. *Link to add a task*

Begin by adding an Azure PowerShell script task, as shown in Figure 19-26. This will be used to stop the Data Factory triggers.

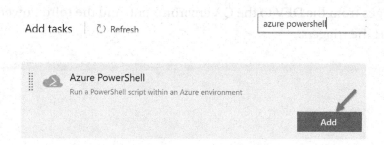

Figure 19-26. *Azure PowerShell script task*

Also add an ARM template deployment task, as shown in Figure 19-27. This will be used to deploy the Data Factory artifacts and parameters to the desired environment.

Figure 19-27. *ARM template deployment task*

Finally, also add another Azure PowerShell task, as shown in Figure 19-28. This will be used to restart the Data Factory triggers.

Figure 19-28. *Azure PowerShell task to restart ADF triggers*

Ensure that all three tasks are organized in the following order shown in Figure 19-29 prior to configuring them. The first PowerShell script task will stop the ADF triggers that you had created in the DEV ADF. The second ARM template deployment task will deploy the ADF changes from the DEV to the QA environment, and the third PowerShell script task will restart the triggers.

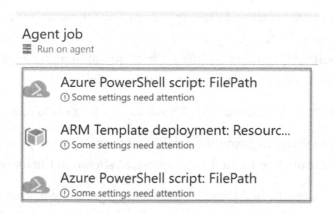

Figure 19-29. *Organize Agent job tasks*

Azure PowerShell Task to Stop Triggers

Begin configuring the Azure PowerShell script to stop the Data Factory triggers in the
QA environment, as shown in Figure 19-30. Ensure that the script path is pointing to the
cicd.ps1 file that you added to the GitHub repo in the prerequisite step.

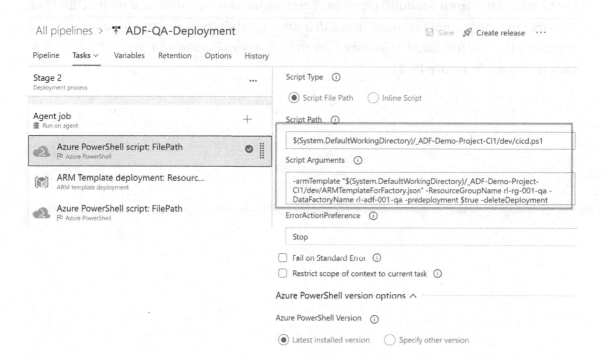

Figure 19-30. *Configure Azure PowerShell script to stop ADF triggers*

Be sure to add a script path and the arguments. The following are the values that you see illustrated in Figure 19-30:

Pre-deployment script path: Should be a fully qualified path or relative to the default working directory, as shown in the following code:

```
$(System.DefaultWorkingDirectory)/_ADF-Demo-Project-CI1/dev/cicd.ps1
```

Pre-deployment script arguments: Additional parameters to pass to PowerShell, which can be either ordinal or named parameters, as shown in the following code:

```
-armTemplate "$(System.DefaultWorkingDirectory)/_ADF-Demo-Project-CI1/dev/
ARMTemplateForFactory.json" -ResourceGroupName rl-rg-001-dev -DataFactoryName
rl-adf-001-dev -predeployment $true -deleteDeployment $false
```

ARM Template Deployment Task

This release pipeline task will incrementally update the QA resource group with the template and template parameters of the DEV Data Factory that you published to the DevOps artifacts from the build pipeline. These artifacts were published in the DEV Data Factory and committed to the *adf_publish* branch GitHub repo. Be sure to select the appropriate *ARMTemplateForFactory.json* and *ARMTemplateParametersForFactory.json* files, as shown in Figure 19-31.

Agent job
🖥 Run on agent

☁ Azure PowerShell script: FilePath
   ⚡ Azure PowerShell

🔲 ARM Template deployment: Resource Group s... ✓
   ARM template deployment

☁ Azure PowerShell script: FilePath
   ⚡ Azure PowerShell

Azure Details ⌃

Deployment scope * ⓘ

| Resource Group | ⌄ |

Azure Resource Manager connection * ⓘ | Manage ⤤

| MSDN Platforms Subscription | ⌄ | ↻ |

ⓘ Scoped to subscription 'MSDN Platforms Subscription'

Subscription * ⓘ

| MSDN Platforms Subscription (| ⌄ | ↻ |

Action * ⓘ

| Create or update resource group | ⌄ |

Resource group * ⓘ

| rl-rg-001-qa | ⌄ | ↻ |

Location * ⓘ

| Central US | ⌄ | ↻ |

Template ⌃

Template location *

| Linked artifact | ⌄ |

Template * ⓘ

| $(System.DefaultWorkingDirectory)/_ADF-Demo-Project-CI1/dev/ARMTemplateForFactory.json | ⋯ |

Template parameters ⓘ

| $(System.DefaultWorkingDirectory)/_ADF-Demo-Project-CI1/dev/ARMTemplateParametersForFactory.json | ⋯ |

Override template parameters ⓘ

| -factoryName "rl-adf-001-qa" -LS_AzureSqlDatabase_connectionString "Integrated Security=False;Encrypt=True;Connection Timeout=30;Data Source=rl-sqlserver-001-qa.database.windows.net;Initial Catalog=rl-sql-001- | ⋯ |

Deployment mode * ⓘ

| Incremental | ⌄ |

Figure 19-31. *Task to create or update resource group*

Once all of the other Azure details in Figure 19-31 are completely configured, choose the override template parameters by clicking the ... icon, as shown in Figure 19-32.

Override template parameters ⓘ

| -factoryName "rl-adf-001-qa" -LS_AzureSqlDatabase_connectionString "Integrated Security=False;Encrypt=True;Connection Timeout=30;Data Source=rl-sqlserver-001-qa.database.windows.net;Initial Catalog=rl-sql-001- | ⋯ |

Figure 19-32. *Override template parameters*

The following template and parameters are defined in the deployment task, as shown in Figures 19-31 and 19-32:

Deployment template: Specify the path or a pattern pointing to the Azure Resource Manager template. More information about the templates can be found at the following URL (`https://aka.ms/azuretemplates`), and to get started with a sample template, please see the following URL: `https://aka.ms/sampletemplate`.

```
$(System.DefaultWorkingDirectory)/_ADF-Demo-Project-CI1/dev/
ARMTemplateForFactory.json.
```

Template parameters: Ensure that you select the path or a pattern pointing to the parameters file for the Azure Resource Manager template:

```
$(System.DefaultWorkingDirectory)/_ADF-Demo-Project-CI1/dev/
ARMTemplateParametersForFactory.json
```

Override template parameters: The "…" next to the Override template parameters textbox will show the template parameters. You can also type the template parameters to override in the textbox, for example, `-storageName fabrikam -adminUsername $(vmusername) -adminPassword $(password) –azureKeyVaultName $(fabrikamFibre)`. If the parameter value that you will be using has multiple words, be sure to enclose them in quotes, even if you're passing them using variables, for example, -name "parameter value" -name2 "$(var)". To override object type parameters, use stringified JSON objects, for example, -options ["option1"] -map {"key1": "value1" }.

```
-factoryName "rl-adf-001-qa" -LS_AzureSqlDatabase_connectionString
"Integrated Security=False;Encrypt=True;Connection Timeout=30;Data
Source=rl-sqlserver-001-qa.database.windows.net;Initial Catalog=rl-sql-001-
qa;User ID=qaadmin" -LS_AzureDataLakeStorage_properties_typeProperties_url
"https://adls2qa.dfs.core.windows.net" -LS_AzureKeyVault_properties_
typeProperties_baseUrl "https://rl-kv-001-qa.vault.azure.net/"
```

Change the override template parameters to the QA resources and connection properties shown in Figure 19-33 and click OK to save the configuration settings.

Override template parameters

Name	Value
factoryName	"rl-adf-001-qa"
LS_AzureSqlDatabase_conn...	"Integrated Security=False;...
LS_AzureDataLakeStorage_...	"https://adls2qa.dfs.core.wi...
LS_AzureKeyVault_properti...	"https://rl-kv-001-qa.vault....

OK   Cancel

Figure 19-33. *Change the override template parameters to the QA*

Azure PowerShell Task to Start Triggers

Finally, configure the Azure PowerShell script to start the Data Factory triggers in the QA environment. Ensure that the script path is pointing to the cicd.ps1 file that you added to the GitHub repo in the prerequisite step. This is shown in Figure 19-34.

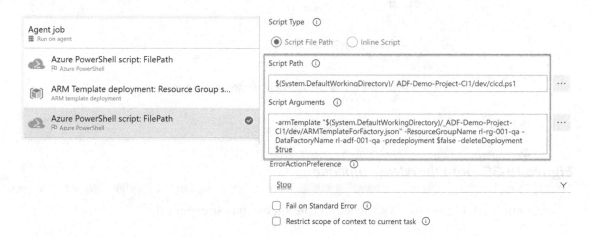

Figure 19-34. *PowerShell task to restart ADF triggers*

Also, add the following post-deployment script path to the task configuration, which is shown in Figure 19-34:

```
$(System.DefaultWorkingDirectory)/_ADF-Demo-Project-CI1/dev/cicd.ps1
```

You'll also need to add the following post-deployment script arguments to the task configuration, which is shown in Figure 19-34:

```
-armTemplate "$(System.DefaultWorkingDirectory)/_ADF-Demo-Project-CI1/dev/
ARMTemplateForFactory.json" -ResourceGroupName rl-rg-001-dev -DataFactory
Name rl-adf-001-dev -predeployment $false -deleteDeployment $true
```

Run the Release Pipeline

After adding and configuring the continuous deployment (CD) release pipeline tasks shown in Figure 19-35, run the release pipeline.

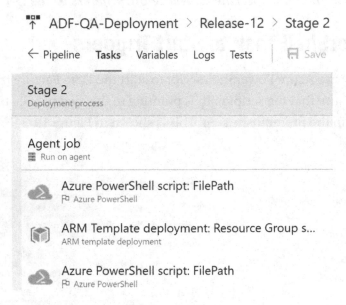

Figure 19-35. *Run the release pipeline*

As expected, Figure 19-36 shows that the release has succeeded.

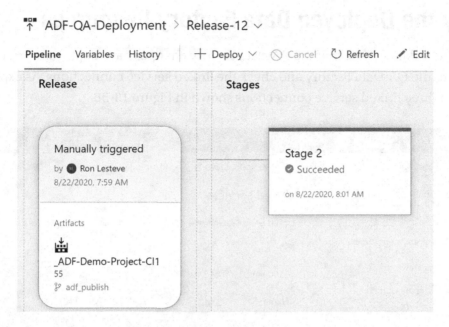

Figure 19-36. *Release has succeeded*

Upon navigating to the logs shown in Figure 19-37, verify that all steps in the release pipeline have successfully completed.

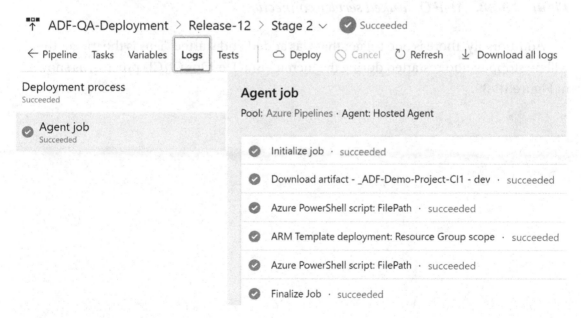

Figure 19-37. *Agent job deployment process success*

Verify the Deployed Data Factory Resources

As a final check to ensure that the QA Data Factory changes have deployed correctly, navigate to the QA Data Factory and check the linked service connections. As expected, notice the three linked service connections shown in Figure 19-38.

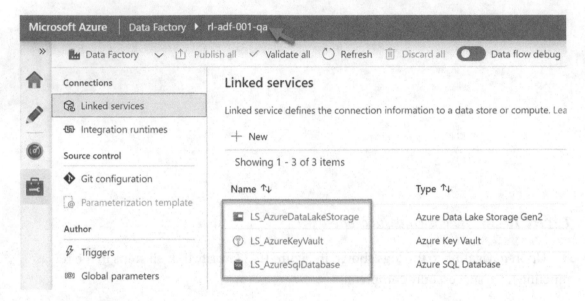

Figure 19-38. *ADF QA linked service connections*

Additionally, there is one trigger that was added and started. This is the trigger that will be stopped and restarted during the incremental DevOps CI/CD process, as shown in Figure 19-39.

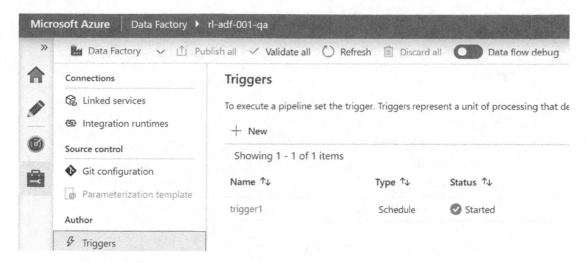

Figure 19-39. *ADF QA trigger in "Started" status*

Upon drilling in further, notice that the connection strings for the data lake have been changed and that the connection has been successfully tested, which is shown in Figure 19-40.

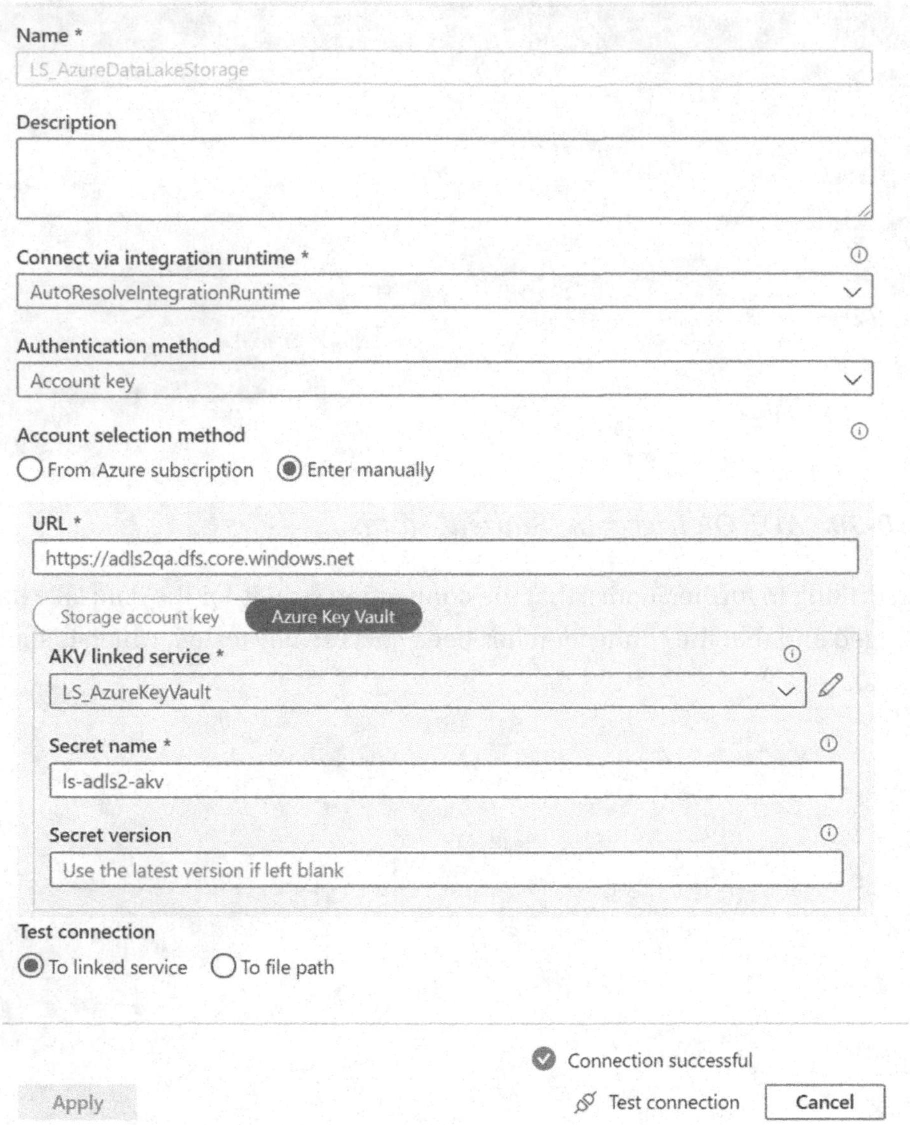

Name *

LS_AzureDataLakeStorage

Description

Connect via integration runtime *   ⓘ

AutoResolveIntegrationRuntime   ⌄

Authentication method

Account key   ⌄

Account selection method   ⓘ

◯ From Azure subscription   ⦿ Enter manually

URL *

https://adls2qa.dfs.core.windows.net

Storage account key   **Azure Key Vault**

AKV linked service *   ⓘ

LS_AzureKeyVault   ⌄   ✎

Secret name *   ⓘ

ls-adls2-akv

Secret version   ⓘ

Use the latest version if left blank

Test connection

⦿ To linked service   ◯ To file path

✓ Connection successful

Apply   ✐ Test connection   Cancel

Figure 19-40.   *ADF QA connection to ADLS Gen2*

Figure 19-41 shows that the Key Vault connection was also changed to QA and has been successfully tested.

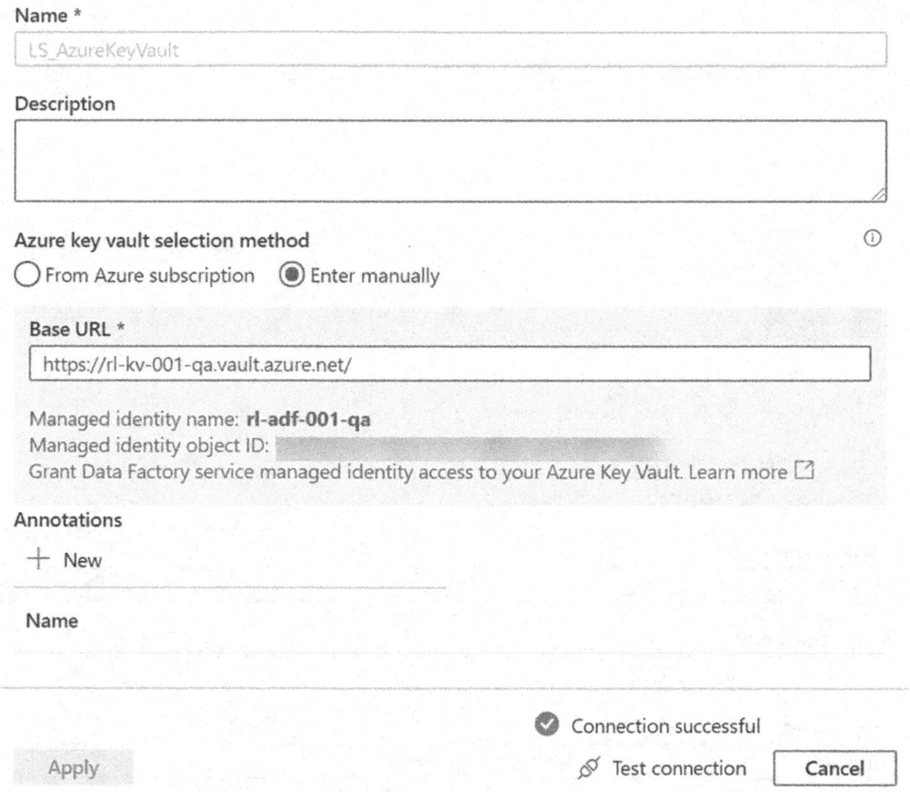

Name *

LS_AzureKeyVault

Description

Azure key vault selection method   ⓘ

○ From Azure subscription   ◉ Enter manually

Base URL *

https://rl-kv-001-qa.vault.azure.net/

Managed identity name: **rl-adf-001-qa**
Managed identity object ID:
Grant Data Factory service managed identity access to your Azure Key Vault. Learn more ⧉

Annotations

➕ New

Name

✅ Connection successful

Apply   ⋫ Test connection   Cancel

Figure 19-41. *ADF QA connection to Key Vault updated*

Finally, notice that the SQL Server and database connections were also changed to QA and the test connection have been successfully validated, as shown in Figure 19-42.

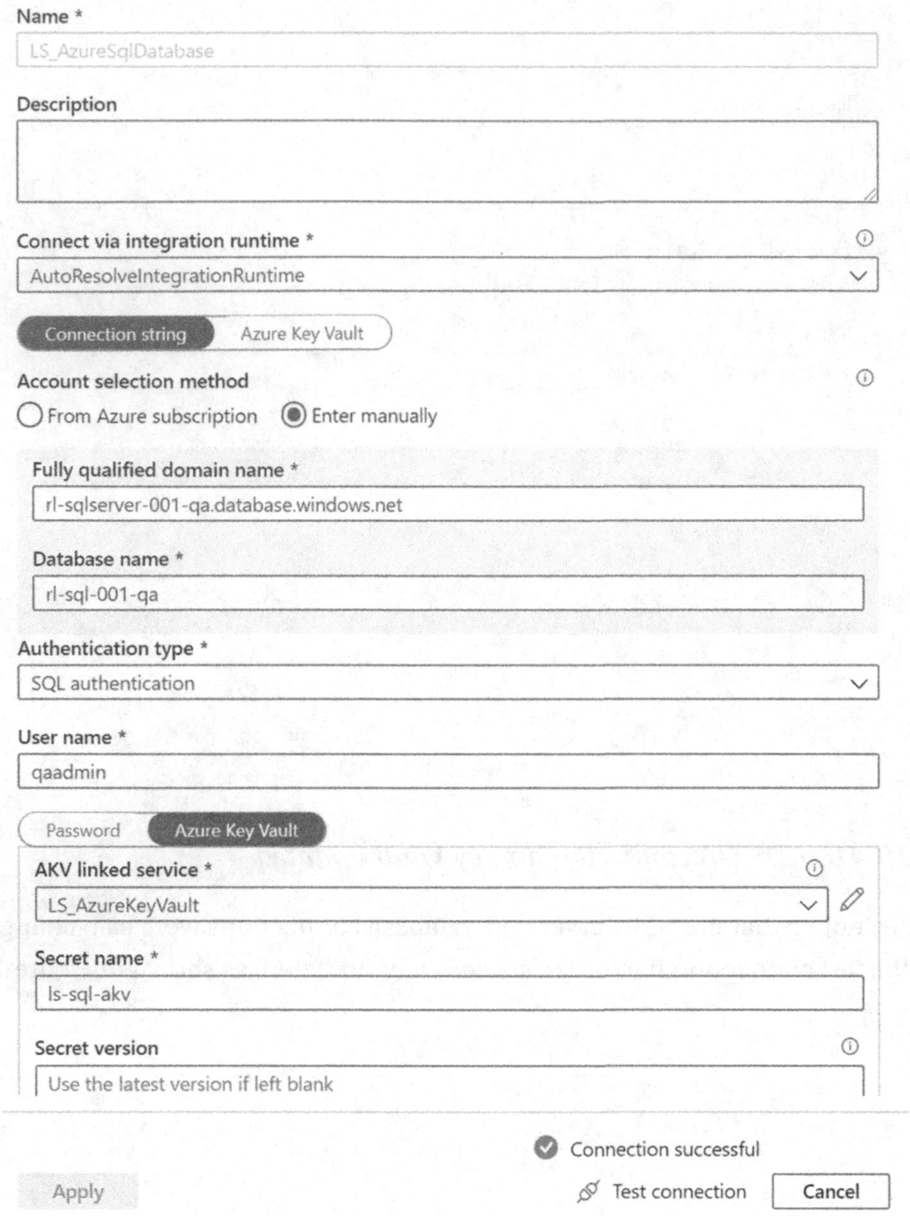

Name *

LS_AzureSqlDatabase

Description

Connect via integration runtime *   ⓘ

AutoResolveIntegrationRuntime   ∨

(Connection string)   (Azure Key Vault)

Account selection method   ⓘ

◯ From Azure subscription   ◉ Enter manually

Fully qualified domain name *

rl-sqlserver-001-qa.database.windows.net

Database name *

rl-sql-001-qa

Authentication type *

SQL authentication   ∨

User name *

qaadmin

(Password)   (Azure Key Vault)

AKV linked service *   ⓘ

LS_AzureKeyVault   ∨   🖉

Secret name *

ls-sql-akv

Secret version   ⓘ

Use the latest version if left blank

✓ Connection successful

Apply   🔌 Test connection   Cancel

Figure 19-42. *ADF QA connection to SQL Server and database updated*

Summary

In this chapter, I demonstrated how to use PowerShell scripts along with the Azure Resource Group Deployment task to start/stop ADF triggers and change ADF environment connection configuration properties through an end-to-end deployment of Azure Data Factory CI/CD changes with Azure DevOps. For more information related to this exercise, specifically to find samples covered in this chapter, please see my GitHub repo at the following URL: `https://github.com/ronlesteve/demo-dev-adf/tree/adf_publish`. For alternative methods of deploying the JSON for ADF pipelines, other than ARM template deployments, explore the Publish ADF task (`https://marketplace.visualstudio.com/items?itemName=SQLPlayer.DataFactoryTools`), which can be found in the Visual Studio marketplace and can be added to the Azure DevOps CI/CD pipeline as a task.

## CHAPTER 20

Deploy a SQL Database

Infrastructure differences and inconsistencies between a DevTest environment that an application was developed and tested in and the production environment are common scenarios that IT professionals and software developers might encounter. Despite an application initially being deployed the right way, configuration changes made to the production environment might cause discrepancies. Customers often ask if there is a seamless way of automating the creation of Azure resources while ensuring consistency across multiple environments.

Infrastructure as code is the process of creating a template that defines and then deploys the environment along with the application to ensure consistency. Azure Resource Manager (ARM) is the deployment and management service for Azure. It provides a consistent management layer that enables you to create, update, and delete resources in your Azure subscription. You can use its access control, auditing, and tagging features to secure and organize your resources after deployment.

By using ARM templates, you can manage your infrastructure through declarative templates rather than scripts and deploy, manage, and monitor all the resources for your solution as a group, rather than handling these resources individually. Additionally, you can repeatedly deploy your solution throughout the development lifecycle and have confidence your resources are deployed in a consistent state.

There are a few source control options within Visual Studio. GitHub is one of these source control options and offers a number of benefits including advanced security options. Integrating multiple applications such as Visual Studio, GitHub, Azure DevOps, and Azure SQL Database for a seamless CI/CD process is a growing need for many enterprises that are on a journey to modernize their data and infrastructure platform. In this chapter, I will demonstrate an end-to-end solution for the architectural flow shown in Figure 20-1 to deploy an AdventureWorksLT2019 database from Visual Studio to an Azure SQL Database through a GitHub repository and then deployed by both an Azure DevOps build (CI) and release (CD) pipeline.

© Ron C. L'Esteve 2021
R. C. L'Esteve, *The Definitive Guide to Azure Data Engineering*, https://doi.org/10.1007/978-1-4842-7182-7_20

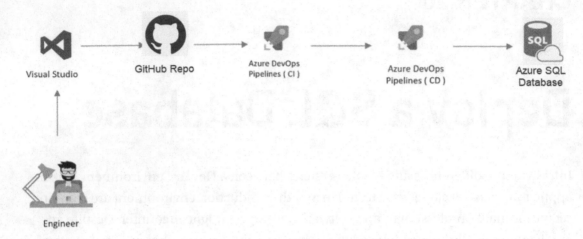

Figure 20-1. *Azure DevOps CI/CD using GitHub repo and Visual Studio Azure SQL Database project*

Pre-Requisites

Before you get started with the development and deployment process for the architectural pattern shown in Figure 20-1, you will need to create the following prerequisite resources:

1) Visual Studio 2019 with SSDT: You'll be using Visual Studio 2019 along with SSDT for Visual Studio to deploy this database project solution. You can download the latest version of Visual Studio from this URL: `https://visualstudio.microsoft.com/downloads/`. Additionally, you could download SQL Server Data Tools (SSDT) for Visual Studio from the following URL: `https://docs.microsoft.com/en-us/sql/ssdt/download-sql-server-data-tools-ssdt?view=sql-server-ver15`.

2) GitHub account and repo: Since you will be associating your Visual Studio project with a GitHub source control account, you can use this URL to learn more about how to create an account in GitHub (`www.wikihow.com/Create-an-Account-on-GitHub`). Additionally, you could read more about how to create a GitHub repo from this URL: `https://docs.github.com/en/enterprise/2.16/user/github/getting-started-with-github/create-a-repo`.

3)   Azure DevOps account and project: Finally, you'll need an
      Azure DevOps account and project that you will use to build the
      CI/CD process. You can use the following URL as reference to
      create an organization and project: `https://docs.microsoft.`
      `com/en-us/azure/devops/organizations/accounts/create-`
      `organization?view=azure-devops`.

Create a Visual Studio SQL Database Project

Begin by creating a new SQL database project in Visual Studio 2019, as shown in
Figure 20-2. Within the project, you will import the sample AdventureWorks database.

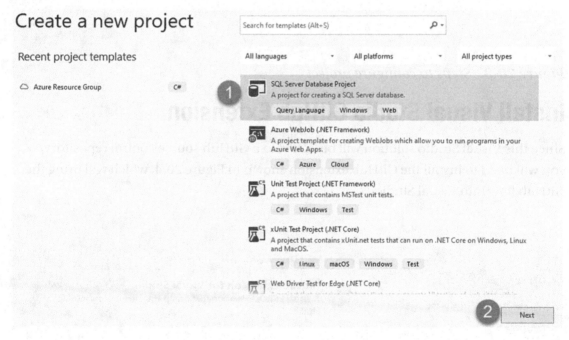

Figure 20-2. *Steps to create a VS project*

Next, Figure 20-3 shows how you should configure and create the new SQL database
project by entering the project name, location, and solution.

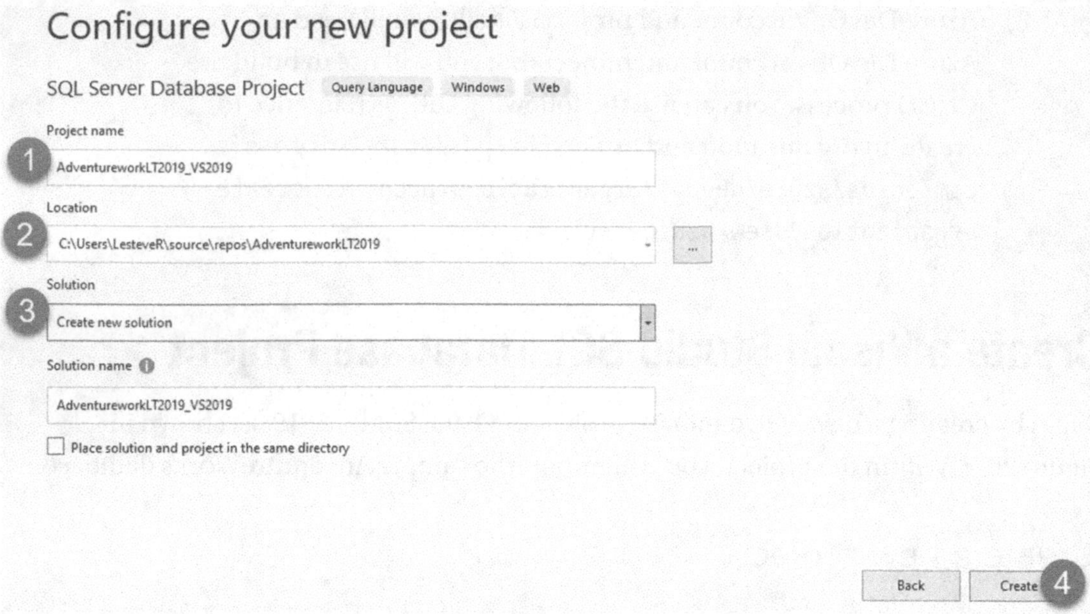

Configure your new project

SQL Server Database Project   Query Language   Windows   Web

Project name

1 AdventureworkLT2019_VS2019

Location

2 C:\Users\LesteveR\source\repos\AdventureworkLT2019   ...

Solution

3 Create new solution

Solution name ⓘ

AdventureworkLT2019_VS2019

☐ Place solution and project in the same directory

Back   Create **4**

Figure 20-3. *Steps to configure project*

Install Visual Studio GitHub Extension

Since this Visual Studio solution will be linked to a GitHub source control repository,
you will need to install the GitHub extension shown in Figure 20-4, which will bring the
GitHub flow into Visual Studio.

Manage Extensions

▷ Installed   Sort by: Most Downloads

▲ Online   **GitHub Extension for Visual Studio**   Download
                                     A Visual Studio Extension that brings the GitHub Flow into Visual
  ▷ Visual Studio Marketplace   Studio.

▷ Updates (2)

Figure 20-4. *Steps to install GitHub extension*

At this point, you should close Visual Studio to complete the GitHub extension
installation, which will launch the install shown in Figure 20-5.

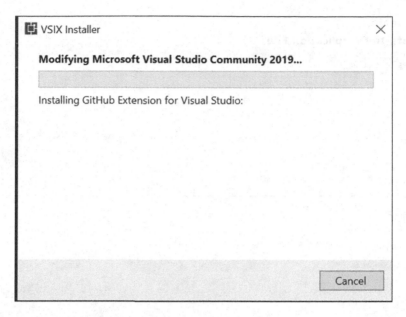

Figure 20-5. *Image showing extension installer*

Import AdventureWorks Database

Reopen the Visual Studio project and import the AdventureWorksLT2019 Dacpac file, as shown in Figure 20-6. You can always download AdventureWorks sample databases to be used for the import process from the following URL: `https://docs.microsoft.com/en-us/sql/samples/adventureworks-install-configure?view=sql-server-ver15&tabs=ssms`.

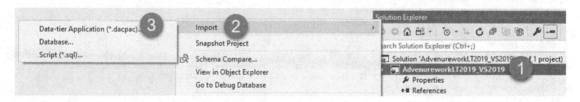

Figure 20-6. *Steps to import data tier*

After clicking the Data-tier Application option, the Import Data-tier Application File UI shown in Figure 20-7 will appear, and you will need to select the appropriate Dacpac file and click Start.

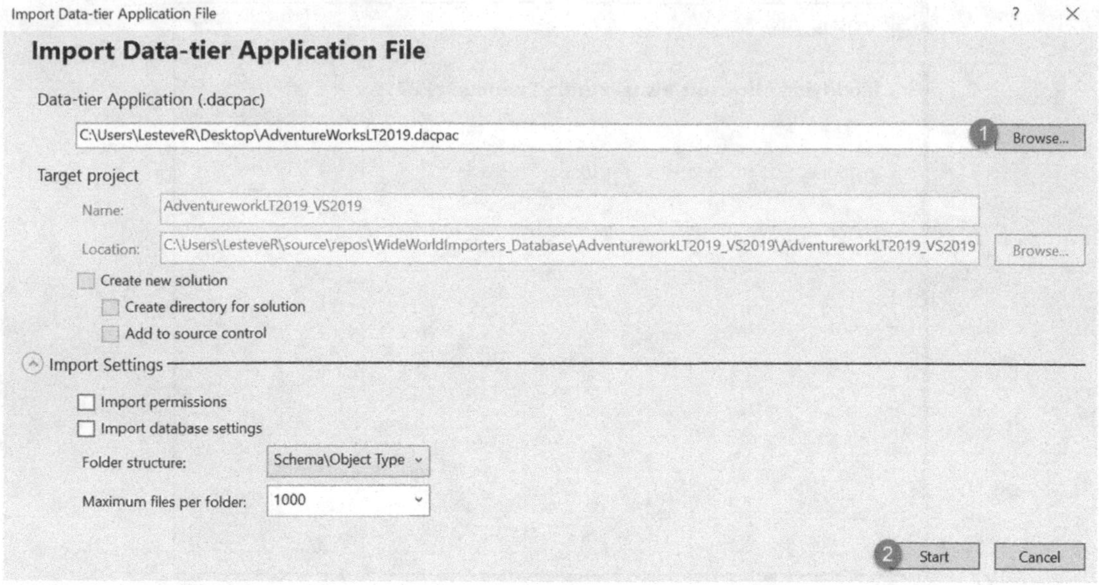

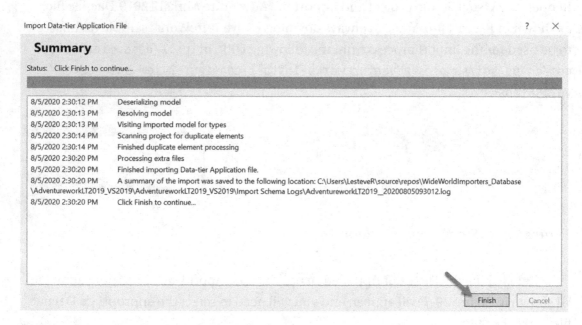

Figure 20-7. *Step 2 to import data tier Dacpac*

Click Finish to continue the import process, as shown in Figure 20-8. Notice that the status and import steps are displayed in this Summary UI.

Figure 20-8. *Image of Summary import progress*

Once the data-tier application is imported, the schemas will be listed as folders in the SQL Server database project shown in Figure 20-9.

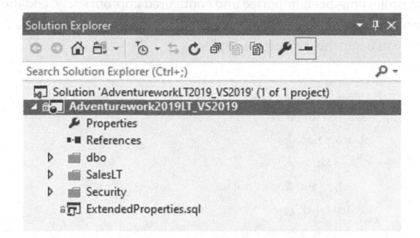

Figure 20-9. *Image of AdventureWorks solution project*

Ensure the database project target platform is set to Microsoft Azure SQL Database from the Database Project Properties GUI in Figure 20-10.

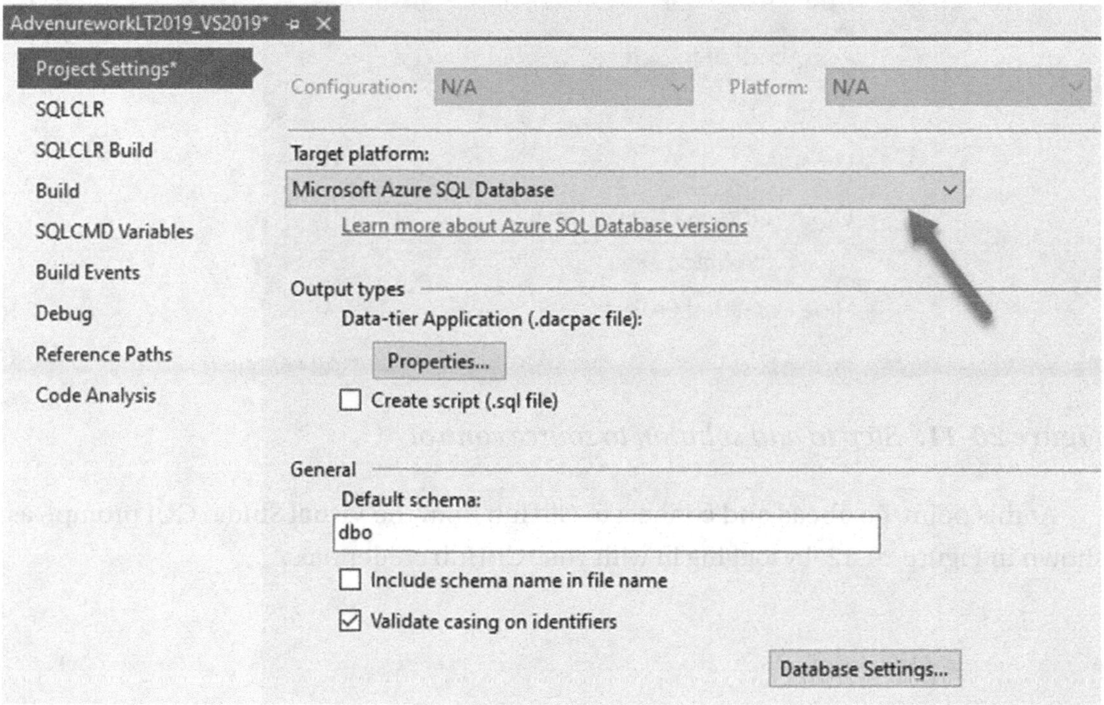

Figure 20-10. *Image of project properties and target platform*

Connect to GitHub Repo Source Control

Now that the project has been imported and configured appropriately, add the solution to the GitHub source control, as shown in Figure 20-11. You'll be asked to sign into GitHub when you begin the connection process.

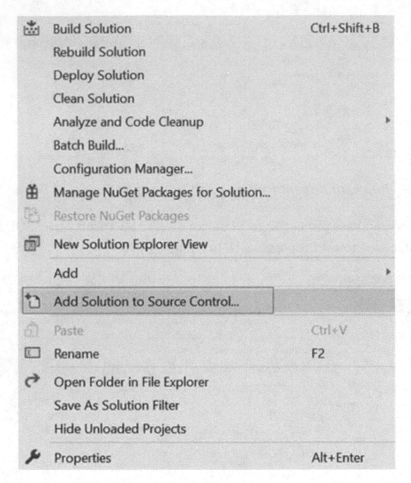

Figure 20-11. *Step to add solution to source control*

At this point, go ahead and connect to GitHub from the Visual Studio GUI prompt, as shown in Figure 20-12, by logging in with your GitHub credentials.

Figure 20-12. *Steps to sign into GitHub*

A prompt will appear to authorize Visual Studio and GitHub to integrate together. Review the configuration settings and then go ahead and click Authorize github, which is shown in Figure 20-13.

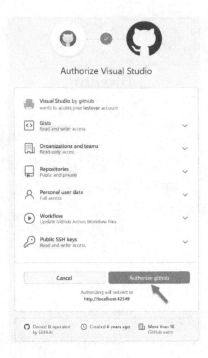

Figure 20-13. *Step to authorize Visual Studio*

Check In Visual Studio Solution to GitHub Repo

Now that the solution has been added to a GitHub repository, click home and then
sync to check in and sync the solution to the GitHub repository, which is illustrated in
Figure 20-14.

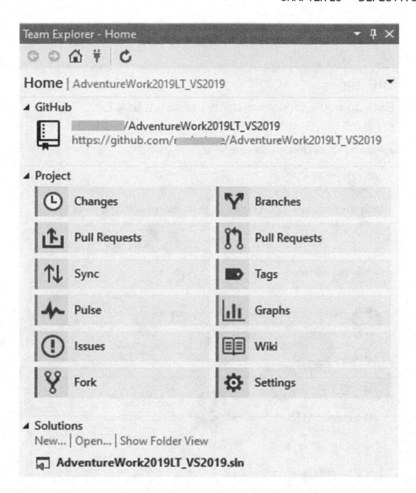

Figure 20-14. *Step to sync to repo*

At this point, verify the synchronization details and click Publish, which is shown in Figure 20-15, to sync the solution to the desired GitHub repo.

Figure 20-15. *Step to publish to repo*

After the sync settings are successfully published, verify that the Visual Studio solution has been checked into the selected GitHub repo and branch. In Figure 20-16, notice that the solution and database have been checked into the repository.

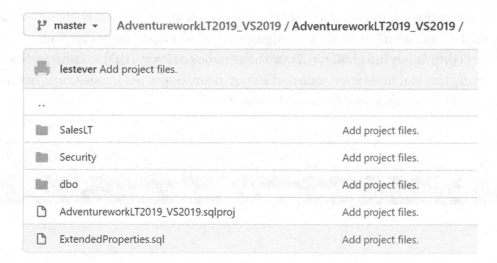

Figure 20-16. *Image of GitHub repo*

Install Azure Pipelines from GitHub

Now that you have integrated Azure GitHub with Visual Studio, it's time to install Azure Pipelines from GitHub Marketplace to integrate GitHub with Azure DevOps. Head over to the marketplace and search for and click Azure Pipelines, shown in Figure 20-17, which will navigate to the Azure Pipelines setup page.

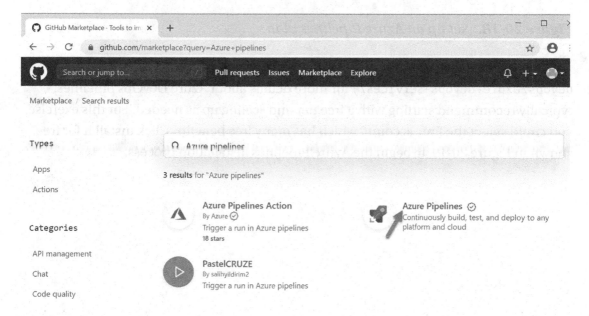

Figure 20-17. *Add Azure Pipelines*

Azure Pipelines automatically builds and tests code projects to make them available to others. It works with just about any language or project type. Azure Pipelines combines continuous integration (CI) and continuous delivery (CD) to constantly and consistently test and build your code and ship it to any target. Set up an Azure Pipelines plan by clicking Set up a plan shown in Figure 20-18.

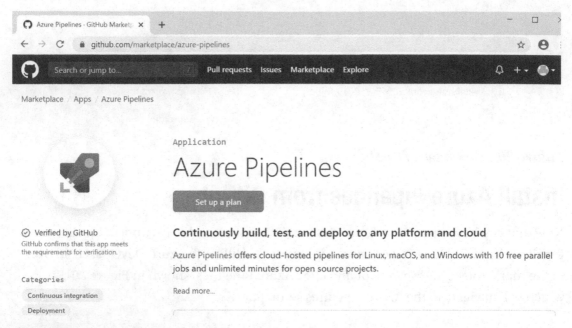

Figure 20-18. *Set up an Azure Pipelines plan*

See pricing details (https://azure.microsoft.com/en-us/pricing/details/ devops/azure-devops-services/) for more details about Azure DevOps pipelines. I typically recommend starting with a free tier and scaling up as needed. For this exercise, you could select the Free account, which has many free benefits. Click Install it for free, shown in Figure 20-19, to begin the Azure Pipelines installation process.

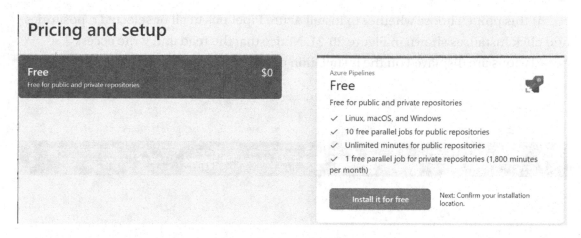

Figure 20-19. *Steps to install Azure Pipelines plan*

You'll be asked to review your order and then complete the order to begin the installation. Notice from Figure 20-20 that with the Free plan, you will not be charged anything on a monthly basis.

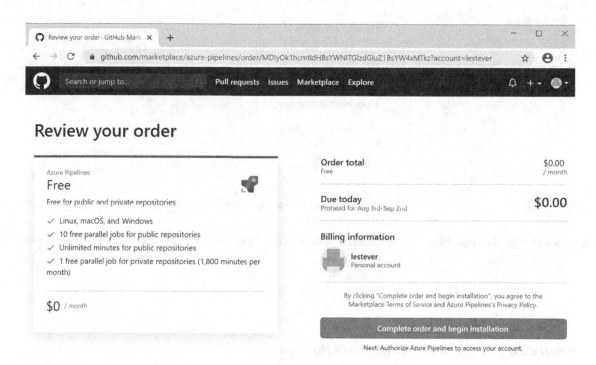

Figure 20-20. *Steps to complete the Azure Pipelines installation plan*

At this point, choose whether to install Azure Pipelines to all or selected repositories and click Install, as shown in Figure 20-21. Notice that the read and write access permissions are displayed on the installation page.

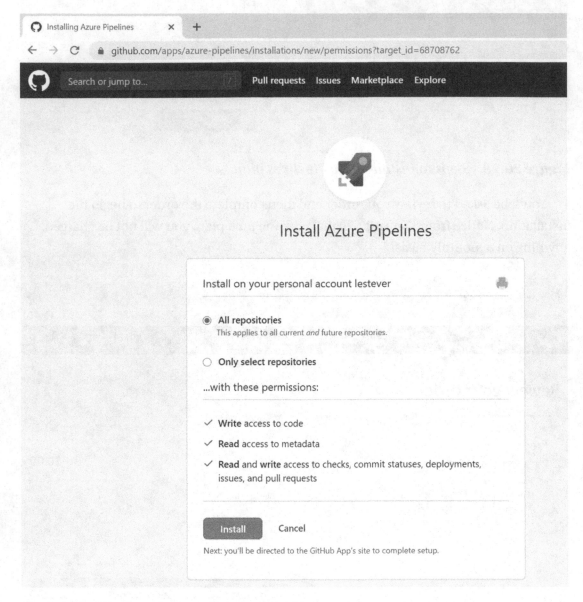

Figure 20-21. *Steps to install Azure Pipelines*

When prompted, select your Azure DevOps pipeline project and organization and click Continue, as shown in Figure 20-22.

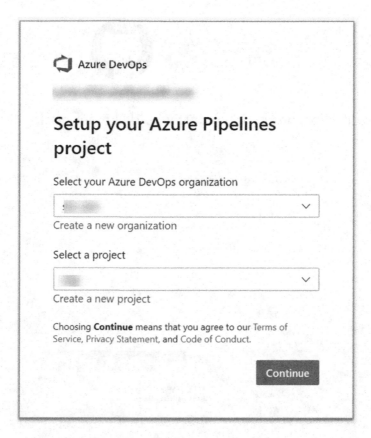

Figure 20-22. *Set up Azure pipeline project*

When prompted, authorize the integration between Azure Pipelines and GitHub, which is shown in Figure 20-23.

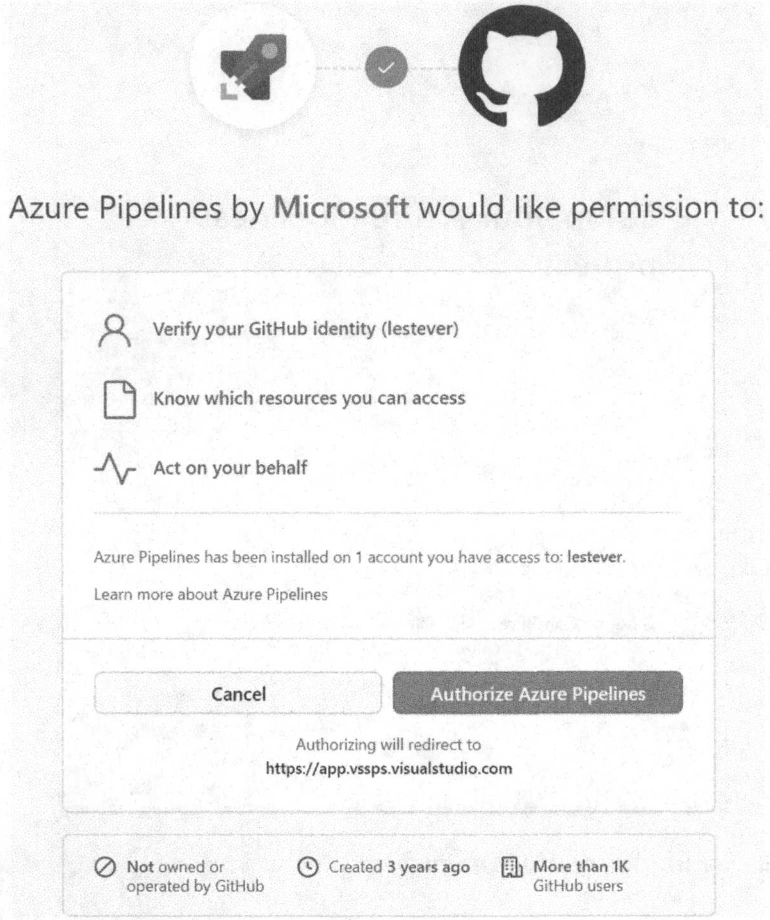

Azure Pipelines by **Microsoft** would like permission to:

🧍 Verify your GitHub identity (lestever)

📄 Know which resources you can access

〰️ Act on your behalf

Azure Pipelines has been installed on 1 account you have access to: **lestever**.

Learn more about Azure Pipelines

| Cancel | Authorize Azure Pipelines |

Authorizing will redirect to
https://app.vssps.visualstudio.com

⊘ **Not** owned or   🕐 Created **3 years ago**   🗂️ **More than 1K**
operated by GitHub   GitHub users

Figure 20-23. *Step to authorize Azure Pipelines*

Build CI Pipeline from GitHub Repo

You are now ready to build a continuous integration (CI) pipeline from the GitHub repo. Click Use the classic editor, shown in Figure 20-24, to create the build pipeline without YAML.

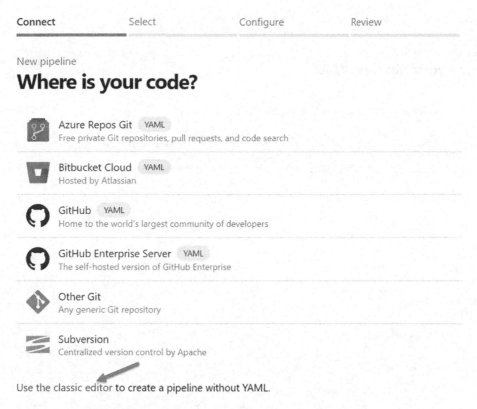

Figure 20-24. Step to select where the code is

Note that there is also an option to create the pipeline with YAML that will require YAML code. For more information on YAML, I would recommend reading this article to get started with YAML in minutes `www.cloudbees.com/blog/yaml-tutorial-everything-you-need-get-started/`. Figure 20-25 shows a sample illustration of what a YAML code block might look like and how you are able to write and configure it to run as a new pipeline option. For this exercise, use the classic editor to create the pipeline.

Figure 20-25. *Image showing YAML code option*

For the classic option that does not use YAML, start with Empty job, shown in Figure 20-26.

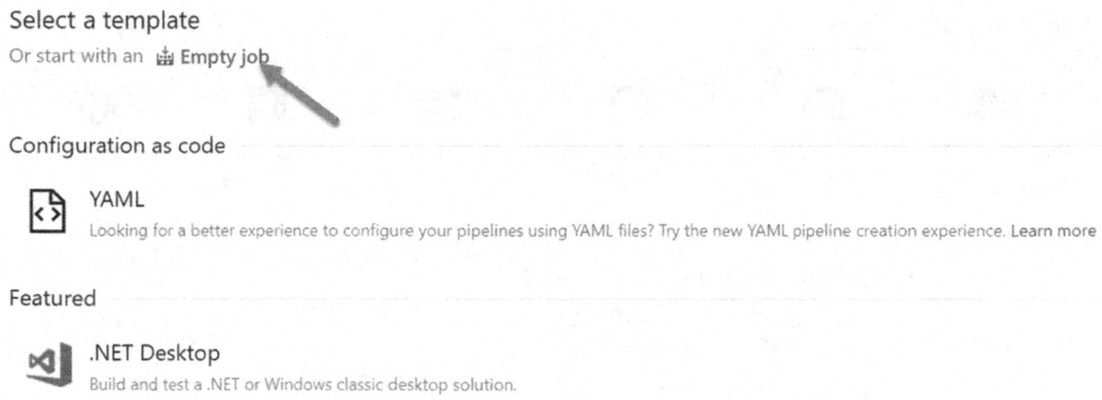

Select a template

Or start with an Empty job

Configuration as code

 YAML

Looking for a better experience to configure your pipelines using YAML files? Try the new YAML pipeline creation experience. **Learn more**

Featured

.NET Desktop

Build and test a .NET or Windows classic desktop solution.

Figure 20-26. *Select an Empty job template*

Next, select GitHub as the source. Notice from Figure 20-27 that there are a number of available source options to choose from. Also, select the desired repo and branch.

There are a few additional available settings that are typically set to the defaults listed in Figure 20-27. These options include the following:

- **Clean**: Allows you to perform various cleaning of the working directory of your private agent before the build is run.

- **Tag sources**: Allows you to tag your source code files after the build completes to identify which version of each file is included in the completed build.

- **Report build status**: Displays a badge on the source repository to indicate whether the build succeeded or failed.

- **Checkout submodules**: Checks out your Git submodules if they are in the same repository or in a public repository.

- **Checkout files from LFS**: Configures local working directory but skips sync sources.

- **Don't sync sources**: Limits fetching to the specified number of commits from the tip of each remote branch history and allows you to specify the number of commits in the Fetch depth option.

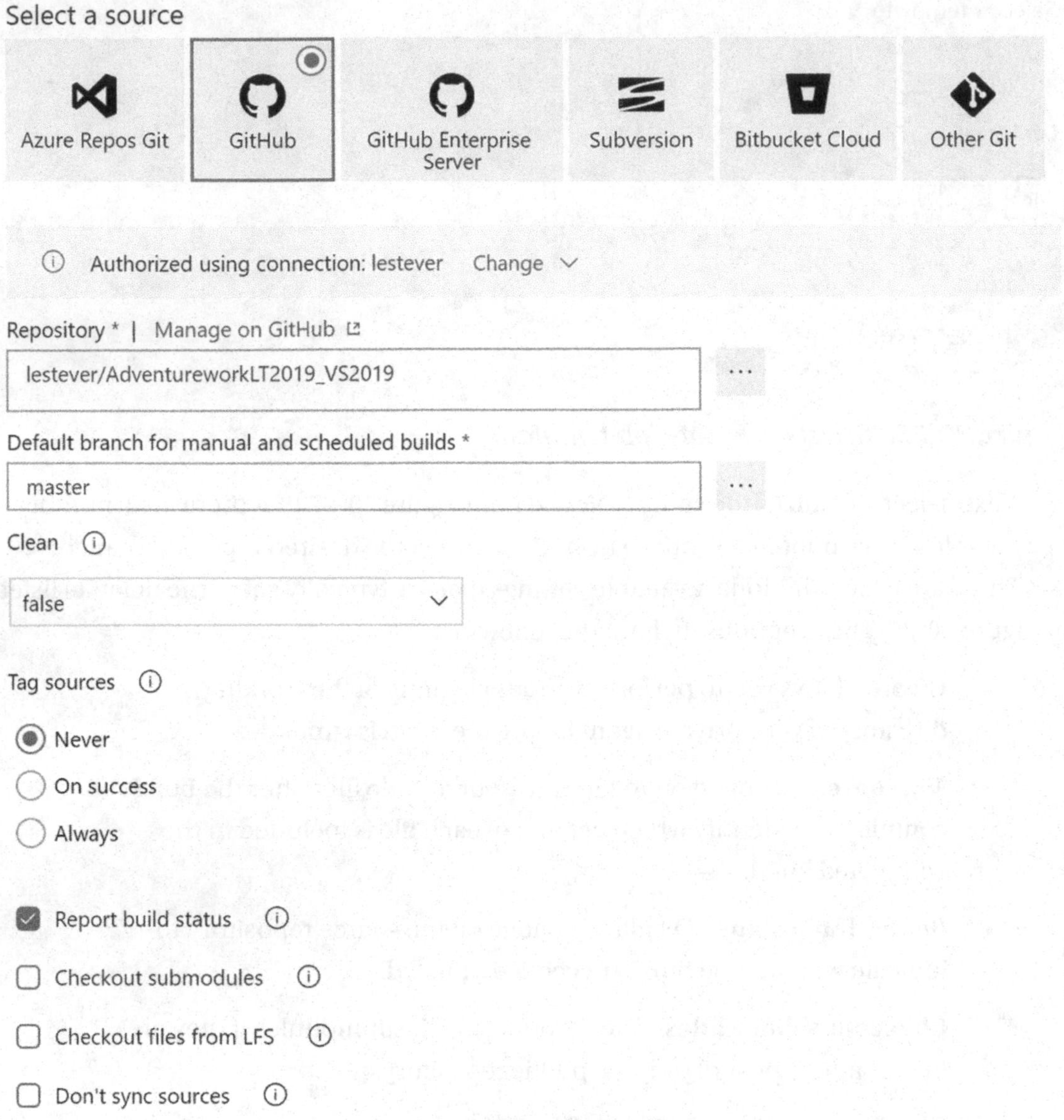

Select a source

| Azure Repos Git | GitHub | GitHub Enterprise Server | Subversion | Bitbucket Cloud | Other Git |

ⓘ Authorized using connection: lestever   Change ⌄

Repository * | Manage on GitHub ↗

lestever/AdventureworkLT2019_VS2019   ...

Default branch for manual and scheduled builds *

master   ...

Clean ⓘ

false   ⌄

Tag sources ⓘ

◉ Never

◯ On success

◯ Always

☑ Report build status ⓘ

☐ Checkout submodules ⓘ

☐ Checkout files from LFS ⓘ

☐ Don't sync sources ⓘ

Figure 20-27. *Select the GitHub source*

Search for and add the *MSBuild* task to the pipeline, as shown in Figure 20-28. This task will build the Visual Studio solution.

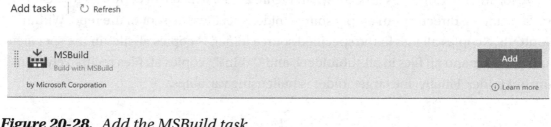

Figure 20-28. *Add the MSBuild task*

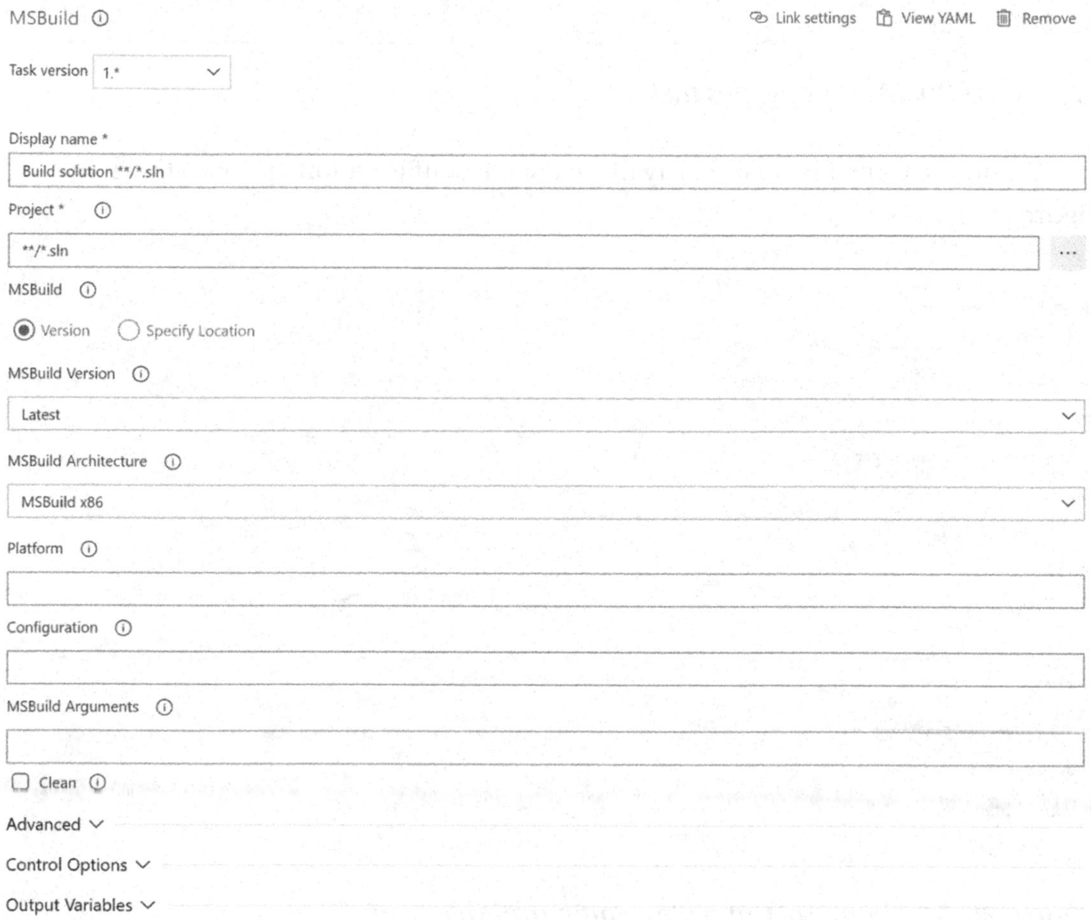

MSBuild ⓘ   ⮂ Link settings   📋 View YAML   🗑 Remove

Task version   1.*   ⌄

Display name *

Build solution **/*.sln

Project *   ⓘ

**/*.sln   ...

MSBuild   ⓘ

⦿ Version   ○ Specify Location

MSBuild Version   ⓘ

Latest   ⌄

MSBuild Architecture   ⓘ

MSBuild x86   ⌄

Platform   ⓘ

Configuration   ⓘ

MSBuild Arguments   ⓘ

☐ Clean   ⓘ

Advanced ⌄

Control Options ⌄

Output Variables ⌄

Figure 20-29. *Image of MSBuildConfig*

Verify the MSBuild configurations shown in Figure 20-29. The relative wildcard path of **/*.sln is sufficient for identifying the solution file.

Also, add the Copy files task shown in Figure 20-30, which will copy files to the artifact staging directory. An empty source folder specifies the root of the repo. Within Contents, * copies all files in the specified source folder, ** copies all files in the specified source folder and all files in all subfolders, and **\bin** copies all files recursively from any bin folder. Finally, the target folder is built using variables.

Figure 20-30. *Add Copy files task*

Within the Copy files task, verify the available configuration options shown in Figure 20-31.

Copy files ⓘ   ⬡ Link settings   ⬡ View YAML   🗑 Remove

Task version 2.*   ⌄

Display name *

Copy Files to: $(build.artifactstagingdirectory)

Source Folder   ⓘ

$(agent.builddirectory)\s   ...

Contents *   ⓘ

**

Target Folder *   ⓘ

$(build.artifactstagingdirectory)

Advanced ⌄

Control Options ⌄

Output Variables ⌄

Figure 20-31. *Image of Copy files configurations*

Finally, add the Publish build artifacts task shown in Figure 20-32 to publish the staging directory artifacts to Azure Pipelines.

Figure 20-32. *Step to add Publish build artifacts*

Verify the Publish build artifacts configurations shown in Figure 20-33. Notice that the path to publish is specified by a variable.

Publish build artifacts ⓘ   🔗 Link settings   📋 View YAML   🗑 Remove

Task version   1.*

Display name *

Publish Artifact: drop

Path to publish *   ⓘ

$(Build.ArtifactStagingDirectory)   ...

Artifact name *   ⓘ

drop

Artifact publish location *   ⓘ

Azure Pipelines

Advanced ⌄

Control Options ⌄

Output Variables ⌄

Figure 20-33. *Step showing Publish build artifacts configurations*

Verify that the build pipeline tasks are in the correct order, which is shown in Figure 20-34. Click Save & queue.

Figure 20-34.   *Summary of build tasks*

Verify the run pipeline parameters shown in Figure 20-35 and click Save and run.

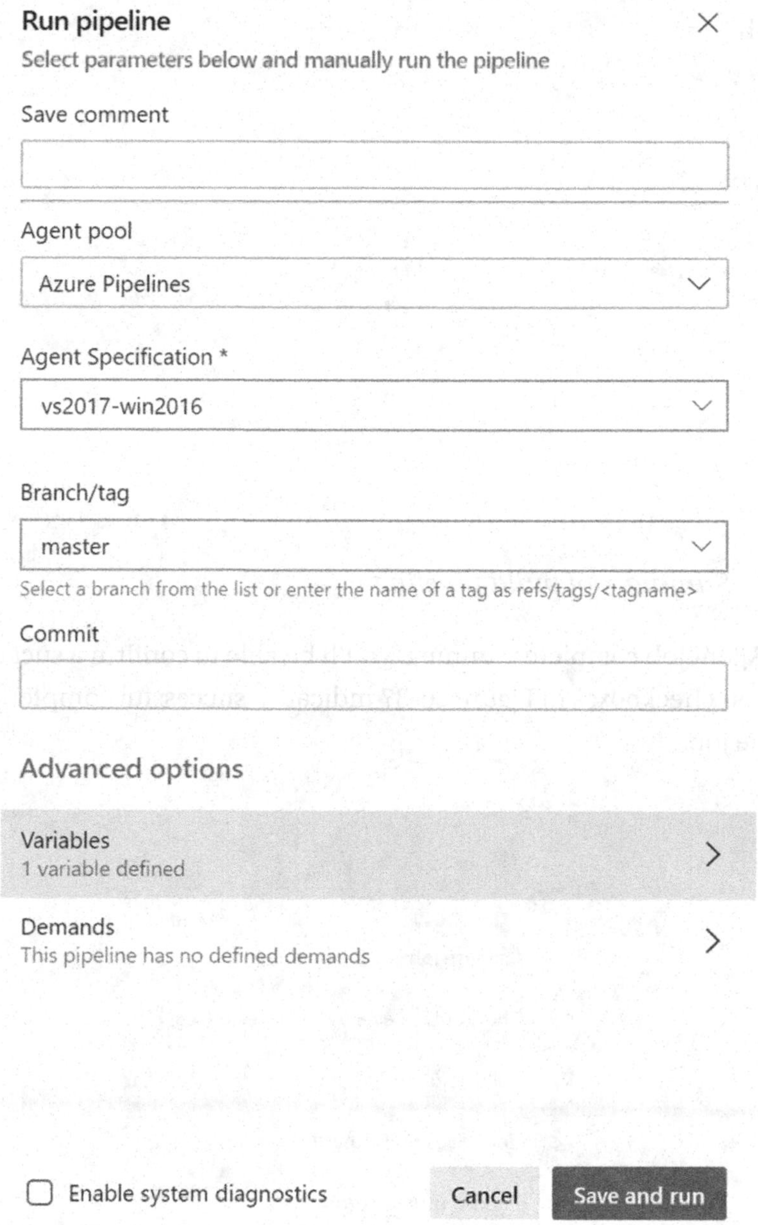

Figure 20-35. *Image of run pipeline task configurations*

Notice the build pipeline summary in Figure 20-36, which contains details and the status of the job.

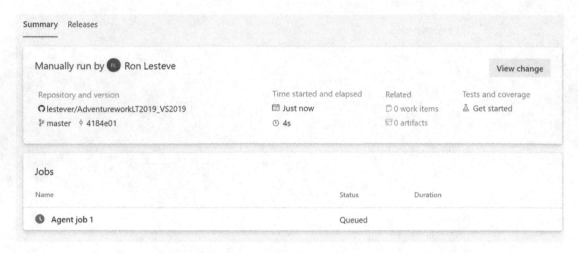

Figure 20-36. *Summary of build pipeline*

Once the Agent job completes running, you'll be able to confirm a successful job status. The green checkboxes in Figure 20-37 indicate a successful completion of all task steps within the job.

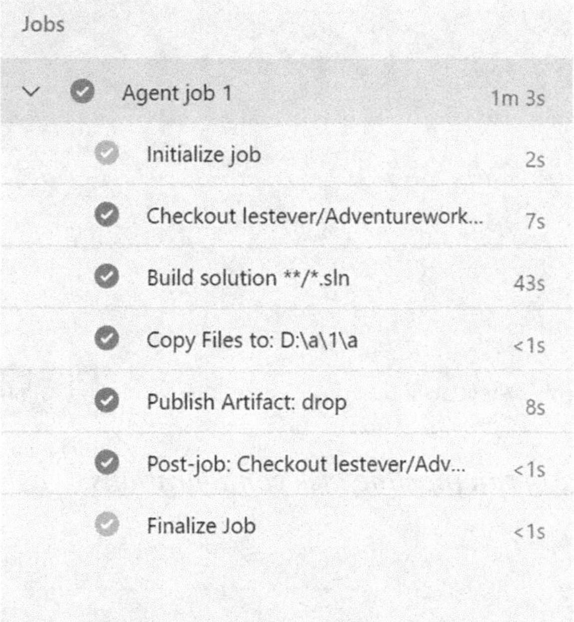

Figure 20-37. *Image of build job summary*

Release CD Pipeline from DevOps Artifact Repo

Now that the build pipeline has been created and successfully deployed, it's time to create a release pipeline. Begin by adding a release pipeline artifact, which is shown in Figure 20-38.

Figure 20-38. *Step to add release artifact*

Select the Build artifact as the source type shown in Figure 20-39. Notice the other available source options, which include a variety of source repos. Complete the configuration of the other required details and click Add.

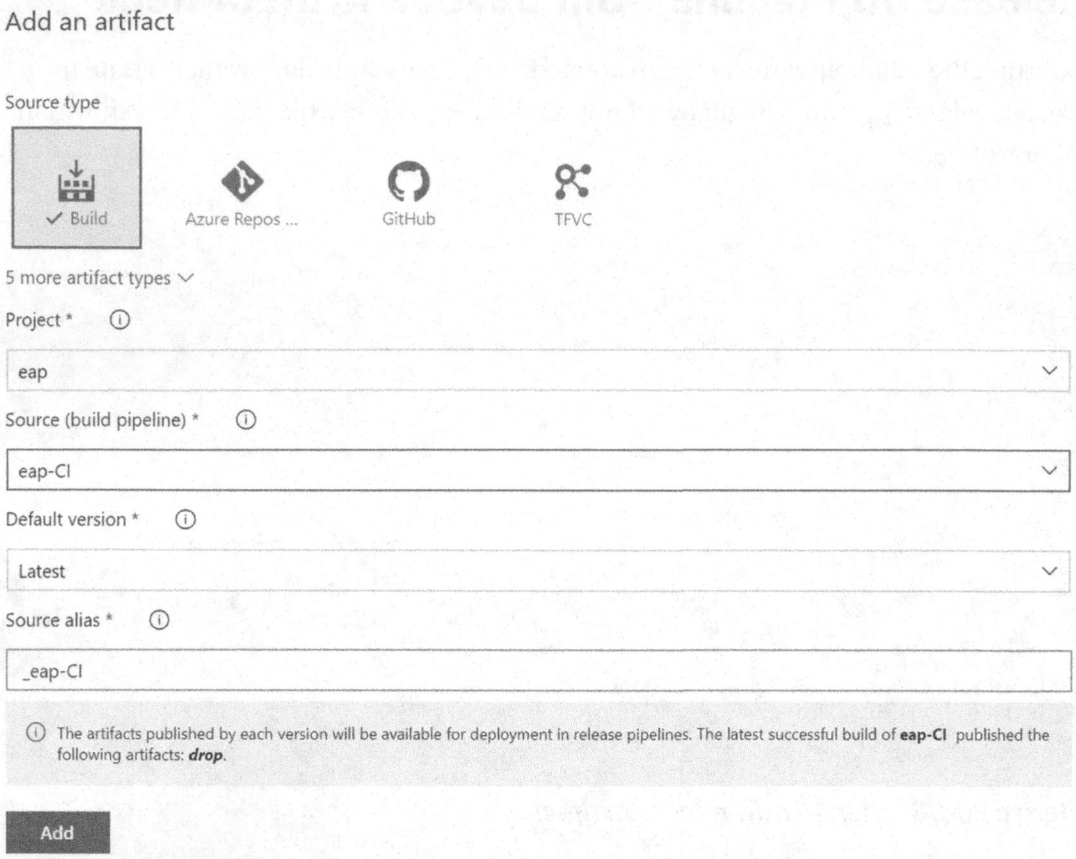

Add an artifact

Source type

5 more artifact types ⌄

Project *   ⓘ

eap	⌄

Source (build pipeline) *   ⓘ

eap-CI	⌄

Default version *   ⓘ

Latest	⌄

Source alias *   ⓘ

_eap-CI

ⓘ The artifacts published by each version will be available for deployment in release pipelines. The latest successful build of **eap-CI** published the following artifacts: **drop**.

Add

Figure 20-39. *Add the Build artifact to the release*

When prompted to select a template, add an Empty job for the stage. This can be found by clicking the blue link that says Empty job, as shown in Figure 20-40.

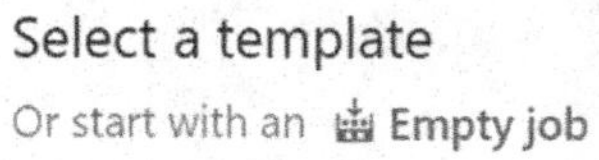

Select a template

Or start with an ▦ **Empty job**

Figure 20-40. *Select the Empty job template*

Verify the stage properties shown in Figure 20-41.

Stage   🗑 Delete   ⇕ Move ∨   ⋯

Stage 1

🗜 Properties ∧

Name and owners of the stage

Stage name

Stage 1

Stage owner

🧑 Ron Lesteve   ✕

Figure 20-41. *Verify the stage properties*

Add a task within the stage, as shown in Figure 20-42.

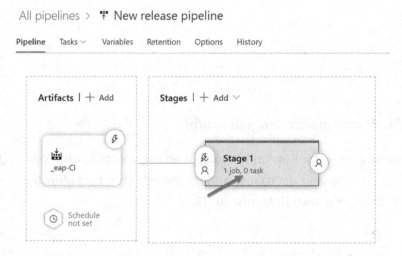

Figure 20-42. *Add a stage task*

Verify the Agent job details within the new release pipeline, as shown in Figure 20-43.

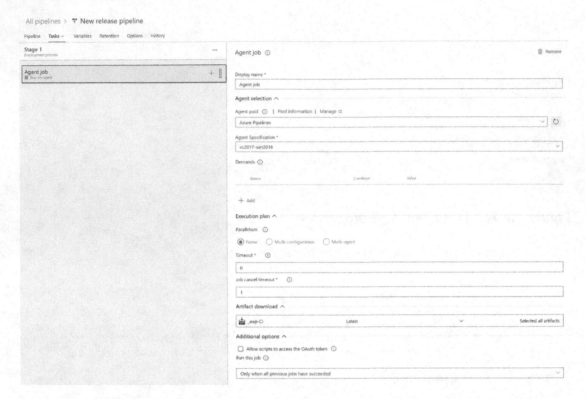

Figure 20-43. *Image of release agent config*

Start by adding an Azure Key Vault task as a secure way of storing your SQL database credentials that include your username and password, which are stored as secrets within the Azure Key Vault, as shown in Figure 20-44.

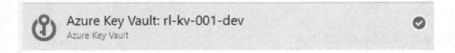

Figure 20-44. *Add Azure Key Vault task*

The steps required to achieve this, as shown in Figure 20-45, would be to first add your Azure subscription by clicking Manage and then logging into your Azure account, select your Key Vault account from your Azure subscription, and finally select the secrets, which in this case would be the *LoginPassword* and *LoginUser*.

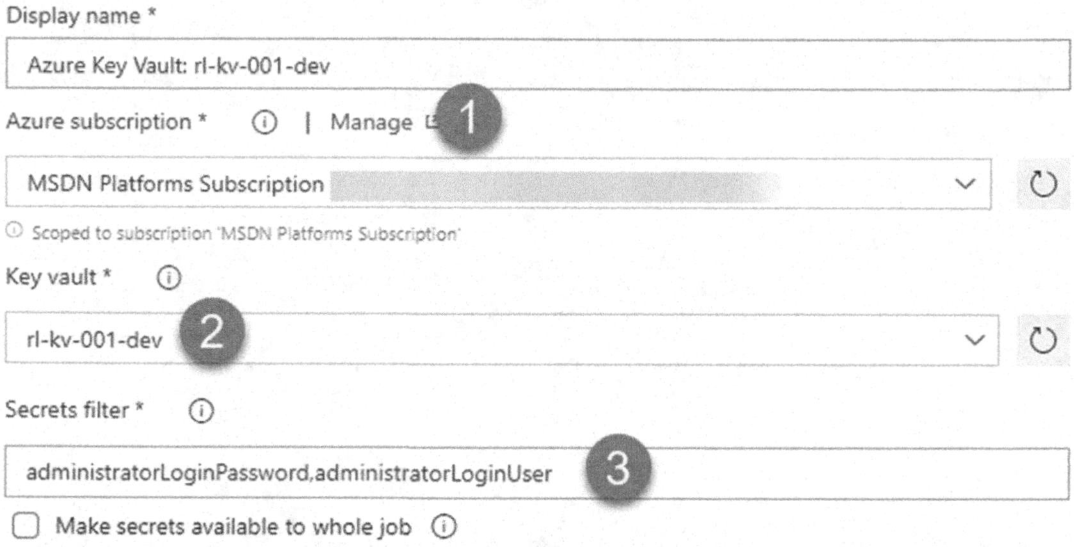

Display name *

Azure Key Vault: rl-kv-001-dev

Azure subscription *   ⓘ | Manage ⏚   1

MSDN Platforms Subscription   ⌄   ↻

ⓘ Scoped to subscription 'MSDN Platforms Subscription'

Key vault *   ⓘ

rl-kv-001-dev   2   ⌄   ↻

Secrets filter *   ⓘ

administratorLoginPassword,administratorLoginUser   3

☐ Make secrets available to whole job ⓘ

Figure 20-45. *Configure Azure Key Vault task*

Next, you'll need to add an Azure SQL Database deployment task, shown in
Figure 20-46, which will deploy the Azure SQL Database using a Dacpac file.

Figure 20-46. *Add the Azure SQL Database deployment task*

Populate the necessary Azure SQL Database deployment parameters, as shown in
Figure 20-47.

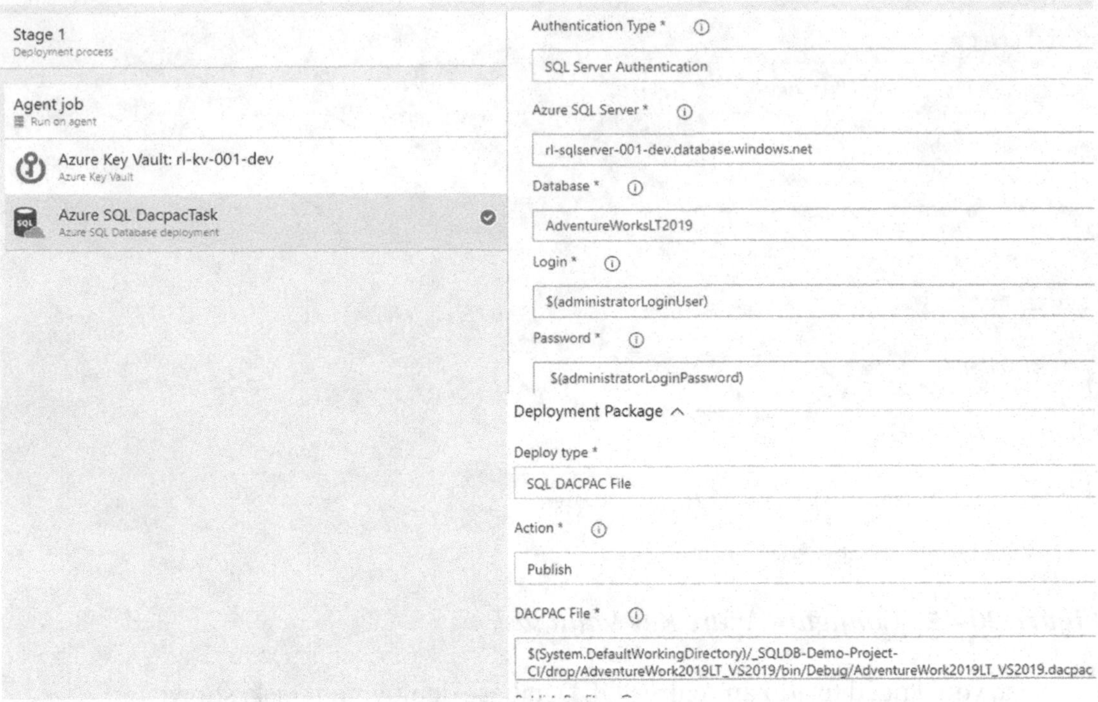

Figure 20-47. *Populate the necessary Azure SQL Database deployment parameters*

Also, within the Azure SQL Dacpac task, select the AdventureWorksLT2019 Dacpac file that was built and stored in the artifact drop directory from the previous build pipeline, as shown in Figure 20-48, and click OK.

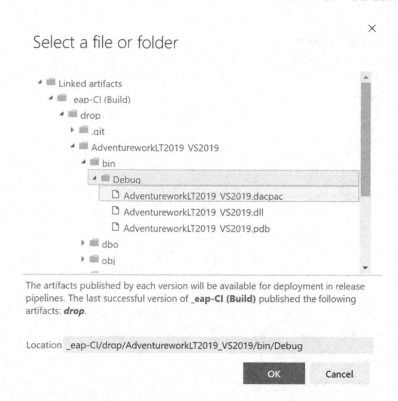

Figure 20-48. *Select the Dacpac file*

Proceed to create the new release pipeline and click Create, which is shown in Figure 20-49.

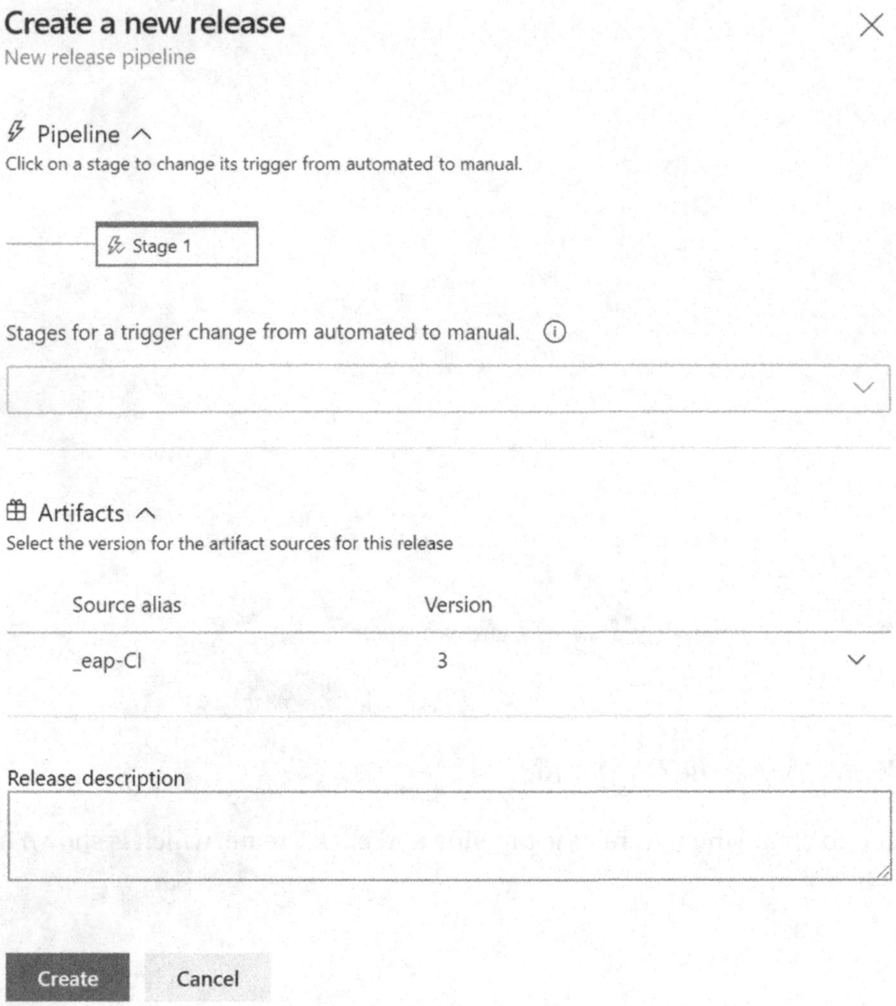

Create a new release

New release pipeline

✕

⚡ Pipeline ∧

Click on a stage to change its trigger from automated to manual.

⚡ Stage 1

Stages for a trigger change from automated to manual.   ⓘ

⊞ Artifacts ∧

Select the version for the artifact sources for this release

Source alias	Version	
_eap-CI	3	∨

Release description

Create   Cancel

Figure 20-49. *Create a new release*

Verify that the release pipeline has succeeded, which is shown in the Stages section of Figure 20-50.

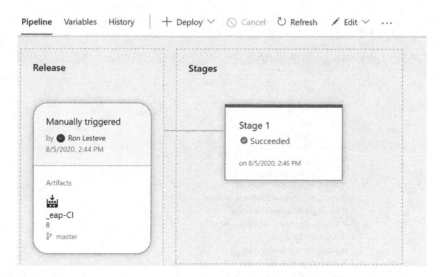

Figure 20-50. *Image showing the release pipeline run status*

Also, verify that the Agent job steps to initialize the job, download the artifact, deploy the Azure SQL Dacpac task, and finalize the job complete successfully. These successfully completed Agent job steps can be seen in Figure 20-51.

Figure 20-51. *Image showing the release agent job summary*

Verify Deployed Azure SQL AdventureWorks Database

In this final step of the process, log into the Azure SQL Database through SSMS to verify that the AdventureWorksLT2019 database exists in the specified deployment location. Notice from Figure 20-52 that all of the database objects have been successfully deployed to the Azure SQL Database.

Figure 20-52. *Verify Azure SQL Database exists through SSMS*

Summary

In this chapter, I have demonstrated how to design and implement an end-to-end solution to deploy AdventureWorksLT2019 database changes by using the database Dacpac file in Visual Studio. This Visual Studio solution was connected to a GitHub repository for source control and was linked to an Azure DevOps pipeline. The exercise concluded by showing you how to use an Azure DevOps build (CI) and release (CD) pipeline to deploy this AdventureWorksLT2019 database to an Azure SQL Database through a continuous integration and deployment methodology.

PART V

Advanced Analytics

CHAPTER 21

Graph Analytics Using Apache Spark's GraphFrame API

Graph technology enables users to store, manage, and query data in the form of a graph in which entities are called vertices or nodes and the relationships between the entities are called edges. Graph analytics enables the capabilities for analyzing deeply grained relationships through queries and algorithms.

Apache Spark's GraphFrame API is an Apache Spark package that provides data frame–based graphs through high-level APIs in Java, Python, and Scala and includes extended functionality for motif finding, data frame–based serialization, and highly expressive graph queries. With GraphFrames, you can easily search for patterns within graphs, find important vertices, and more.

A graph is a set of points, called nodes or *vertices*, which are interconnected by a set of lines called *edges*, as shown in Figure 21-1.

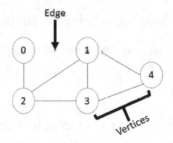

Figure 21-1. *Depicting edges and vertices*

© Ron C. L'Esteve 2021
R. C. L'Esteve, *The Definitive Guide to Azure Data Engineering*, https://doi.org/10.1007/978-1-4842-7182-7_21

Directed graphs have edges with direction. The edges indicate a *one-way* relationship in that each edge can only be traversed in a single direction, which is illustrated in Figure 21-2. A common directed graph is a genealogical tree, which maps the relationship between parents and children.

Undirected graphs have edges with no direction, as shown in Figure 21-2. The edges indicate a two-way relationship in that each edge can be traversed in both directions. A common undirected graph is the topology of connections in a computer network. The graph is undirected because we can assume that if one device is connected to another, then the second one is also connected to the first. It is worth mentioning that an *isolated vertex* is a vertex that is not an endpoint of any edge.

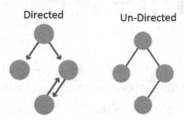

Figure 21-2. *Directed vs. undirected*

In this chapter, you will learn about a practical example of using the GraphFrame API by achieving the following:

1. Installing the JAR library

2. Loading new data tables

3. Loading the data to data frames in a Databricks notebook

4. Running queries and algorithms using the GraphFrame API

Install JAR Library

Now that you have a basic understanding of graphs, let's get started with this exercise on graph analytics using the Apache Spark GraphFrame API.

For this chapter's exercise, I have chosen to utilize a Databricks notebook for running the graph analysis. However, you could just as easily use a Synapse Workspace notebook by adding and configuring an Apache Spark job definition in Synapse Analytics (https://docs.microsoft.com/en-us/azure/synapse-analytics/spark/apache-spark-job-definitions) and adding the GraphFrame API JAR file to the Spark definition.

A complete list of compatible GraphFrame version releases can be found here: https://spark-packages.org/package/graphframes/graphframes. For this exercise, use a JAR file for release version 0.8.1-spark2.4-s_2.12 and install it within the library of a Standard Databricks Cluster, as shown in Figure 21-3.

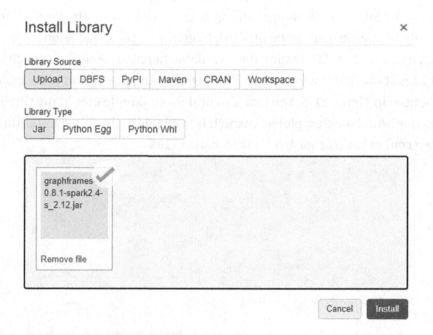

Figure 21-3. *Steps to install JAR library*

Once the installation is complete, notice in Figure 21-4 that the JAR is installed on the cluster, which will be used to run the notebook code in the subsequent sections.

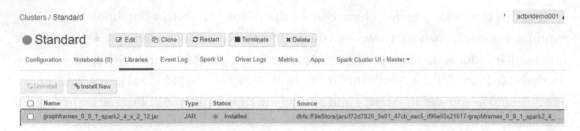

Figure 21-4. *Cluster with library JAR*

Load New Data Tables

Since the GraphFrame API JAR and cluster are ready, upload some data via DBFS. For this exercise, CTA Station data along with trip data is used. There are plenty of free datasets available online that can be utilized. Kaggle, a website that allows users to find and publish datasets, has CTA Station data available here: `www.kaggle.com/chicago/chicago-transit-authority-cta-data`. Start by creating the *station_data* table using the UI presented in Figure 21-5. You can also find these sample files in my GitHub repository GraphAnalyticsSampleFiles, which is available at the following URL: `https://github.com/ronlesteve/GraphAnalyticsSampleFiles`.

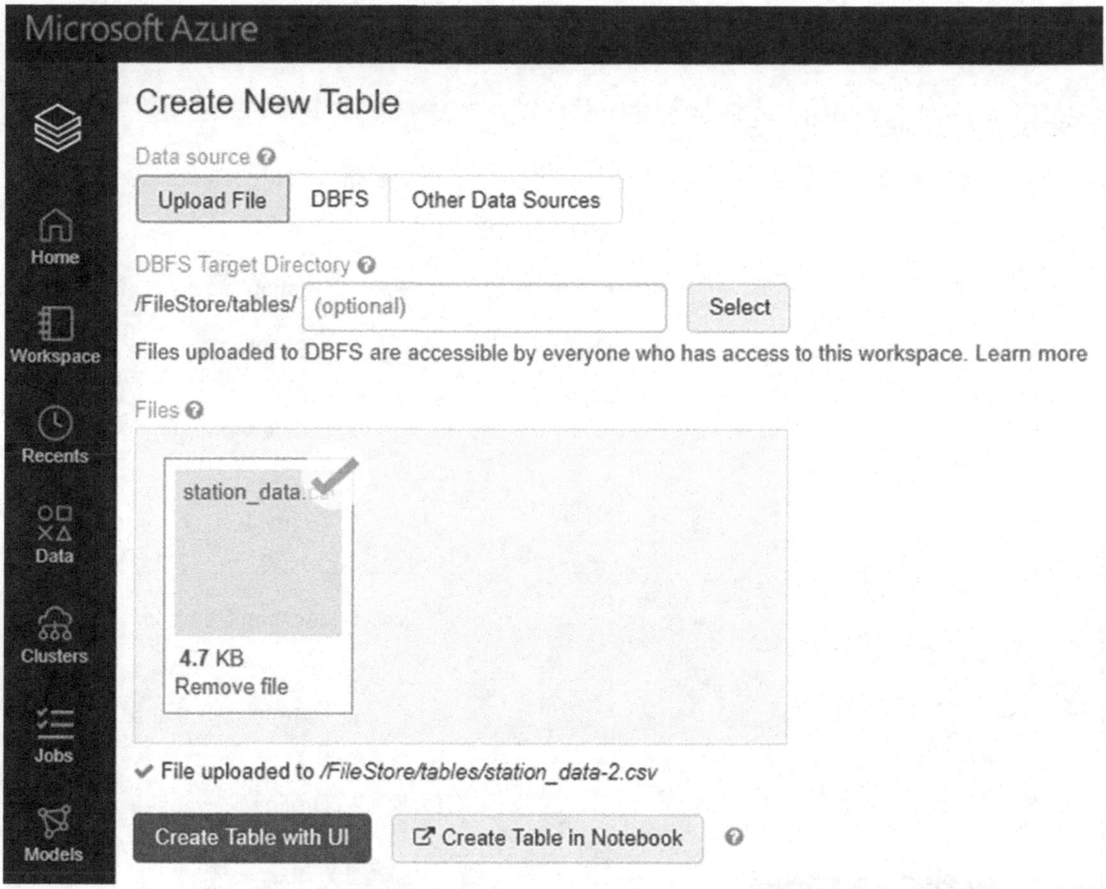

Figure 21-5. *Create station data table*

Validate the schema, ensure that "First row is header" is checked, and click "Create Table," as shown in Figure 21-6.

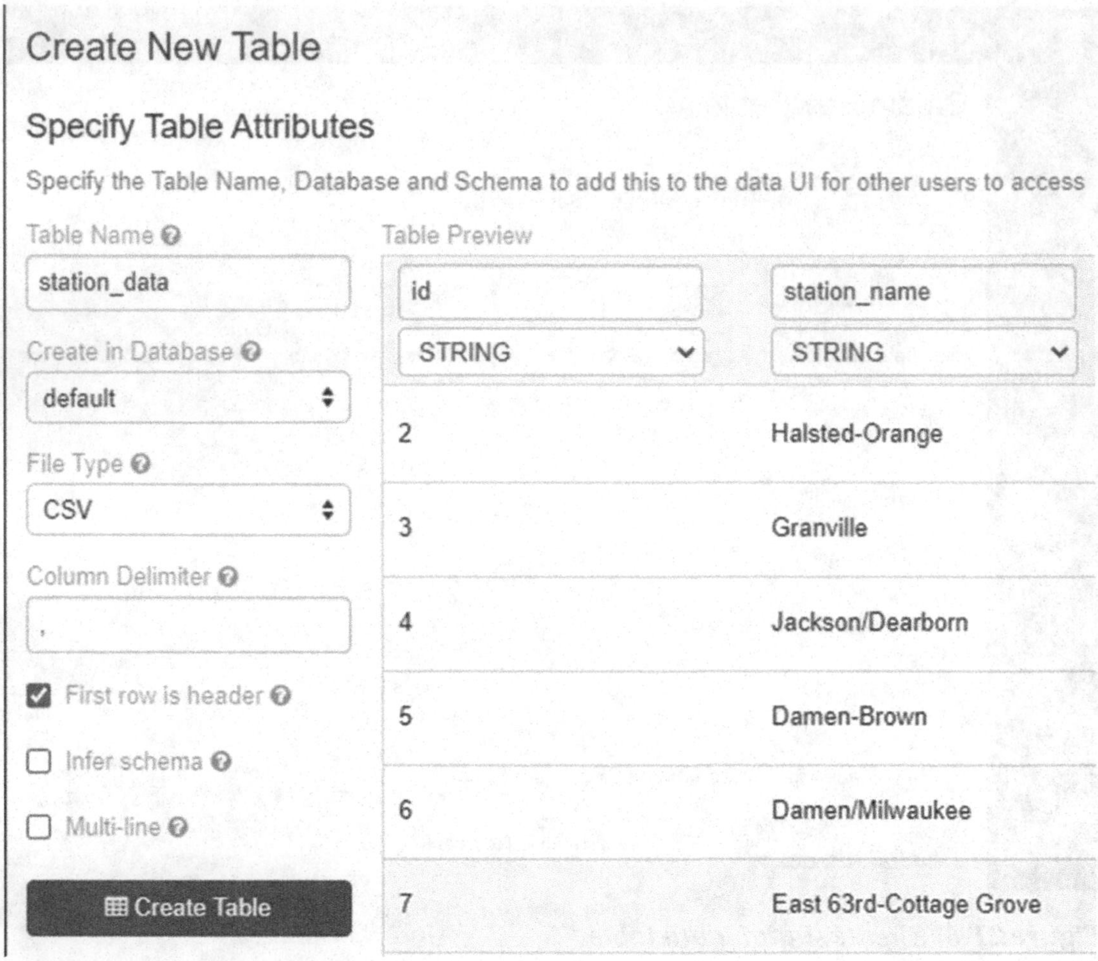

Figure 21-6. *Review table attributes*

Figure 21-7 shows how to follow this same process for the *trip_data* and create the table using the Create New Table UI.

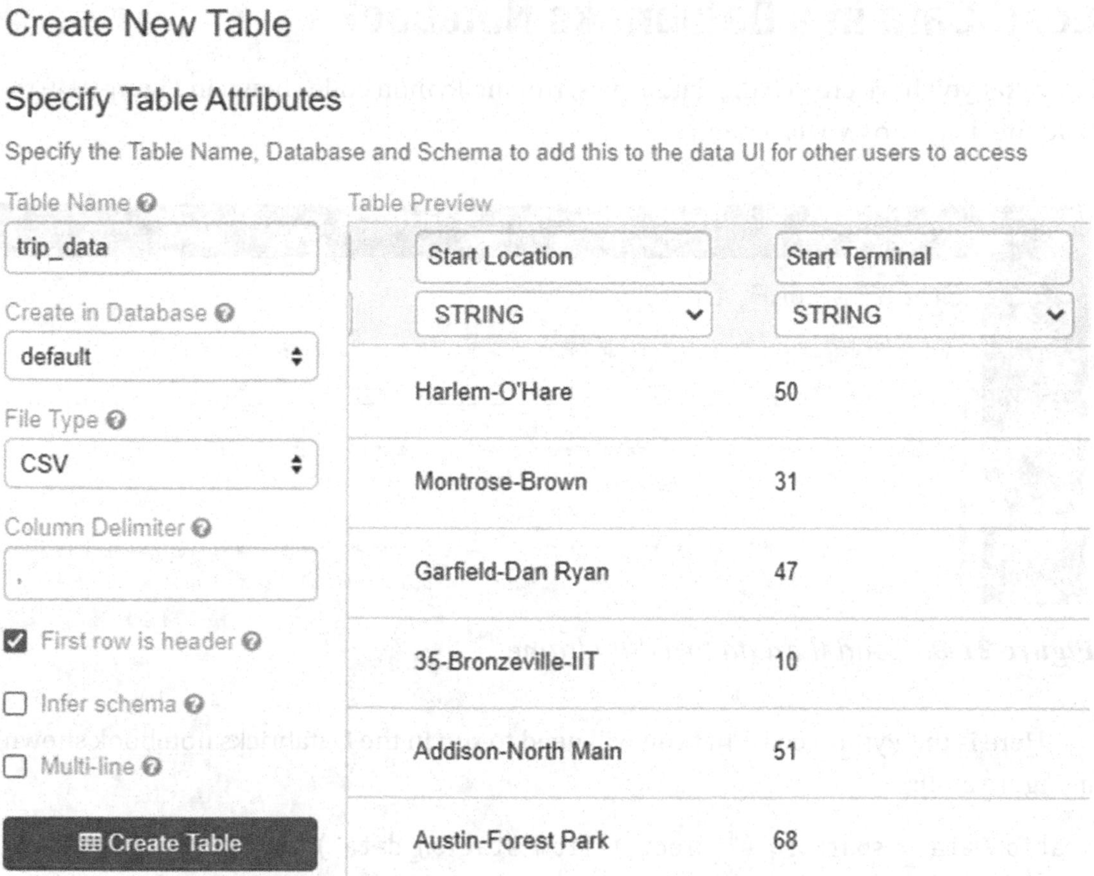

Create New Table

Specify Table Attributes

Specify the Table Name, Database and Schema to add this to the data UI for other users to access

Table Name ❓

| trip_data |

Create in Database ❓

| default ⬍ |

File Type ❓

| CSV ⬍ |

Column Delimiter ❓

| , |

☑ First row is header ❓

☐ Infer schema ❓

☐ Multi-line ❓

| ▦ Create Table |

Table Preview

Start Location	Start Terminal
STRING ⌄	STRING ⌄
Harlem-O'Hare	50
Montrose-Brown	31
Garfield-Dan Ryan	47
35-Bronzeville-IIT	10
Addison-North Main	51
Austin-Forest Park	68

Figure 21-7.   *Create table for trip data*

Once the tables are created, they will both appear in the Tables section, as shown in Figure 21-8.

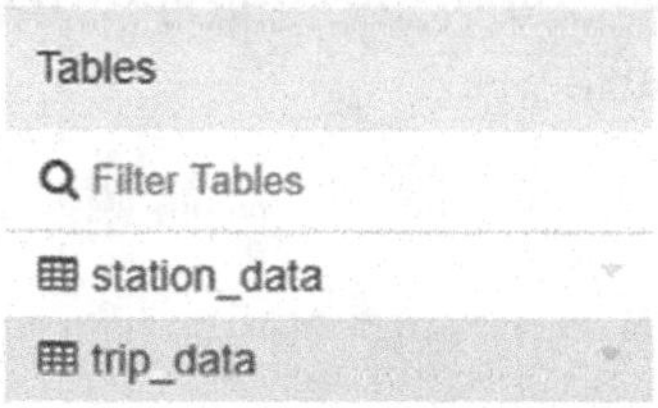

Tables

🔍 Filter Tables

▦ station_data

▦ trip_data

Figure 21-8.   *Verify that tables are created*

Load Data in a Databricks Notebook

Now that you have created the data tables, run the Python code shown in Figure 21-9 to load the data into two data frames.

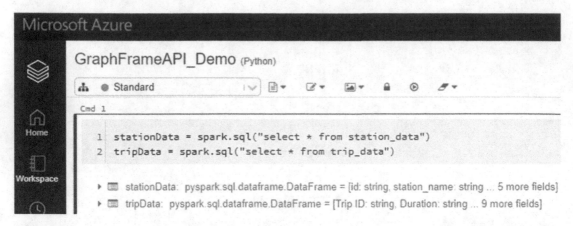

Figure 21-9.   *Load the data to a data frame*

Here is the Python code that you will need to run in the Databricks notebook shown in Figure 21-9:

```
stationData = spark.sql("select * from station_data")
tripData = spark.sql("select * from trip_data")
```

Build a Graph with Vertices and Edges

Now that the data is loaded into data frames, it's time to define the edges and vertices. Since you'll be creating a directed graph, the code shown in Figure 21-10 defines the trips' beginning and end locations.

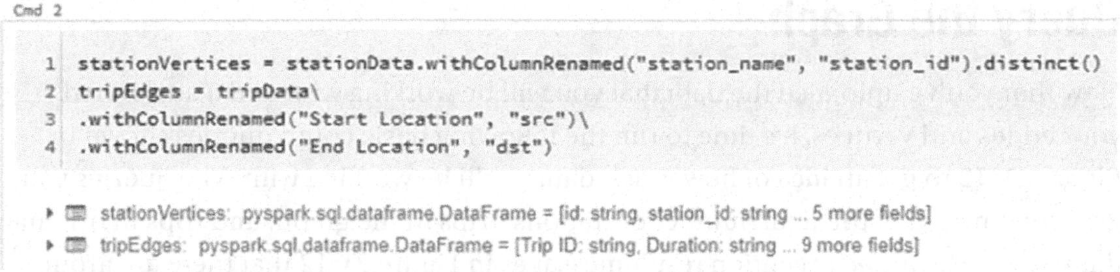

Cmd 2

```
1  stationVertices = stationData.withColumnRenamed("station_name", "station_id").distinct()
2  tripEdges = tripData\
3  .withColumnRenamed("Start Location", "src")\
4  .withColumnRenamed("End Location", "dst")
```

▸ ▦ stationVertices: pyspark.sql.dataframe.DataFrame = [id: string, station_id: string ... 5 more fields]
▸ ▦ tripEdges: pyspark.sql.dataframe.DataFrame = [Trip ID: string, Duration: string ... 9 more fields]

Figure 21-10. *Create data frame for vertices and edges*

Here is the Python code that you will need to run in the Databricks notebook shown in Figure 21-10:

```
stationVertices = stationData.withColumnRenamed("station_name", "station_
id").distinct()
tripEdges = tripData\
.withColumnRenamed("Start Location", "src")\
.withColumnRenamed("End Location", "dst")
```

Next, you will need to build a GraphFrame object, which will represent your graph using the edge and vertex data frames that were defined earlier. Also, you should cache the data for quicker access later. The Python code block that you'll need to run to achieve this is shown in Figure 21-11.

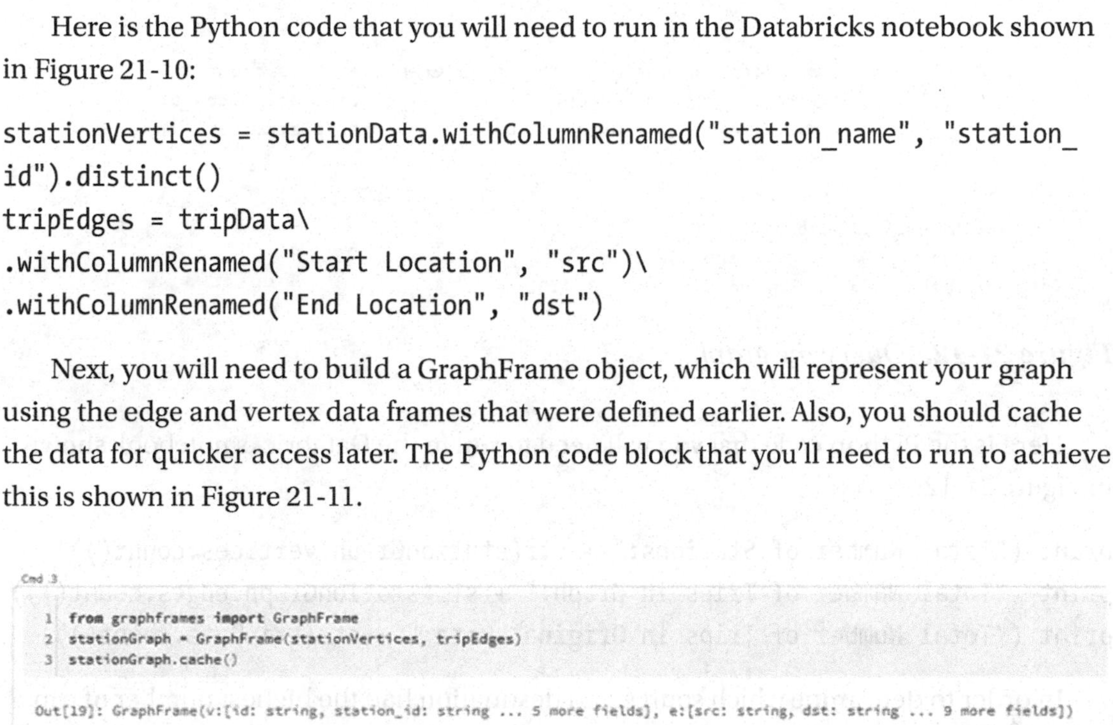

Cmd 3

```
1  from graphframes import GraphFrame
2  stationGraph = GraphFrame(stationVertices, tripEdges)
3  stationGraph.cache()
```

Out[19]: GraphFrame(v:[id: string, station_id: string ... 5 more fields], e:[src: string, dst: string ... 9 more fields])

Figure 21-11. *Build the GraphFrame*

Here is the Python code that you will need to run in the Databricks notebook shown in Figure 21-11:

```
from graphframes import GraphFrame
stationGraph = GraphFrame(stationVertices, tripEdges)
stationGraph.cache()
```

Query the Graph

Now that you've uploaded the data that you will be working with and also defined your edges and vertices, it's time to run the following basic count queries shown in Figure 21-12 to get an idea of how much data you'll be working with. The queries will give you an idea of the total number of stations, trips in the graph, and trips in original data. Notice from the execution result messages in Figure 21-12 that there are around 354K trips spanning 71 stations.

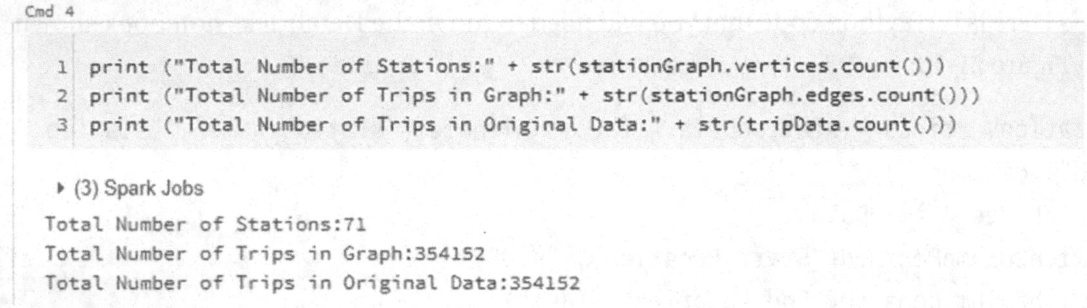

```
Cmd 4

1  print ("Total Number of Stations:" + str(stationGraph.vertices.count()))
2  print ("Total Number of Trips in Graph:" + str(stationGraph.edges.count()))
3  print ("Total Number of Trips in Original Data:" + str(tripData.count()))

  ▶ (3) Spark Jobs
Total Number of Stations:71
Total Number of Trips in Graph:354152
Total Number of Trips in Original Data:354152
```

Figure 21-12. *Query the graph*

Here is the Python code that you will need to run in the Databricks notebook shown in Figure 21-12:

```
print ("Total Number of Stations:" + str(stationGraph.vertices.count()))
print ("Total Number of Trips in Graph:" + str(stationGraph.edges.count()))
print ("Total Number of Trips in Original Data:" + str(tripData.count()))
```

In order to determine which source and destination had the highest number of trips, run the queries shown in Figure 21-13.

Cmd 5

```
1  from pyspark.sql.functions import desc
2  stationGraph.edges.groupBy("src", "dst").count().orderBy(desc("count")).show(10)
```

▸ (1) Spark Jobs

```
+--------------------+--------------------+-----+
|                 src|                 dst|count|
+--------------------+--------------------+-----+
|      Western-Orange|            Harrison| 3748|
|       Harlem-O'Hare|  Cicero-Forest Park| 3145|
|           Southport|       Harlem-O'Hare| 2973|
|            Harrison|      Western-Orange| 2734|
|       Harlem-O'Hare|           Southport| 2640|
|  Addison-North Main|  Belmont-North Main| 2439|
|California/Milwaukee|           Southport| 2356|
|  Cicero-Forest Park|California/Milwaukee| 2330|
|            Harrison|  Belmont-North Main| 2192|
|Kedzie-Homan-Fore...|  Belmont-North Main| 2184|
+--------------------+--------------------+-----+
only showing top 10 rows
```

Figure 21-13. *Run graph queries*

Here is the Python code that you will need to run in the Databricks notebook shown in Figure 21-13:

```
from pyspark.sql.functions import desc
stationGraph.edges.groupBy("src", "dst").count().orderBy(desc("count")).
show(10)
```

Similarly, also run the following query shown in Figure 21-14 to find the number of trips in and out of a particular station.

Cmd 6

```
1  stationGraph.edges\
2  .where("src = 'Cicero-Forest Park' OR dst = 'Cicero-Forest Park'")\
3  .groupBy("src", "dst").count()\
4  .orderBy(desc("count"))\
5  .show(10)
```

▶ (1) Spark Jobs

```
+--------------------+--------------------+-----+
|                 src|                 dst|count|
+--------------------+--------------------+-----+
|      Harlem-O'Hare|  Cicero-Forest Park| 3145|
|  Cicero-Forest Park|California/Milwaukee| 2330|
|  Cicero-Forest Park|       Harlem-O'Hare| 1798|
|California/Milwaukee|  Cicero-Forest Park| 1614|
|  Cicero-Forest Park|  Cicero-Forest Park|  850|
|  Belmont-North Main|  Cicero-Forest Park|  754|
|  Cicero-Forest Park|  Belmont-North Main|  741|
|  Cicero-Forest Park|Kedzie-Homan-Fore...|  668|
|   Irving Park-Brown|  Cicero-Forest Park|  627|
|  Cicero-Forest Park|   Western/Milwaukee|  598|
+--------------------+--------------------+-----+
only showing top 10 rows
```

Figure 21-14. *Run more graph queries*

Here is the Python code that you will need to run in the Databricks notebook shown in Figure 21-14:

```
stationGraph.edges\
.where("src = 'Cicero-Forest Park' OR dst = 'Cicero-Forest Park'")\
.groupBy("src", "dst").count()\
.orderBy(desc("count"))\
.show(10)
```

Find Patterns with Motifs

After running a few basic queries on the graph data, let's get a little more advanced as a next step by finding patterns with something called motifs, which are a way of expressing structural patterns in a graph and querying for patterns in the data instead of actual data.

For example, if you are interested in finding all trips in your dataset that form a triangle between three stations, you would use the following: "(a)-[ab]->(b); (b)-[bc]->(c); (c)-[ca]->(a)." Basically (a), (b), and (c) represent your vertices, and [ab], [bc], and [ca] represent your edges that are excluded from the query by using the following operators as an example: (a)-[ab]->(b), as shown in Figure 21-15.

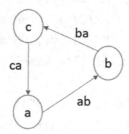

Figure 21-15. Sample pattern with motif

A motif query can be run using a custom pattern like the following code block shown in Figure 21-16.

```
Cmd  7

1  motifs = stationGraph.find("(a)-[ab]->(b); (b)-[bc]->(c); (c)-[ca]->(a)")
2  motifs.show()

▸ (5) Spark Jobs
  ▸ 🗔  motifs:  pyspark.sql.dataframe.DataFrame = [a: struct, ab: struct ... 4 more fields]
+---+---+---+---+---+---+
|  a| ab|  b| bc|  c| ca|
```

Figure 21-16. Pattern algorithm for motif

Here is the Python code that you will need to run in the Databricks notebook shown in Figure 21-16:

```
motifs = stationGraph.find("(a)-[ab]->(b); (b)-[bc]->(c); (c)-[ca]->(a)")
motifs.show()
```

Once the motif is added to a data frame, it can be used in the following query, for example, to find the shortest time from station (a) to (b) to (c) back to (a) by leveraging the timestamps:

```
from pyspark.sql.functions import expr
motifs.selectExpr("*",
"to_timestamp(ab.`Start Date`, 'MM/dd/yyyy HH:mm') as abStart",
"to_timestamp(bc.`Start Date`, 'MM/dd/yyyy HH:mm') as bcStart",
"to_timestamp(ca.`Start Date`, 'MM/dd/yyyy HH:mm') as caStart")\
.where("ca.`Station_Name` = bc.`Station_Name`").where("ab.`Station_Name` =
bc.`Station_Name`")\
.where("a.id != b.id").where("b.id != c.id")\
.where("abStart < bcStart").where("bcStart < caStart")\
.orderBy(expr("cast(caStart as long) - cast(abStart as long)"))\
.selectExpr("a.id", "b.id", "c.id", "ab.`Start Date`", "ca.`End Date`")
.limit(1).show(1, False)
```

Discover Importance with PageRank

The GraphFrames API also leverages graph theory and algorithms to analyze the data. *PageRank* is one such graph algorithm that works by counting the number and quality of links to a page to determine a rough estimate of how important the website is, as shown in Figure 21-17. The underlying assumption is that more important websites are likely to receive more links from other websites.

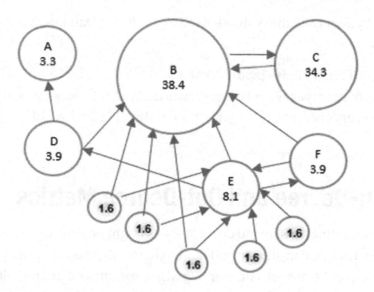

Figure 21-17. *PageRank sample image*

The concept of PageRank can be applied to your data to get an understanding of train stations that receive a lot of bike traffic. In this example shown in Figure 21-18, important stations will be assigned large PageRank values.

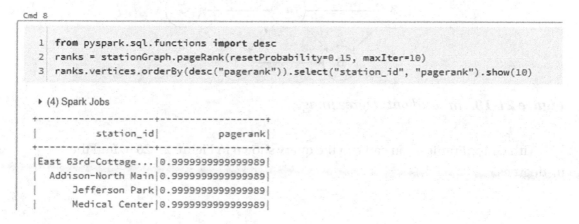

```
Cmd 8

1  from pyspark.sql.functions import desc
2  ranks = stationGraph.pageRank(resetProbability=0.15, maxIter=10)
3  ranks.vertices.orderBy(desc("pagerank")).select("station_id", "pagerank").show(10)

▶ (4) Spark Jobs

+--------------------+-------------------+
|          station_id|           pagerank|
+--------------------+-------------------+
|East 63rd-Cottage...|0.9999999999999989|
|  Addison-North Main|0.9999999999999989|
|      Jefferson Park|0.9999999999999989|
|      Medical Center|0.9999999999999989|
```

Figure 21-18. *Code for PageRank*

Here is the Python code that you will need to run in the Databricks notebook shown in Figure 21-18:

```
from pyspark.sql.functions import desc
ranks = stationGraph.pageRank(resetProbability=0.15, maxIter=10)
ranks.vertices.orderBy(desc("pagerank")).select("station_id", "pagerank").
show(10)
```

Explore In-Degree and Out-Degree Metrics

Measuring and counting trips in and out of stations might be a necessary task, and you can use metrics called in-degree and out-degree for this task. This may be more applicable in the context of social networking where we can find people with more followers (in-degree) than people whom they follow (out-degree). This concept of in- and out-degree is illustrated in Figure 21-19.

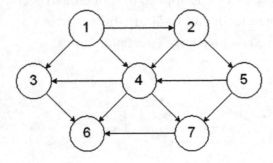

Figure 21-19. *In- and out-degree image*

With GraphFrames, you can run the query shown in Figure 21-20 to find the in-degrees.

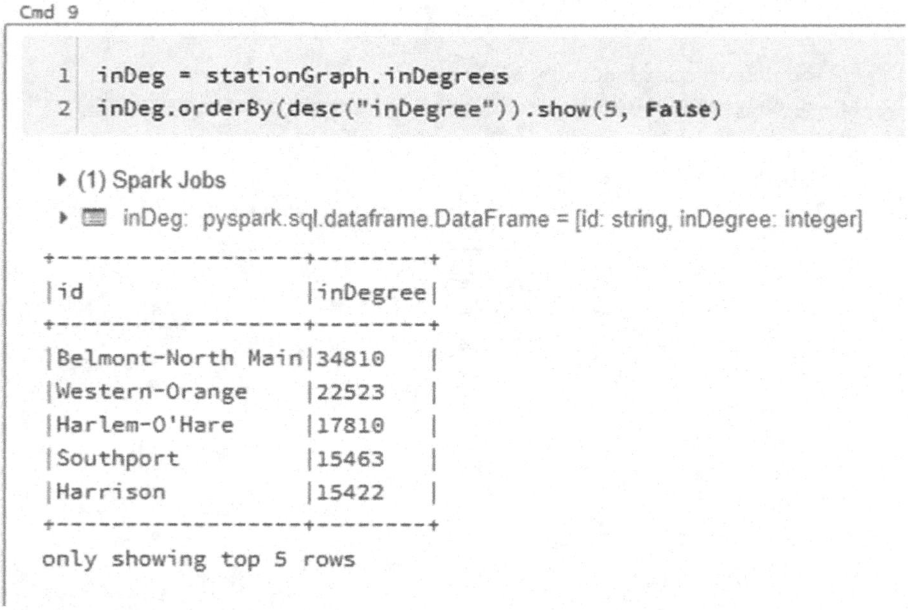

Cmd 9

```
1  inDeg = stationGraph.inDegrees
2  inDeg.orderBy(desc("inDegree")).show(5, False)
```

▸ (1) Spark Jobs
▸ 🔳 inDeg: pyspark.sql.dataframe.DataFrame = [id: string, inDegree: integer]

```
+------------------+--------+
|id                |inDegree|
+------------------+--------+
|Belmont-North Main|34810   |
|Western-Orange    |22523   |
|Harlem-O'Hare     |17810   |
|Southport         |15463   |
|Harrison          |15422   |
+------------------+--------+
only showing top 5 rows
```

Figure 21-20. *Code for in-degree*

Here is the Python code that you will need to run in the Databricks notebook shown in Figure 21-20:

```
inDeg = stationGraph.inDegrees
inDeg.orderBy(desc("inDegree")).show(5, False)
```

Similarly, run the code block shown in Figure 21-21 to find the out-degrees and also notice the results.

Cmd 10

```
1   outDeg = stationGraph.outDegrees
2   outDeg.orderBy(desc("outDegree")).show(5, False)
```

▶ (1) Spark Jobs
▶ 🖾 outDeg: pyspark.sql.dataframe.DataFrame = [id: string, outDegree: integer]

```
+---------------------------+----------+
|id                         |outDegree|
+---------------------------+----------+
|Belmont-North Main         |26304    |
|Western-Orange             |21758    |
|Harlem-O'Hare              |17255    |
|Kedzie-Homan-Forest Park|14436    |
|Cicero-Forest Park         |14158    |
+---------------------------+----------+
only showing top 5 rows
```

Figure 21-21. *Code for out-degree*

Here is the Python code that you will need to run in the Databricks notebook shown in Figure 21-21:

```
outDeg = stationGraph.outDegrees
outDeg.orderBy(desc("outDegree")).show(5, False)
```

Finally, run the code shown in Figure 21-22 to find the ratio between in- and out-degrees. A higher ratio value will indicate where many trips end (but rarely begin), while a lower value tells us where trips often begin (but infrequently end).

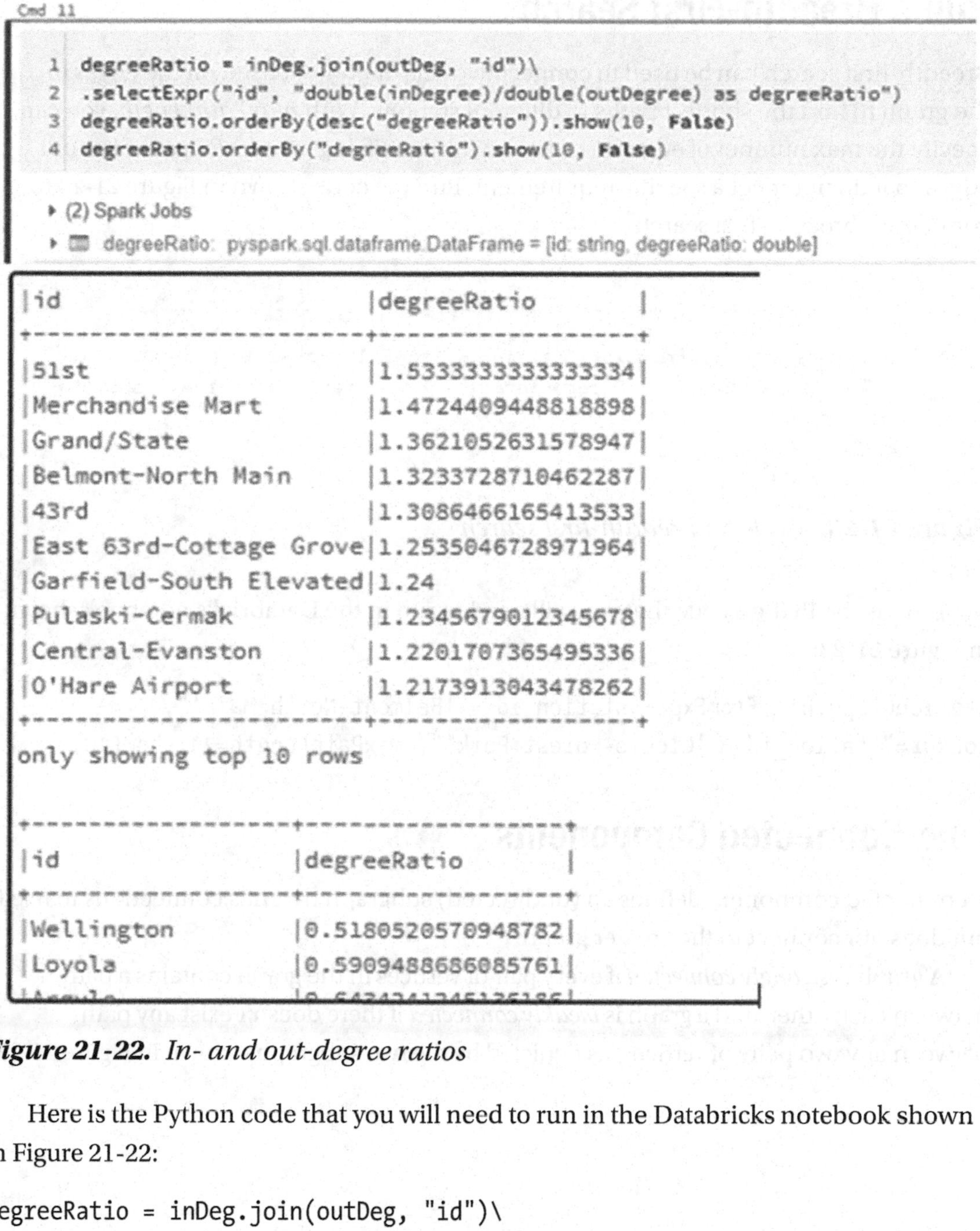

```
Cmd 11
1  degreeRatio = inDeg.join(outDeg, "id")\
2  .selectExpr("id", "double(inDegree)/double(outDegree) as degreeRatio")
3  degreeRatio.orderBy(desc("degreeRatio")).show(10, False)
4  degreeRatio.orderBy("degreeRatio").show(10, False)
```

▶ (2) Spark Jobs

▶ 🔲 degreeRatio: pyspark.sql.dataframe.DataFrame = [id: string, degreeRatio: double]

```
|id                      |degreeRatio         |
+------------------------+--------------------+
|51st                    |1.5333333333333334  |
|Merchandise Mart        |1.4724409448818898  |
|Grand/State             |1.3621052631578947  |
|Belmont-North Main      |1.3233728710462287  |
|43rd                    |1.3086466165413533  |
|East 63rd-Cottage Grove |1.2535046728971964  |
|Garfield-South Elevated |1.24                |
|Pulaski-Cermak          |1.2345679012345678  |
|Central-Evanston        |1.2201707365495336  |
|O'Hare Airport          |1.2173913043478262  |
+------------------------+--------------------+
only showing top 10 rows


+------------------------+--------------------+
|id                      |degreeRatio         |
+------------------------+--------------------+
|Wellington              |0.5180520570948782  |
|Loyola                  |0.5909488686085761  |
```

Figure 21-22. *In- and out-degree ratios*

Here is the Python code that you will need to run in the Databricks notebook shown in Figure 21-22:

```
degreeRatio = inDeg.join(outDeg, "id")\
.selectExpr("id", "double(inDegree)/double(outDegree) as degreeRatio")
degreeRatio.orderBy(desc("degreeRatio")).show(10, False)
degreeRatio.orderBy("degreeRatio").show(10, False)
```

Run a Breadth-First Search

Breadth-first search can be used to connect two sets of nodes, based on the edges in the graph to find the shortest paths to different stations. With *maxPathLength*, you can specify the max number of edges to connect with specifying an *edgeFilter* to filter out edges that do not meet a specific requirement. Run the code shown in Figure 21-23 to complete a breadth-first search.

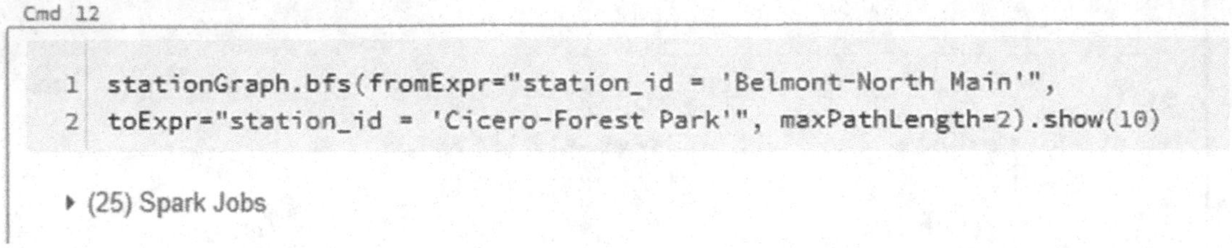

```
Cmd 12

1  stationGraph.bfs(fromExpr="station_id = 'Belmont-North Main'",
2  toExpr="station_id = 'Cicero-Forest Park'", maxPathLength=2).show(10)


▶ (25) Spark Jobs
```

Figure 21-23. *Code for breadth-first search*

Here is the Python code that you will need to run in the Databricks notebook shown in Figure 21-23:

```
stationGraph.bfs(fromExpr="station_id = 'Belmont-North Main'",
toExpr="station_id = 'Cicero-Forest Park'", maxPathLength=2).show(10)
```

Find Connected Components

A connected component defines an (undirected) subgraph that has connections to itself but does not connect to the greater graph.

A graph is *strongly connected* if every pair of vertices in the graph contains a path between each other, and a graph is *weakly connected* if there doesn't exist any path between any two pairs of vertices, as depicted by the two subgraphs shown in Figure 21-24.

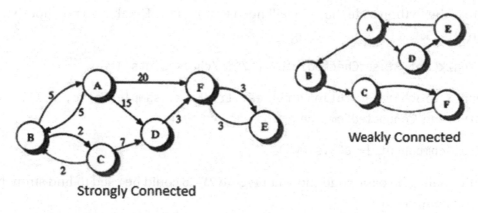

Figure 21-24. *Strongly vs. weakly connected components*

The code shown in Figure 21-25 will generate the connected components. Notice the results.

```
1   spark.sparkContext.setCheckpointDir("/tmp/checkpoints")
2
3
4   minGraph = GraphFrame(stationVertices, tripEdges.sample(False, 0.1))
5   cc = minGraph.connectedComponents()
6
7   cc.where("component != 0").show()
8
```

▸ (9) Spark Jobs

▸ 🔲 cc: pyspark.sql.dataframe.DataFrame = [id: string, station_id: string ... 6 more fields]

id	station_id	lat	long	dockcount	landmark	installation	component
2	Halsted-Orange	37.329732	-121.901782	27	San Jose	8/6/2013	1623497637888
84	Randolph/Wabash	37.342725	-121.895617	15	San Jose	4/9/2014	1159641169920
33	47th-South Elevated	37.400241	-122.099076	15	Mountain View	8/16/2013	1056561954816
59	Paulina	37.781332	-122.418603	23	San Francisco	8/21/2013	197568495616
28	Chicago/Franklin	37.394358	-122.076713	23	Mountain View	8/15/2013	249108103169
39	Roosevelt	37.783871	-122.408433	19	San Francisco	8/25/2013	1434519076864
75	Ashland-Orange	37.7913	-122.399051	19	San Francisco	8/25/2013	506806140928

Figure 21-25. *Code to run for finding connected components*

Here is the Python code that you will need to run in the Databricks notebook shown in Figure 21-25:

```
spark.sparkContext.setCheckpointDir("/tmp/checkpoints")

minGraph = GraphFrame(stationVertices, tripEdges.sample(False, 0.1))
cc = minGraph.connectedComponents()

cc.where("component != 0").show()
```

Additionally, the basic code shown in Figure 21-26 could be used to find strongly connected components.

```
1  scc = minGraph.stronglyConnectedComponents(maxIter=3)
2  scc.groupBy("component").count().show()
3
```

▸ (8) Spark Jobs
▸ 🔲 scc: pyspark.sql.dataframe.DataFrame = [id: string, station_id: string ... 6 more fields]

```
+--------------+-----+
|     component|count|
+--------------+-----+
|  506806140928|    1|
|  128849018880|    1|
|1451698946048|    1|
```

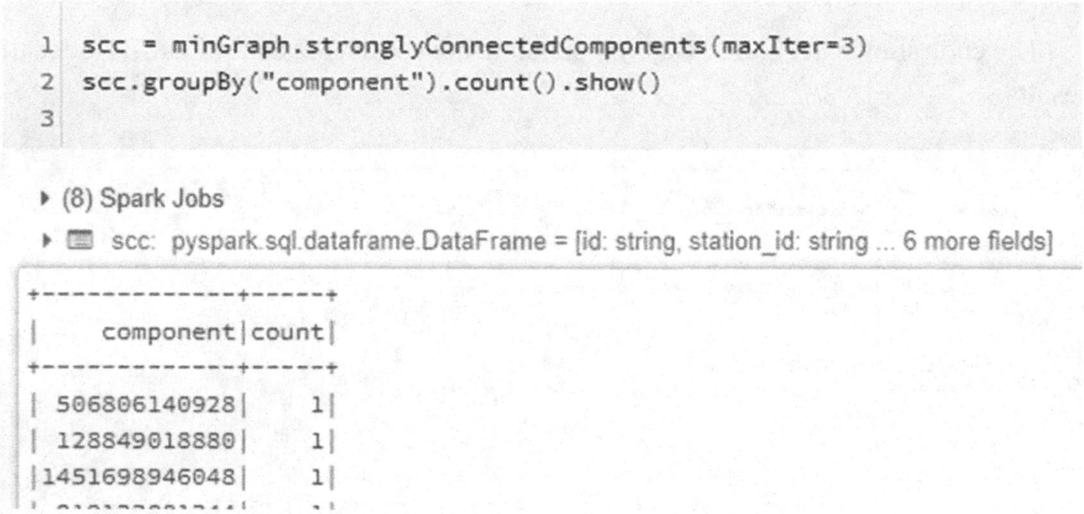

Figure 21-26. *Code for finding strongly connected components*

Here is the Python code that you will need to run in the Databricks notebook shown in Figure 21-26:

```
scc = minGraph.stronglyConnectedComponents(maxIter=3)
scc.groupBy("component").count().show()
```

Summary

In this chapter, I demonstrated a few practical examples of how to use the Databricks GraphFrame API to easily search for patterns within graphs and vertices and then query them to gain insights into the data. Within Azure, graph database services are available through the Cosmos DB Gremlin API. Cosmos DB is a multi-model, globally distributed NoSQL database, and its Gremlin API supports a structure that is composed of vertices and edges. Other Microsoft technologies such as SQL Server have also added support for graph databases to handle data that contains complex entity relationships. All of these various graph technologies show the value of having the capability to perform analytics on extremely large graph datasets such as social media feeds with Apache Spark.

CHAPTER 22

Synapse Analytics Workspaces

The numerous new additions to Microsoft's Azure Data Platform have been creating both excitement and confusion around a number of similar offerings and their purpose in the modern Azure Data Platform. Azure Synapse Analytics workspaces are one such offering, which is not to be confused with Azure Synapse Analytics DW, which has traditionally been the next-generation Azure data warehouse, which has been rebranded as Synapse Analytics dedicated SQL pools (Synapse Analytics DW). Azure Synapse Analytics workspaces are the evolution of the unified modern data and analytics platform, which brings together a variety of Azure services into one centralized location.

Azure Synapse Analytics workspaces are a web-native experience that unifies end-to-end analytics solutions for data engineers to empower and enable them to ingest, explore, prepare, orchestrate, and visualize their data through one experience by utilizing either SQL or Spark pools. Additionally, it brings along with it the capabilities for debugging, performance optimization, data governance, and integration with CI/CD. In this chapter, I will cover how to get started with Synapse Analytics workspaces by leveraging some practical examples, code, and use cases provided by the Synapse Analytics Workspaces Knowledge Center.

© Ron C. L'Esteve 2021
R. C. L'Esteve, *The Definitive Guide to Azure Data Engineering*, https://doi.org/10.1007/978-1-4842-7182-7_22

Create a Synapse Analytics Workspace

Begin by creating a new Azure Synapse Analytics workspace, as shown in Figure 22-1.

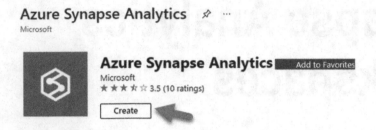

Figure 22-1. *Synapse Analytics workspace creation*

Next, configure and complete the "Create Synapse workspace" section. Note that you will need to create and use an Azure Data Lake Storage Gen2 in the workspace creation process, as shown in Figure 22-2.

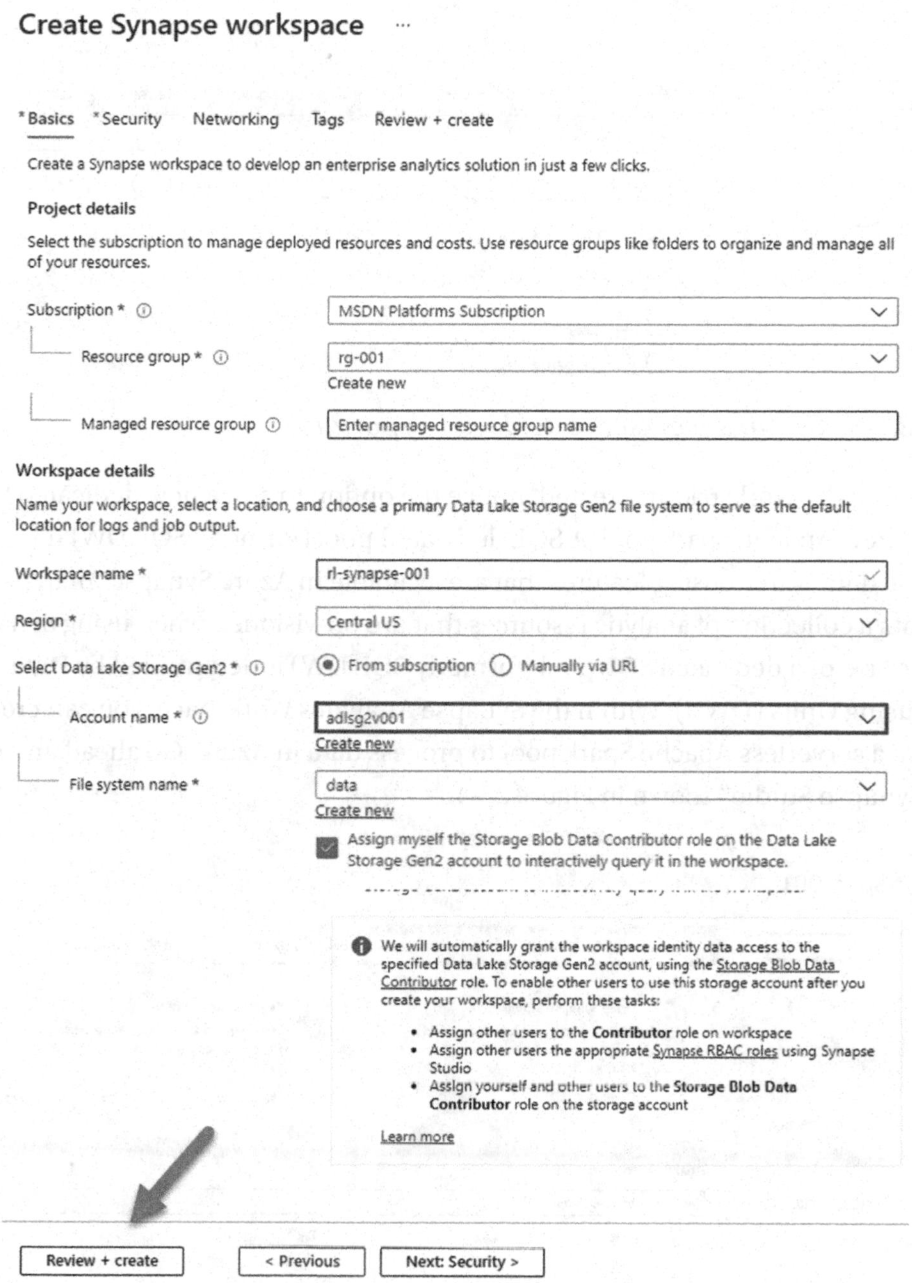

Create Synapse workspace ...

*Basics   *Security   Networking   Tags   Review + create

Create a Synapse workspace to develop an enterprise analytics solution in just a few clicks.

Project details

Select the subscription to manage deployed resources and costs. Use resource groups like folders to organize and manage all of your resources.

Subscription * ⓘ | MSDN Platforms Subscription   ∨ |

　　└── Resource group * ⓘ | rg-001   ∨ |
                                         Create new

　　└── Managed resource group ⓘ | Enter managed resource group name |

Workspace details

Name your workspace, select a location, and choose a primary Data Lake Storage Gen2 file system to serve as the default location for logs and job output.

Workspace name * | rl-synapse-001   ✓ |

Region * | Central US   ∨ |

Select Data Lake Storage Gen2 * ⓘ   ⦿ From subscription   ◯ Manually via URL

　　└── Account name * ⓘ | adlsg2v001   ∨ |
                                         Create new

　　└── File system name * | data   ∨ |
                                         Create new

                                         ☑ Assign myself the Storage Blob Data Contributor role on the Data Lake
                                            Storage Gen2 account to interactively query it in the workspace.

                                         ┌──┐
                                         │ ① We will automatically grant the workspace identity data access to the │
                                         │   specified Data Lake Storage Gen2 account, using the Storage Blob Data │
                                         │   Contributor role. To enable other users to use this storage account after you │
                                         │   create your workspace, perform these tasks: │
                                         │   │
                                         │   • Assign other users to the **Contributor** role on workspace │
                                         │   • Assign other users the appropriate Synapse RBAC roles using Synapse │
                                         │   Studio   │
                                         │   • Assign yourself and other users to the **Storage Blob Data** │
                                         │   **Contributor** role on the storage account   │
                                         │   │
                                         │   Learn more   │
                                         └──┘

[Review + create]   [< Previous]   [Next: Security >]

Figure 22-2. *Review and create Synapse workspace*

Once the workspace is created, it will be available in Azure Portal in the chosen resource group and appear as shown in Figure 22-3. Serverless SQL will be available immediately after provisioning the workspace and will include a pay-per-query billing model.

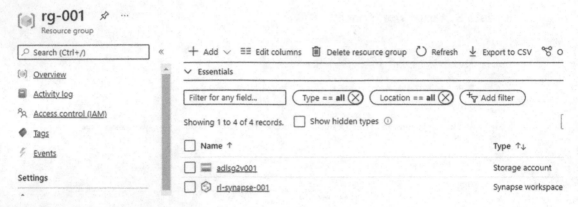

Figure 22-3. *Synapse workspace deployed in portal*

Now that the workspace is created, notice the option to add a new dedicated SQL pool and new Apache Spark pool. A SQL dedicated pool (formerly SQL DW) refers to the enterprise data warehousing features that are available in Azure Synapse Analytics and represents a collection of analytic resources that are provisioned when using Synapse SQL. The size of a dedicated SQL pool (formerly SQL DW) is determined by Data Warehousing Units (DWU). Within the Synapse Analytics Workspace, you can create and configure a serverless Apache Spark pool to process data in Azure. Go ahead and click "Open Synapse Studio" shown in Figure 22-4.

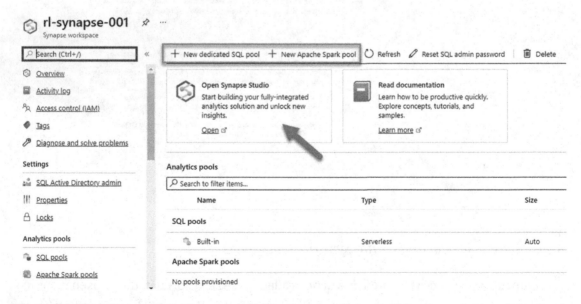

Figure 22-4. *Launch Synapse Studio from portal*

Once the Synapse Analytics Studio workspace launches, navigate to the Knowledge Center and click "Use samples immediately," shown in Figure 22-5, to instantly explore scripts, notebooks, datasets, and pools that are automatically provisioned for you.

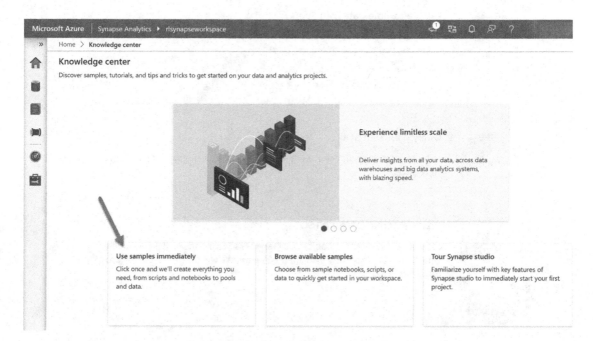

Figure 22-5. *Use samples from Knowledge Center*

Explore Sample Data with Spark

There are three samples included in the Knowledge Center for immediate exploration, as shown in Figure 22-6. The first explores sample data with Spark, includes sample scripts, and even creates a new Spark pool for you, or you can also bring you own.

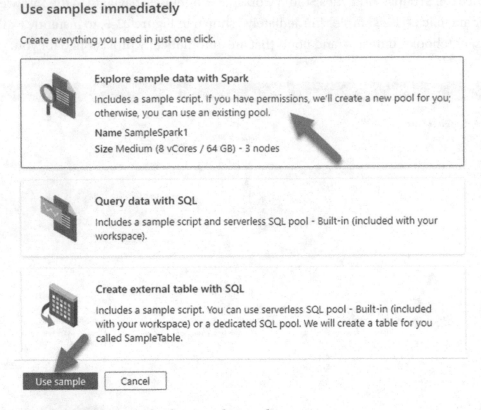

Figure 22-6. *Explore sample data with Spark*

Select the first sample and click "Use sample." A notebook will be created that will contain a number of scripts containing data from the Azure ML OpenDatasets package (https://docs.microsoft.com/en-us/python/api/azureml-opendatasets). Specifically, this notebook uses the NYC Yellow Taxi Dataset. Click "Run all" to execute the notebook and notice from the completed execution results in Figure 22-7 that the job used two Spark executors and eight cores to ingest the data.

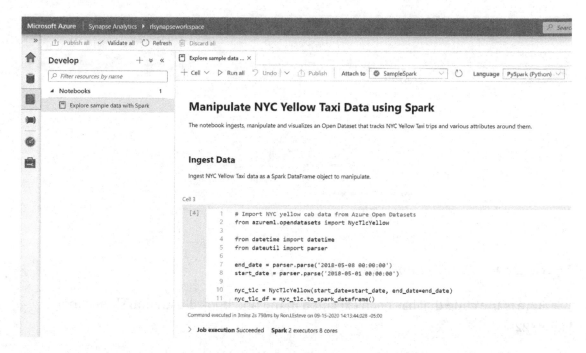

Figure 22-7. *NYC taxi data sample notebook*

Here is the code that has been executed in the notebook illustration shown in Figure 22-7:

```python
# Read NYC yellow cab data from Azure Open Datasets
from azureml.opendatasets import NycTlcYellow

from datetime import datetime
from dateutil import parser

end_date = parser.parse('2018-05-08 00:00:00')
start_date = parser.parse('2018-05-01 00:00:00')

nyc_tlc = NycTlcYellow(start_date=start_date, end_date=end_date)
nyc_tlc_df = nyc_tlc.to_spark_dataframe()
```

The next step in the process demonstrates how to copy the data to the associated Azure Data Lake Store Gen2 account. The script in the notebook, shown in Figure 22-8, will complete this Copy Data task by creating a Spark table, a CSV, and a parquet file with the copied data. You could just work with the data frame; however, this step will demonstrate how to integrate ADLS Gen2 and persist the data.

Figure 22-8. *Sample Copy Data with Spark pool*

Here is the code that has been executed in the notebook illustration shown in Figure 22-8:

```
nyc_tlc_df.createOrReplaceTempView('nyc_tlc_df')
nyc_tlc_df.write.csv('nyc_tlc_df_csv', mode='overwrite')
nyc_tlc_df.write.parquet('nyc_tlc_df_parquet', mode='overwrite')
```

After the job completes running, navigate to the Azure Data Lake Storage Gen2 account to verify that both the CSV and parquet files have been created. Notice the two new folders containing the NYC taxi datasets shown in Figure 22-9.

Figure 22-9. *Folders and files created in ADLS Gen2*

Open the parquet folder and notice the number of snappy compressed parquet files shown in Figure 22-10.

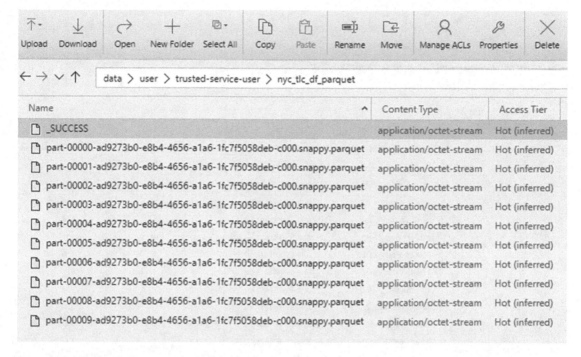

Figure 22-10. *Parquet files created in ADLS Gen2*

Next, navigate back to the Azure Synapse Studio workspace and run the select statement shown in Figure 22-11 to ensure that the data was also ingested into a Spark table.

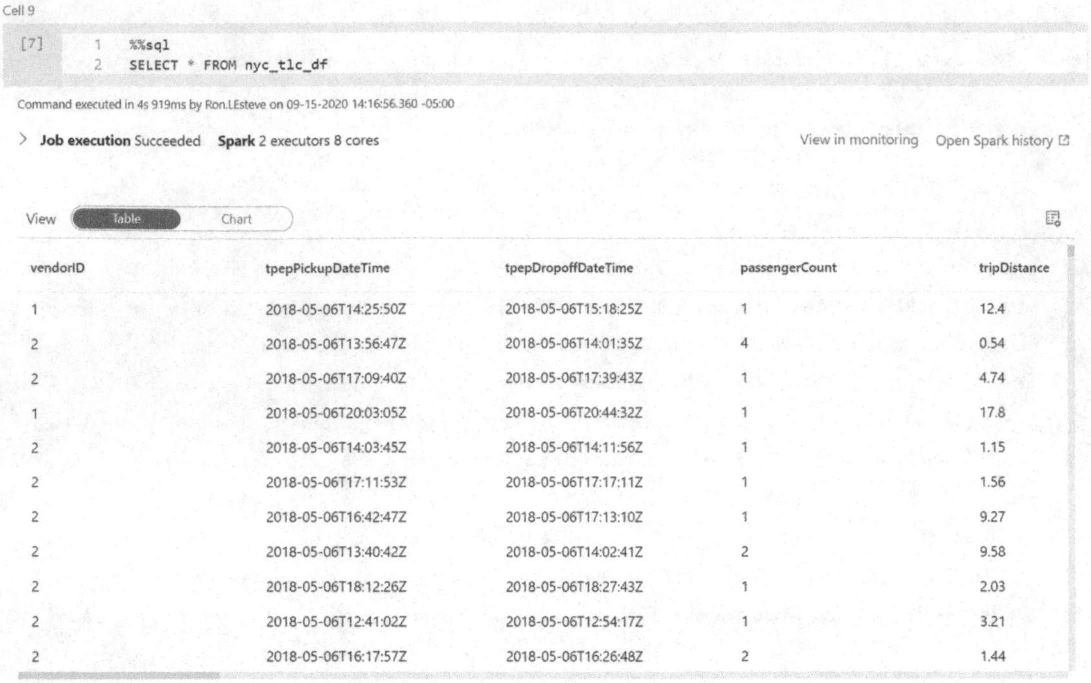

Cell 9

```
[7]    1   %%sql
       2   SELECT * FROM nyc_tlc_df
```

Command executed in 4s 919ms by Ron.LEsteve on 09-15-2020 14:16:56.360 -05:00

> **Job execution** Succeeded   **Spark** 2 executors 8 cores   View in monitoring   Open Spark history ⬚

View   Table   Chart   ⬚

vendorID	tpepPickupDateTime	tpepDropoffDateTime	passengerCount	tripDistance
1	2018-05-06T14:25:50Z	2018-05-06T15:18:25Z	1	12.4
2	2018-05-06T13:56:47Z	2018-05-06T14:01:35Z	4	0.54
2	2018-05-06T17:09:40Z	2018-05-06T17:39:43Z	1	4.74
1	2018-05-06T20:03:05Z	2018-05-06T20:44:32Z	1	17.8
2	2018-05-06T14:03:45Z	2018-05-06T14:11:56Z	1	1.15
2	2018-05-06T17:11:53Z	2018-05-06T17:17:11Z	1	1.56
2	2018-05-06T16:42:47Z	2018-05-06T17:13:10Z	1	9.27
2	2018-05-06T13:40:42Z	2018-05-06T14:02:41Z	2	9.58
2	2018-05-06T18:12:26Z	2018-05-06T18:27:43Z	1	2.03
2	2018-05-06T12:41:02Z	2018-05-06T12:54:17Z	1	3.21
2	2018-05-06T16:17:57Z	2018-05-06T16:26:48Z	2	1.44

Figure 22-11. *Selection of all data within the notebook*

Here is the code that has been executed in the notebook illustration shown in Figure 22-11:

```
%%sql
SELECT * FROM myc_tlc_df
```

From here, you could get even more advanced with the data exploration by beginning to aggregate data, as shown in Figure 22-12.

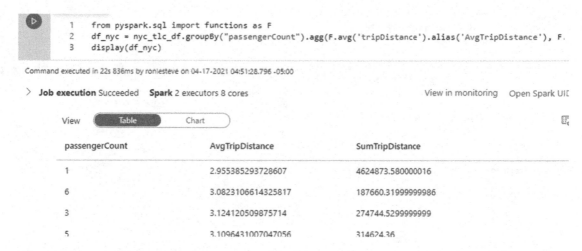

```
1  from pyspark.sql import functions as F
2  df_nyc = nyc_tlc_df.groupBy("passengerCount").agg(F.avg('tripDistance').alias('AvgTripDistance'), F.
3  display(df_nyc)
```

Command executed in 22s 836ms by ronlesteve on 04-17-2021 04:51:28.796 -05:00

> **Job execution** Succeeded   **Spark** 2 executors 8 cores   View in monitoring   Open Spark UI

passengerCount	AvgTripDistance	SumTripDistance
1	2.955385293728607	4624873.580000016
6	3.0823106614325817	187660.31999999986
3	3.124120509875714	274744.5299999999
5	3.1096431007047056	314624.36

Figure 22-12. *Aggregate data in notebook*

Here is the code that has been executed in the notebook illustration shown in Figure 22-12:

```
from pyspark.sql import functions as F
df_nyc = nyc_tlc_df.groupBy("passengerCount").agg(F.avg('tripDistance').
alias('AvgTripDistance'), F.sum('tripDistance').alias('SumTripDistance'))
display(df_nyc)
```

At this point, you could get even more advanced with your analysis by using popular libraries such as matplotlib and seaborn to render line charts, as shown in Figure 22-13.

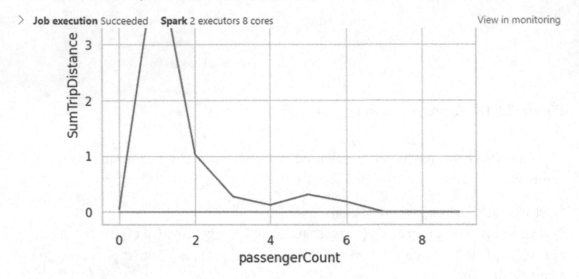

Figure 22-13. Customize data visualization with Spark and notebooks

Here is the code that has been executed in the notebook illustration shown in Figure 22-13:

```
import matplotlib.pyplot
import seaborn

seaborn.set(style = "whitegrid")
pdf_nyc = df_nyc.toPandas()
seaborn.lineplot(x="passengerCount", y="SumTripDistance" , data = pdf_nyc)
seaborn.lineplot(x="passengerCount", y="AvgTripDistance" , data = pdf_nyc)
matplotlib.pyplot.show()
```

Finally, clean up the resources by ending the connected session to ensure that the Spark instance is shut down. Note that the pool shuts down when the idle time specified in the Apache Spark pool is reached. You can also select to end the session from the status bar at the bottom of the notebook.

Query Data with SQL

Next, let's move on to the second sample that includes a sample script and a SQL on-demand pool to query data using SQL, as shown in Figure 22-14. A serverless SQL pool is a query service over the data in your data lake that uses a pay-per-query model. It enables you to access your data through T-SQL syntax and is optimized for querying and analyzing big data in ADLS Gen2. Since it is serverless, there is no infrastructure or clusters to maintain. It can be used by data engineers, data scientists, data analysts, and BI professionals to perform basic data exploration, transformation, and data abstraction on top of raw data or disparate data in the lake.

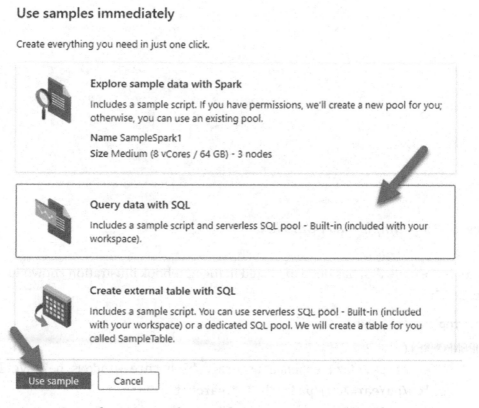

Figure 22-14. *Sample to query data with SQL in Synapse Studio*

Similar to the first sample, a new SQL script is created that demonstrates how to query the Azure Data Lake Storage Gen2 account by using standard SQL commands along with the OPENROWSET function, shown in Figure 22-15. OPENROWSET is a

T-SQL function that allows for reading data from many sources including using SQL Server's BULK import capability. The FORMAT argument within the code allows for PARQUET and CSV types.

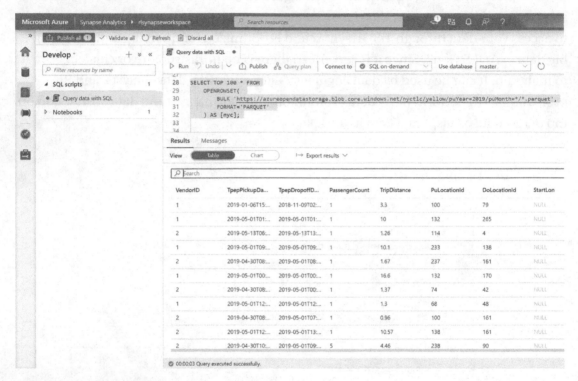

Figure 22-15. *Sample query data lake*

Here is the code that has been executed in the notebook illustration shown in Figure 22-15:

```
SELECT TOP 100 * FROM
    OPENROWSET(
        BULK 'https://azureopendatastorage.blob.core.windows.net/nyctlc/
        yellow/puYear=2019/puMonth=*/*.parquet',
        FORMAT='PARQUET'
    ) AS [nyc];
```

The next query, shown in Figure 22-16, adds more complexity by adding additional filters and clauses and does an excellent job at displaying the capabilities of Synapse Studio workspaces to directly query an Azure Data Lake Storage Gen2 account from a familiar SQL experience. Use SQL on demand to analyze Azure Open Datasets and visualize the results in Azure Synapse Studio. You can find more code samples in the following URL: `https://docs.microsoft.com/en-us/azure/synapse-analytics/sql/tutorial-data-analyst`.

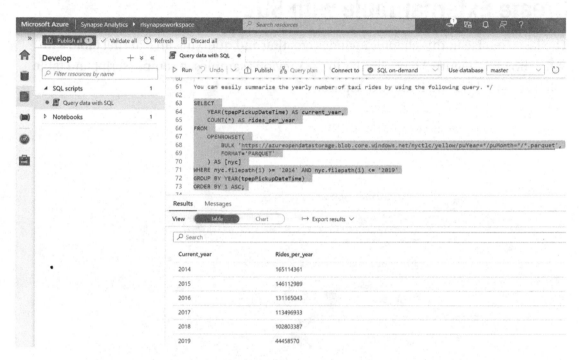

Figure 22-16. *More complex query of data lake*

Here is the code that has been executed in the notebook illustration shown in Figure 22-16:

```
SELECT
    YEAR(tpepPickupDateTime) AS current_year,
    COUNT(*) AS rides_per_year
FROM
    OPENROWSET(
        BULK 'https://azureopendatastorage.blob.core.windows.net/nyctlc/
        yellow/puYear=*/puMonth=*/*.parquet',
        FORMAT='PARQUET'
    ) AS [nyc]
```

```
WHERE nyc.filepath(1) >= '2014' AND nyc.filepath(1) <= '2019'
GROUP BY YEAR(tpepPickupDateTime)
ORDER BY 1 ASC;
```

Create External Table with SQL

The last exercise in this chapter includes creating and querying an external table with SQL, shown in Figure 22-17. Similar to the previous samples, a sample script and table will be created for you in the workspace.

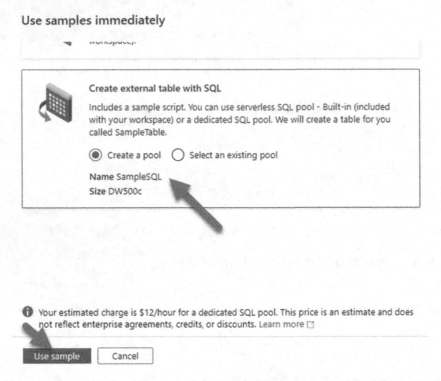

Figure 22-17. *Sample to create an external table in Synapse Studio*

The SQL code block shown in Figure 22-18, which is included in the sample, will demonstrate how to create an external table.

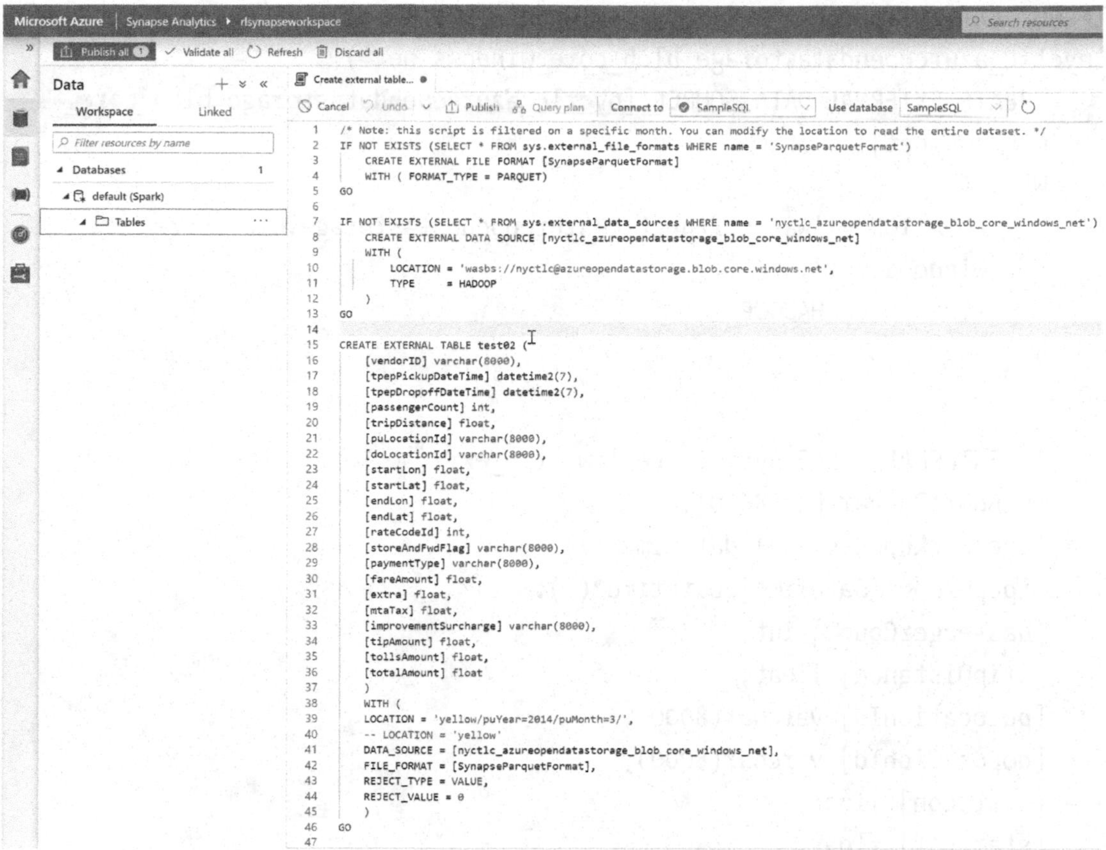

Figure 22-18. *Sample script to create external table*

Here is the code that you will need to run in the notebook illustration shown in Figure 22-18:

```
/* Note: this script is filtered on a specific month. You can modify the
location to read the entire dataset. */
IF NOT EXISTS (SELECT * FROM sys.external_file_formats WHERE name =
'SynapseParquetFormat')
    CREATE EXTERNAL FILE FORMAT [SynapseParquetFormat]
    WITH ( FORMAT_TYPE = PARQUET)
GO
```

```
IF NOT EXISTS (SELECT * FROM sys.external_data_sources WHERE name =
'nyctlc_azureopendatastorage_blob_core_windows_net')
    CREATE EXTERNAL DATA SOURCE [nyctlc_azureopendatastorage_blob_core_
    windows_net]
    WITH (
        LOCATION = 'wasbs://nyctlc@azureopendatastorage.blob.core.
        windows.net',
        TYPE      = HADOOP
    )
GO

CREATE EXTERNAL TABLE nyc_tlc_yellow_trip_ext (
    [vendorID] varchar(8000),
    [tpepPickupDateTime] datetime2(7),
    [tpepDropoffDateTime] datetime2(7),
    [passengerCount] int,
    [tripDistance] float,
    [puLocationId] varchar(8000),
    [doLocationId] varchar(8000),
    [startLon] float,
    [startLat] float,
    [endLon] float,
    [endLat] float,
    [rateCodeId] int,
    [storeAndFwdFlag] varchar(8000),
    [paymentType] varchar(8000),
    [fareAmount] float,
    [extra] float,
    [mtaTax] float,
    [improvementSurcharge] varchar(8000),
    [tipAmount] float,
    [tollsAmount] float,
    [totalAmount] float
    )
    WITH (
```

```
    LOCATION = 'yellow/puYear=2014/puMonth=3/',
    -- LOCATION = 'yellow'
    DATA_SOURCE = [nyctlc_azureopendatastorage_blob_core_windows_net],
    FILE_FORMAT = [SynapseParquetFormat],
    REJECT_TYPE = VALUE,
    REJECT_VALUE = 0
    )
GO

SELECT TOP 100 * FROM nyc_tlc_yellow_trip_ext
GO
```

After you create the table, run a simple SQL select statement, as shown in Figure 22-19, to query the newly created external table. Notice that there is also a "Query plan" button that is available.

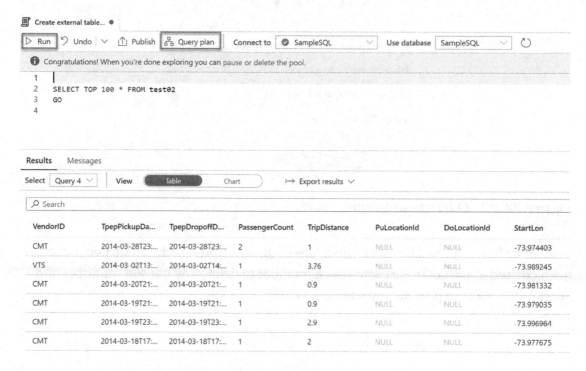

Figure 22-19. Sample to query external table

Here is the code that you will need to execute in the notebook illustration shown in Figure 22-19:

```
SELECT TOP 100 * FROM nyc_tlc_yellow_trip_ext
GO
```

There is the capability of downloading the query plan and opening it in SQL Server Management Studio (SSMS) as shown in Figure 22-20, which could be a useful feature offering when compared to analyzing query plans in the traditional SQL Server experience with SSMS.

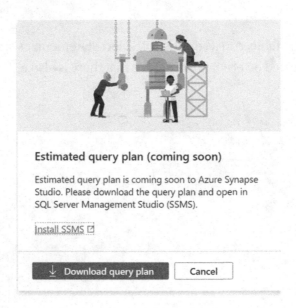

Figure 22-20. *Query plan feature*

Notice from Figure 22-21 that a dedicated SQL pool and Apache Spark pool have been created as part of these exercises. Remember to delete any unused resources and Spark/ SQL pools to prevent any additional costs.

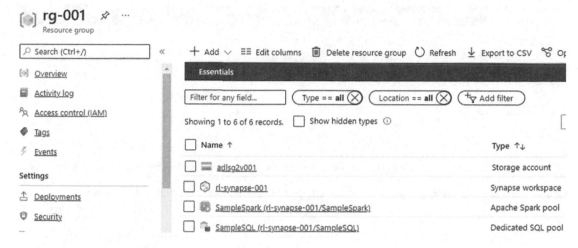

Figure 22-21. *Image of SQL and Spark pools created in Azure Portal from sample demo*

Summary

In this chapter, I demonstrated how to create a new Azure Synapse Analytics Studio workspace and then showed you how to create three samples from the Knowledge Center to achieve the following:

1. Explore data with Spark.

2. Query data with SQL.

3. Create an external table with SQL.

There are numerous other benefits of an Azure Synapse workspace that we have not explored in this chapter such as the capability to connect to Power BI accounts to visualize and build interactive reports; create, schedule, and monitor ETL jobs much like you would do in the Azure Data Factory experience; and search for your data assets by integrating with Azure Purview, a new data governance offering from Microsoft. Getting started on your journey with Azure Synapse Analytics couldn't be easier with all of the available resources and easy-to-use navigation UI shown in Figure 22-22. Azure Synapse Analytics workspaces truly have the potential of being the unified modern data and analytics go-to platform for a variety of business and technology professionals.

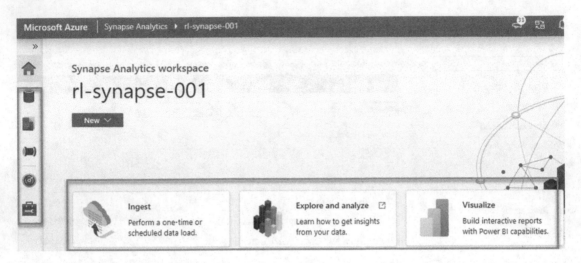

Figure 22-22. *Azure Synapse Analytics homepage*

CHAPTER 23

Machine Learning in Databricks

Organizations and developers that are seeking to leverage the power of machine learning (ML) and AI spend a significant amount of time building ML models and are seeking a method for streamlining their machine learning development lifecycle to track experiments, package code into reproducible runs, as well as build, share, and deploy ML models.

MLflow is an open source platform for managing the end-to-end machine learning lifecycle, as shown in Figure 23-1. It tackles four primary functions:

1. Tracks experiments to compare and record parameters and results

2. Packages ML code to share with other data scientists or transfer to production.

3. Manages and deploys ML models using a variety of available libraries

4. Registers models for easy model management and serves models to host them as REST endpoints

In this chapter, I will go over how to get started with MLflow using Azure Databricks. Databricks offers an integrated MLflow experience for tracking and securing ML model training runs and for running ML projects. In addition to providing a fully managed version of MLflow, Databricks integrates enterprise security features, high availability, and experiment and run management and notebook revision capture.

© Ron C. L'Esteve 2021
R. C. L'Esteve, *The Definitive Guide to Azure Data Engineering*, https://doi.org/10.1007/978-1-4842-7182-7_23

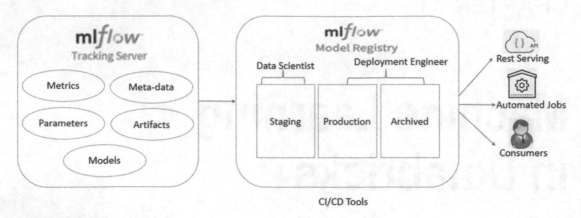

Figure 23-1. *Image depicting the capabilities of MLflow*

Create an MLflow Experiment

Let's begin by creating an MLflow experiment in Azure Databricks. Start by navigating to the Home menu and selecting "New MLflow Experiment," as shown in Figure 23-2.

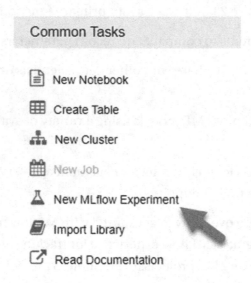

Figure 23-2. *Create a new MLflow experiment from the Home menu*

Select "New MLflow Experiment" to open a new "Create MLflow Experiment" UI. Populate the name of the experiment, and then create it, as shown in Figure 23-3.

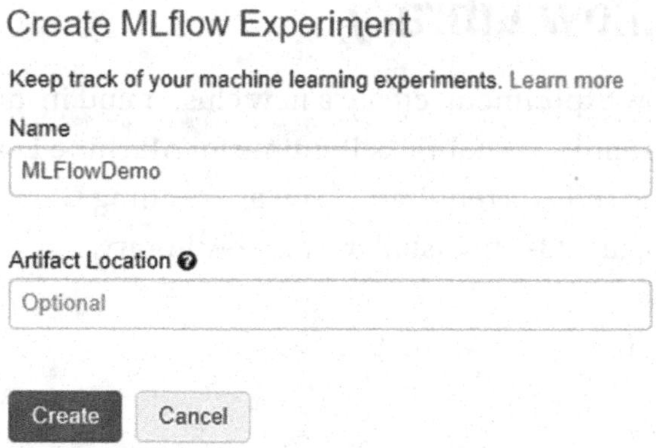

Figure 23-3. *Create experiment for MLflow*

Once the experiment is created, it will have an experiment ID that is associated with it that you will need when you configure your notebook containing the ML model code. Notice the Experiment UI shown in Figure 23-4 contains options to compare multiple models, filter by model status, search by criteria, and download CSV details.

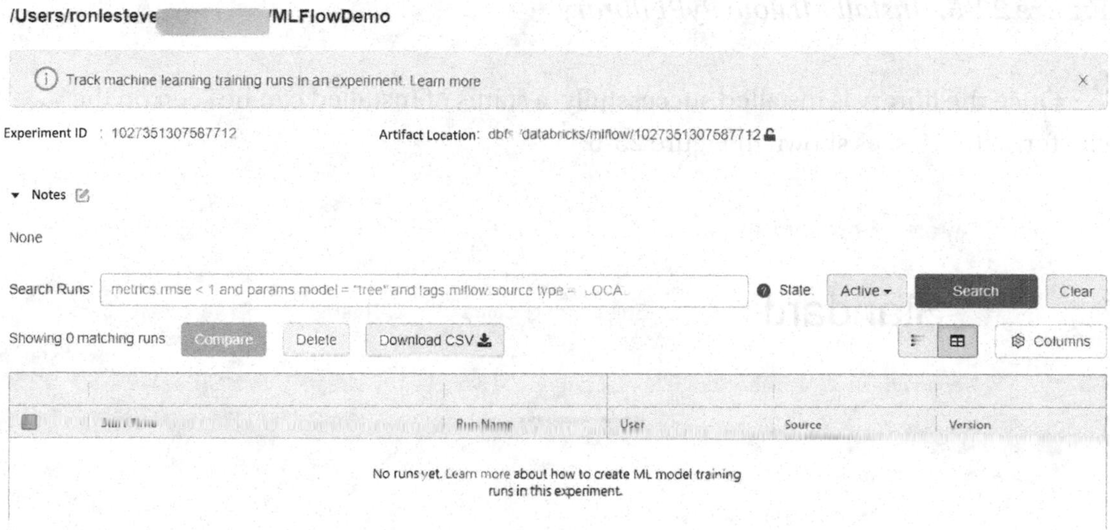

Figure 23-4. *Sample image of the Experiment UI*

Install the MLflow Library

After creating an MLflow experiment, create a new cluster and install the MLflow PyPI library on it. If you are running Databricks Runtime for Machine Learning, MLflow is already installed, and no setup is required. If you are running Databricks Runtime, follow these steps shown in Figure 23-5 to install the MLflow library.

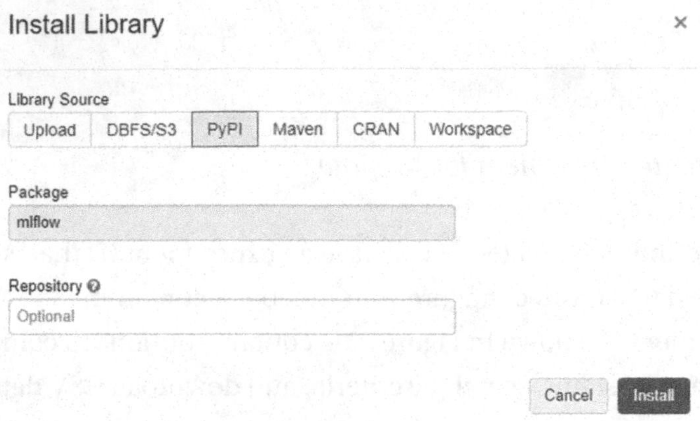

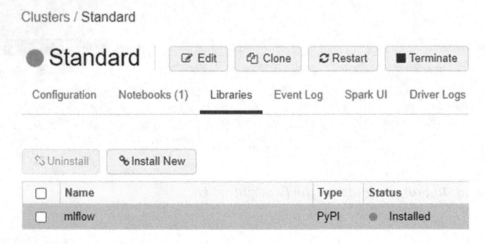

Figure 23-5. *Install MLflow PyPI library*

Once the library is installed successfully, a status of Installed can be seen on the cluster's libraries, as shown in Figure 23-6.

Clusters / Standard

● **Standard**   ☑ Edit   🗐 Clone   ♻ Restart   ■ Terminate

Configuration   Notebooks (1)   Libraries   Event Log   Spark UI   Driver Logs

⤺ Uninstall   ⚲ Install New

	Name	Type	Status
☐	mlflow	PyPI	● Installed

Figure 23-6. *Installed MLflow library*

Create a Notebook

After installing an experiment, a cluster, and the MLflow library, you'll need to create a new notebook that you can use to build the ML model and then associate it with the MLflow experiment. Note that Databricks automatically creates a notebook experiment if there is no active experiment when you start a run.

Use the following code to start the run. Since you have an experiment ID, start your run using

```
mlflow.start_run().
mlflow.start_run(experiment_id=1027351307587712)
```

Alternatively, you can define the experiment name as follows, which will simply use the experiment name rather than ID:

```
experiment_name = "/Demo/MLFlowDemo/"
mlflow.set_experiment(experiment_name)
```

Selective Logging

From a logging perspective, there is the option to auto-log model-specific metrics, parameters, and artifacts. Alternatively, it is possible to define these metrics, parameters, or models by adding the following commands to the notebook code as desired:

Numerical metrics: `mlflow.log_metric("accuracy", 0.9)`
Training parameters: `mlflow.log_param("learning_rate", 0.001)`
Models: `mlflow.sklearn.log_model(model, "myModel")`
Other artifacts (files): `mlflow.log_artifact("/tmp/my-file", "myArtifactPath")`

The following code imports a dataset from scikit-learn and creates the training and test datasets. This code is meant to demonstrate a sample ML model. Add this to a code block in the Databricks notebook:

```
from sklearn.model_selection import train_test_split
from sklearn.datasets import load_diabetes

db = load_diabetes()
X = db.data
```

```
y = db.target
X_train, X_test, y_train, y_test = train_test_split(X, y)
```

This next block of code will import MLflow and sklearn, start the experiment, and log details of the execution run. Simply add this code to your Databricks notebook and run it:

```
import mlflow
import mlflow.sklearn
from sklearn.ensemble import RandomForestRegressor
from sklearn.metrics import mean_squared_error

# In this run, neither the experiment_id nor the experiment_name parameter
  is provided. MLflow automatically creates a notebook experiment and logs
  runs to it.
# Access these runs using the Experiment sidebar. Click Experiment at the
  upper right of this screen.
with mlflow.start_run(experiment_id=1027351307587712):
  n_estimators = 100
  max_depth = 6
  max_features = 3
  # Create and train model
  rf = RandomForestRegressor(n_estimators = n_estimators, max_depth = max_
  depth, max_features = max_features)
  rf.fit(X_train, y_train)
  # Make predictions
  predictions = rf.predict(X_test)

  # Log parameters
  mlflow.log_param("num_trees", n_estimators)
  mlflow.log_param("maxdepth", max_depth)
  mlflow.log_param("max_feat", max_features)

  # Log model
  mlflow.sklearn.log_model(rf, "random-forest-model")
```

```
# Create metrics
mse = mean_squared_error(y_test, predictions)

# Log metrics
mlflow.log_metric("mse", mse)
```

Figure 23-7 shows the defined parameters, metrics, and model that need to be logged.

```
from sklearn.metrics import mean_squared_error

# In this run, neither the experiment_id nor the experiment_name parameter is provided. MLflow automatically
creates a notebook experiment and logs runs to it.
# Access these runs using the Experiment sidebar. Click Experiment at the upper right of this screen.
with mlflow.start_run():
  n_estimators = 100
  max_depth = 6
  max_features = 3
  # Create and train model
  rf = RandomForestRegressor(n_estimators = n_estimators, max_depth = max_depth, max_features = max_features)
  rf.fit(X_train, y_train)
  # Make predictions
  predictions = rf.predict(X_test)

  # Log parameters
  mlflow.log_param("num_trees", n_estimators)
  mlflow.log_param("maxdepth", max_depth)
  mlflow.log_param("max_feat", max_features)

  # Log model
  mlflow.sklearn.log_model(rf, "random-forest-model")

  # Create metrics
  mse = mean_squared_error(y_test, predictions)

  # Log metrics
  mlflow.log_metric("mse", mse)
```

Figure 23-7. *Defined parameters in the code*

Click the icon highlighted in Figure 23-8 to see the experiment run details. Additionally, click Experiment UI at the bottom to open the UI. This is particularly useful if there is no experiment created and a new random experiment will be created for you.

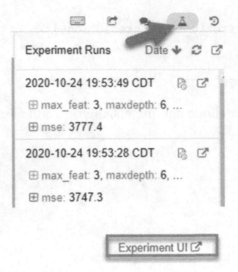

Figure 23-8. *Click to open the UI*

Since there is already an MLflow experiment that was created earlier, open it by clicking MLFlowDemo as shown in Figure 23-9.

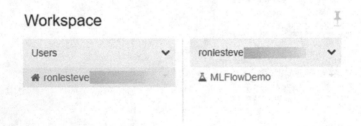

Figure 23-9. *Open the MLflow experiment*

Notice that the experiment ran twice and notice the parameters and metrics associated with it. Additionally, there is now the option to compare the two runs, as shown in Figure 23-10.

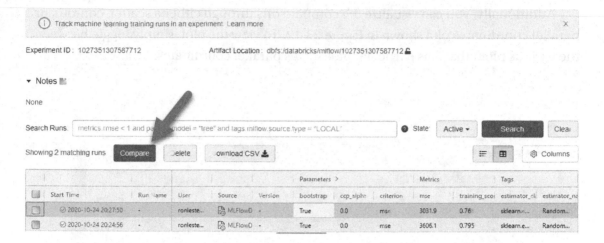

Figure 23-10. *Compare multiple runs*

Click Compare and notice all the associated details between the two runs, which are shown in Figure 23-11.

Run ID:	b90f657741c440c399d4ce0b990ba100	e436b568370f405bb2b2d2a7236cb1a9
Run Name:		
Start Time:	2020-10-24 20:27:50	2020-10-24 20:24:56
Parameters		
bootstrap	True	True
ccp_alpha	0.0	0.0
criterion	mse	mse
max_depth	6	6
max_feat	3	3
max_features	3	3
max_leaf_nodes	None	None
max_samples	None	None
maxdepth	6	6
min_impurity_decrease	0.0	0.0
min_impurity_split	None	None

Figure 23-11. *Comparing parameters from two model runs*

Additionally, you can visualize the comparisons through either a scatter, contour, or parallel coordinates plot shown in Figure 23-12. To view the plot, simply toggle between the various plot tabs. This particular plot shows parallel coordinates.

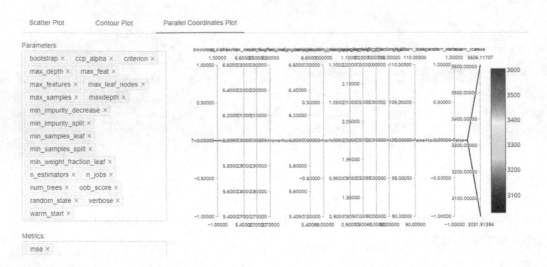

Figure 23-12. *Sample plot 1*

Auto-logging

After testing the defined logging option, also test auto-logging. You can do that using the following variation of the code shown in Figure 23-13, which basically replaces all of the defined metrics, parameters, and so forth with a call to `mlflow.sklearn.autolog()`.

```python
1   import mlflow
2   import mlflow.sklearn
3   from sklearn.ensemble import RandomForestRegressor
4   from sklearn.metrics import mean_squared_error
5
6   mlflow.sklearn.autolog()
7
8   # In this run, neither the experiment_id nor the experiment_name parameter is provided. MLflow automatically
    creates a notebook experiment and logs runs to it.
9   # Access these runs using the Experiment sidebar. Click Experiment at the upper right of this screen.
10  with mlflow.start_run():
11      n_estimators = 100
12      max_depth = 6
13      max_features = 3
14      # Create and train model
15      rf = RandomForestRegressor(n_estimators = n_estimators, max_depth = max_depth, max_features = max_features)
16      rf.fit(X_train, y_train)
17      # Make predictions
18      predictions = rf.predict(X_test)
```

Figure 23-13. *Auto-log in code example*

Here is the code that you will need to run in the Databricks notebook as shown in Figure 23-13:

```
import mlflow
import mlflow.sklearn
from sklearn.ensemble import RandomForestRegressor
from sklearn.metrics import mean_squared_error

mlflow.sklearn.autolog()

# In this run, neither the experiment_id nor the experiment_name parameter
  is provided. MLflow automatically creates a notebook experiment and logs
  runs to it.
# Access these runs using the Experiment sidebar. Click Experiment at the
  upper right of this screen.
with mlflow.start_run():
  n_estimators = 100
  max_depth = 6
  max_features = 3
  # Create and train model
  rf = RandomForestRegressor(n_estimators = n_estimators, max_depth = max_
  depth, max_features = max_features)
  rf.fit(X_train, y_train)
  # Make predictions
  predictions = rf.predict(X_test)
```

After running this new model twice and navigating back to the MLflow experiment, notice the new runs logged in the Experiment UI, and this time there are a few differences in the metrics captured, as expected and shown in Figure 23-14.

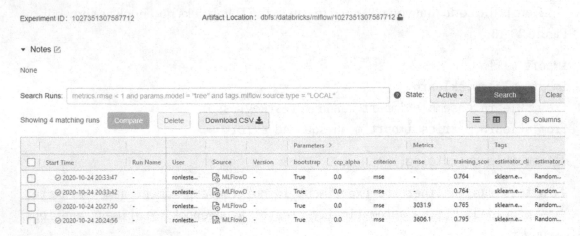

Figure 23-14. *Experiment UI 2*

Click into one of the runs to view a variety of details including plots, which are shown in Figure 23-15.

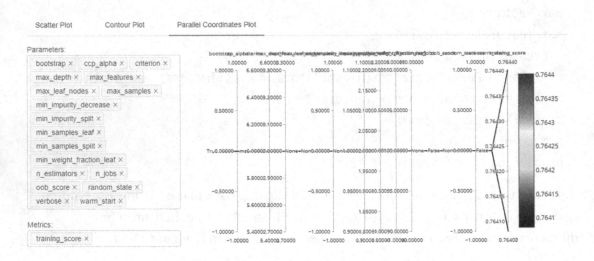

Figure 23-15. *Sample plot 2*

Register a Model

After creating, logging, and comparing models, you can register the models by clicking the desired logged model and scrolling down to the artifacts in the run details, as shown in Figure 23-16.

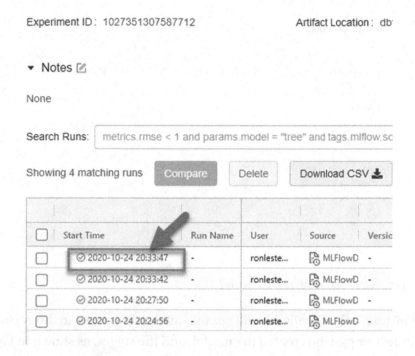

Figure 23-16. *Select model from Experiment UI*

Select the model and click Register Model, as shown in Figure 23-17.

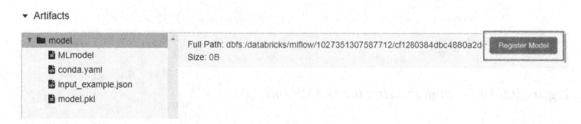

Figure 23-17. *Click to register model from artifacts*

Within the Register Model UI shown in Figure 23-18, create a new model, name it, and register the model.

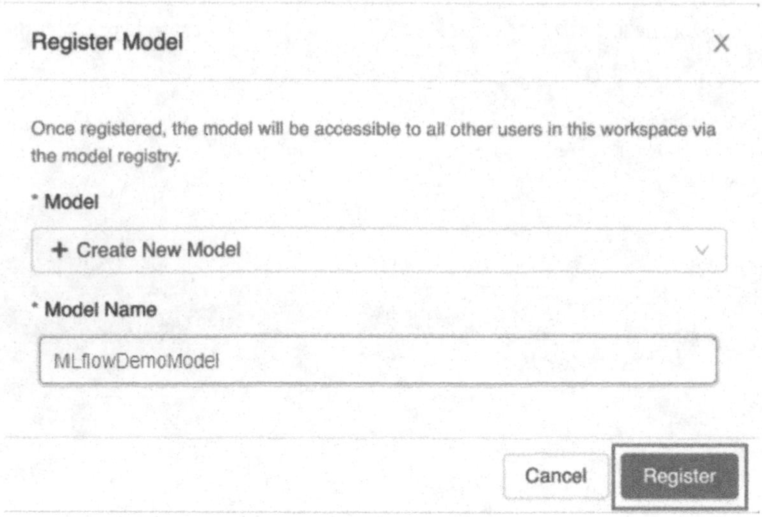

Register Model   ✕

Once registered, the model will be accessible to all other users in this workspace via
the model registry.

* Model

    ＋ Create New Model   ⌄

* Model Name

    MLflowDemoModel

                                              Cancel   Register

Figure 23-18. *Steps to register a model*

Once registered, the Details tab will contain all details related to the versions,
registered timestamps, who created the model, and the stages, as shown in Figure 23-19.

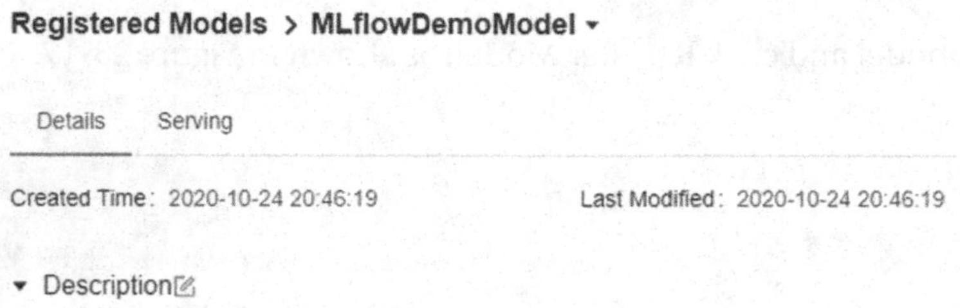

Registered Models > **MLflowDemoModel** ▾

Details   Serving

Created Time : 2020-10-24 20:46:19   Last Modified : 2020-10-24 20:46:19

▾ Description ✎

Figure 23-19. *Image showing the Details tab*

Figure 23-20 shows the various versions along with other details for each stage.

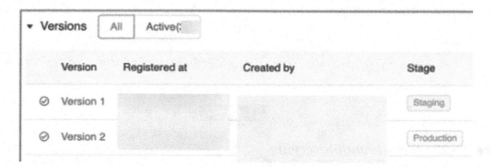

Figure 23-20. *Image showing the version details*

Figure 23-21 shows how stages can be transitioned from Staging to Production to Archived.

Figure 23-21. *Steps to move from one stage to another*

When model serving is enabled, Databricks automatically creates a unique single-node cluster for the model and deploys all non-archived versions of the model on that cluster. Start by clicking Enable Serving shown in Figure 23-22.

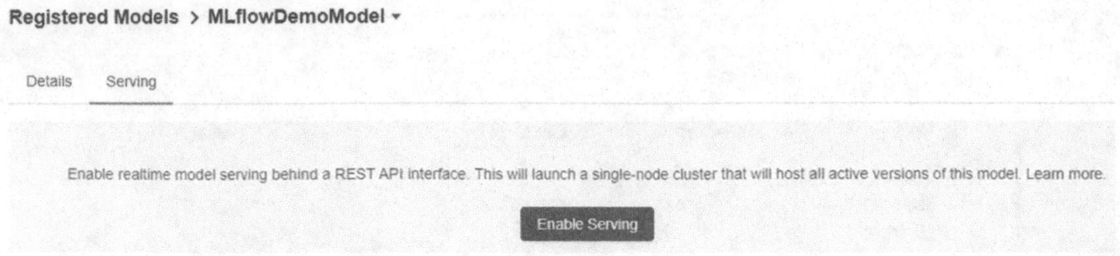

Registered Models > MLflowDemoModel ▾

Details	Serving

Enable realtime model serving behind a REST API interface. This will launch a single-node cluster that will host all active versions of this model. Learn more.

Enable Serving

Figure 23-22. *Steps to enable serving*

At this point, you'll be able to see Serving and Details tabs, as shown in Figure 23-23. MLflow model serving allows hosting registered ML models as REST endpoints that are updated automatically based on the availability of model versions and their stages.

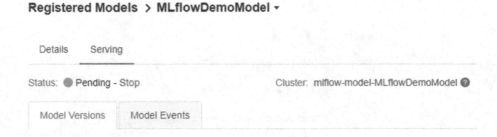

Figure 23-23. *Image showing the new Serving tab properties*

A sample Serving page may look like the illustration shown in Figure 23-24.

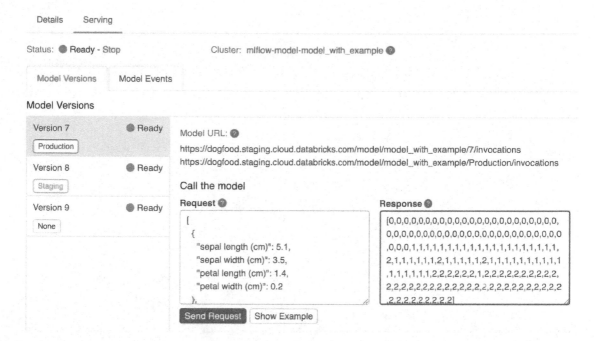

Figure 23-24. *Sample image of Serving tab*

Finally, when a model has been registered, you can find it by navigating to the Models tab in the Databricks UI, as shown in Figure 23-25.

Figure 23-25. *Image of the "Models" tab icon*

Summary

In this chapter, I demonstrated how to create an MLflow experiment, install the MLflow library, and create an MLflow notebook. I also showed you practical examples of selective and auto-logging and finally showed you how to register a model. With MLflow's integration through the Databricks experience, you now have the capability to manage the ML lifecycle, which includes experimentation, reproducibility, deployment, and central model registry.

PART VI

Data Governance

CHAPTER 24

Purview for Data Governance

Numerous organizations are needing to establish data governance processes, standards, and methodologies and have been able to do this with on-premises SQL Server tools such as Master Data Services. However, there has been a major gap in the Azure space for such data governance products. Previously, Microsoft has attempted to bring data governance to Azure through hosting an MDS database on an Azure managed instance or through Azure Data Catalog, which hasn't really been a full-fledged data governance product. Azure Purview is a cloud-based data governance product that is used to centrally manage data governance across your data estate, spanning both cloud and on-premises environments.

Azure Purview's easy-to-use UI and catalog makes data sources easily discoverable and understandable by the users who manage the data. Azure Purview provides a cloud-based service into which users can register data sources while maintaining a copy of the indexed metadata as well as a reference to the source location. Additionally, this metadata can be further enriched in Purview through tags, descriptions, and more. Azure Purview is intended to address some of the challenges for data consumers and producers. In this chapter, I will show you how to get started with Azure Purview and then demonstrate how to explore some of the features and capabilities within Purview Studio.

© Ron C. L'Esteve 2021
R. C. L'Esteve, *The Definitive Guide to Azure Data Engineering*, https://doi.org/10.1007/978-1-4842-7182-7_24

Create Azure Purview Account

Azure Purview offers the capability of automating and managing metadata, accelerating classification at scale, enabling discovery of data assets by data consumers, assigning ownership, managing sensitive data, and much more. In this section, you will learn how to create an Azure Purview account. Azure Purview can be accessed via Azure Portal, as shown in Figure 24-1. Go ahead and click Create to begin the process of creating a new Purview account.

Figure 24-1. Create new Purview account

As with all Azure resources, Purview also requires both project and instance details, as depicted in Figure 24-2.

Create Purview account   ...

Provide Purview account info

* Basics   * Networking   * Configuration   Tags   Review + Create

Create a Purview account to develop a data governance solution in just a few clicks. A storage account and eventhub will be created in a managed resource group in your subscription for catalog ingestion scenarios. Learn more [↗]

Project details

Subscription *	MSDN Platforms Subscription ⌄
Resource group *	rg-001 ⌄
	Create new

Instance details

Purview account name * ⓘ	demo-purview-001 ✓
Location *	East US ⌄

[Review + Create]   [Previous] [Next: Networking >]

Figure 24-2. *New Purview account details*

Additional configuration options are also available for selection, as shown in Figure 24-3:

- Platform size: The following link (`https://azure.microsoft.com/ en-us/pricing/details/azure-purview/`) captures capacity units, which are a provisioned set of resources to keep Purview Data Map up to date and running. A minimum of four units will need to be selected.

- Catalog and data features: Allows you to specify the feature set levels for the Purview account. There are a few available options including business glossary and lineage visualization, source registration, data discovery, automated scanning and classification, and catalog and sensitive data insights.

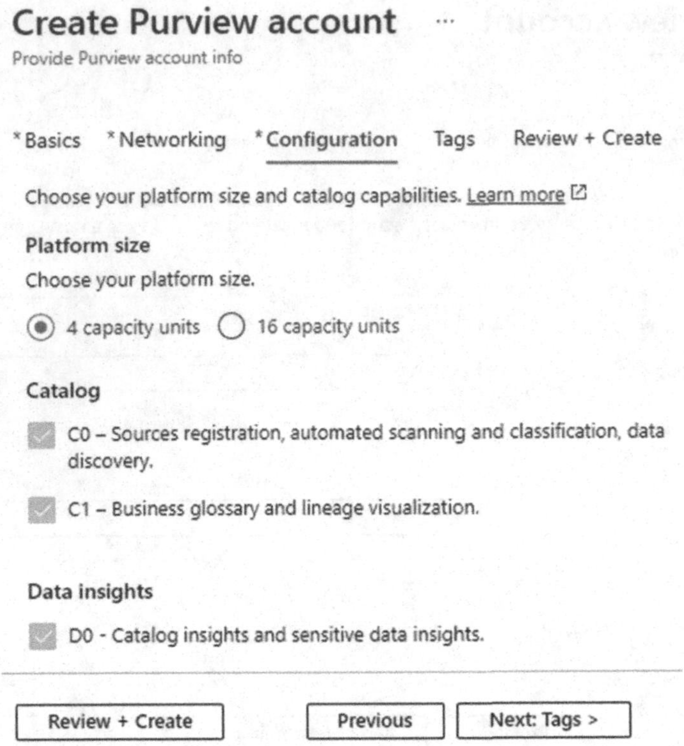

Create Purview account ⋯

Provide Purview account info

* Basics   * Networking   * Configuration   Tags   Review + Create

Choose your platform size and catalog capabilities. Learn more ⧉

Platform size

Choose your platform size.

⦿ 4 capacity units   ◯ 16 capacity units

Catalog

☑ C0 – Sources registration, automated scanning and classification, data discovery.

☑ C1 – Business glossary and lineage visualization.

Data insights

☑ D0 - Catalog insights and sensitive data insights.

[Review + Create]   [Previous]   [Next: Tags >]

Figure 24-3. *New Purview account config details*

Initially, you may run into errors, shown in Figure 24-4, when trying to create Azure Purview instances as Purview will need to be added as a provider at the subscription level. Also, users must have the appropriate Azure Active Directory permissions to Azure Purview. A Purview Data Reader has access to Purview Studio and can read all content except for scan bindings. A Purview Data Curator has the same access as the Data Reader with the added privilege of being able to edit and apply classifications and glossary terms to assets. Finally, the Purview Data Source Administrator does not have access to Purview Studio and does not have read or write access to content in Purview other than scanning.

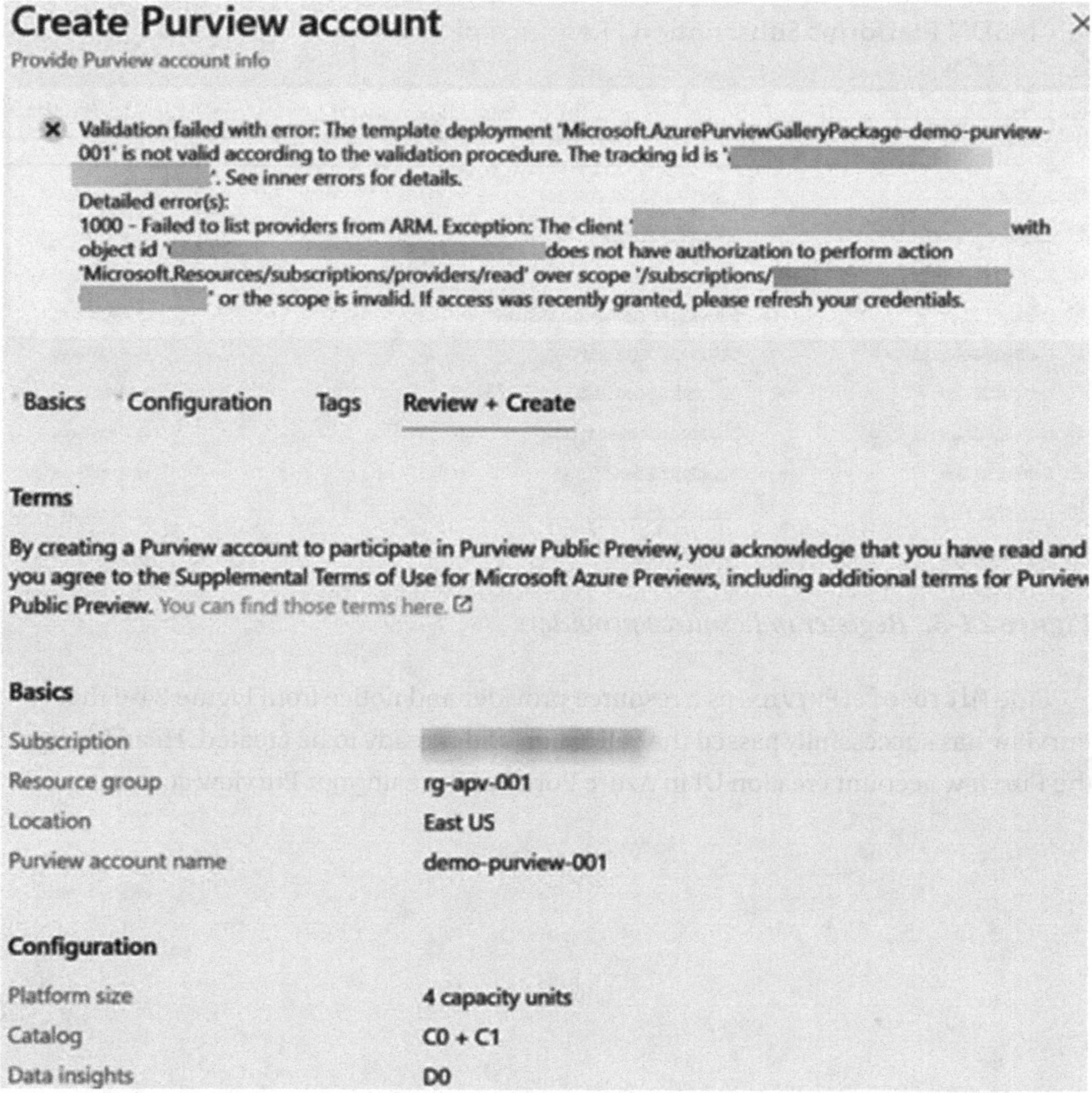

Create Purview account   ✕

Provide Purview account info

> ✕ Validation failed with error: The template deployment 'Microsoft.AzurePurviewGalleryPackage-demo-purview-001' is not valid according to the validation procedure. The tracking id is '▮▮▮▮▮▮▮▮▮▮'. See inner errors for details.
> Detailed error(s):
> 1000 - Failed to list providers from ARM. Exception: The client '▮▮▮▮▮▮▮▮▮▮▮▮' with object id '▮▮▮▮▮▮▮▮▮▮▮▮▮▮' does not have authorization to perform action 'Microsoft.Resources/subscriptions/providers/read' over scope '/subscriptions/▮▮▮▮▮▮▮▮' or the scope is invalid. If access was recently granted, please refresh your credentials.

Basics   Configuration   Tags   **Review + Create**

Terms

By creating a Purview account to participate in Purview Public Preview, you acknowledge that you have read and you agree to the Supplemental Terms of Use for Microsoft Azure Previews, including additional terms for Purview **Public Preview.** You can find those terms here. ☑

Basics

Subscription	▮▮▮▮▮
Resource group	**rg-apv-001**
Location	**East US**
Purview account name	**demo-purview-001**

Configuration

Platform size	**4 capacity units**
Catalog	**C0 + C1**
Data insights	**D0**

Figure 24-4. Error when creating new Purview account

Figure 24-5 illustrates the steps required to add Purview as a resource provider at the subscription level. First, navigate to your subscription where Azure Purview will be deployed. Next, click Resource providers, and finally search for and register `Microsoft.Purview` as a resource provider.

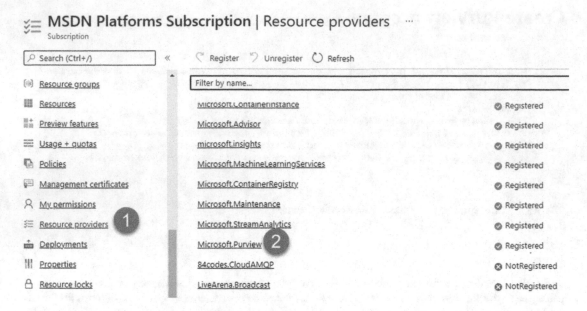

Figure 24-5. *Register in Resource providers*

Add `Microsoft.Purview` as a resource provider and notice from Figure 24-6 that Purview has successfully passed the validation and is ready to be created. Head back to the Purview account creation UI in Azure Portal and create your Purview account.

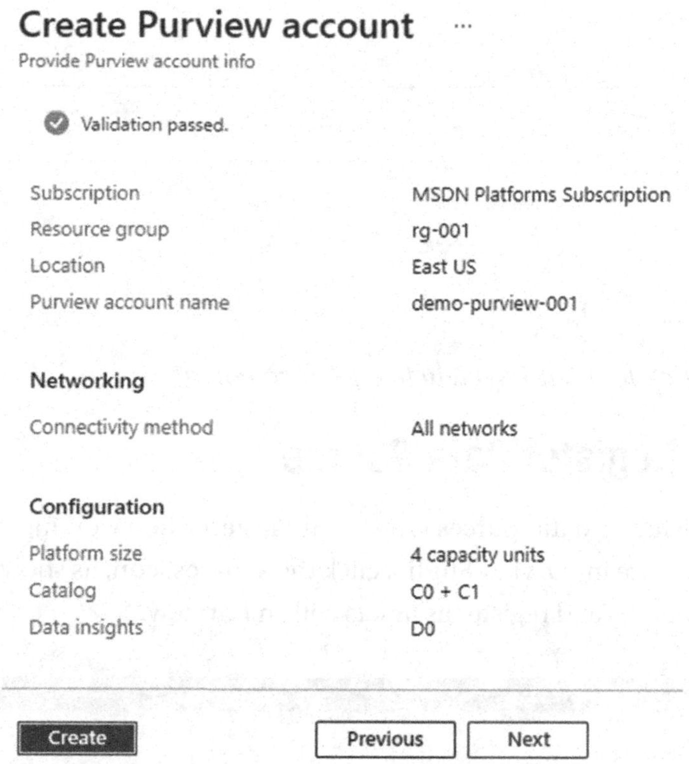

Create Purview account ···

Provide Purview account info

✅ Validation passed.

Subscription	MSDN Platforms Subscription
Resource group	rg-001
Location	East US
Purview account name	demo-purview-001

Networking

Connectivity method	All networks

Configuration

Platform size	4 capacity units
Catalog	C0 + C1
Data insights	D0

[Create]   [Previous] [Next]

Figure 24-6. *Validation passed for creating Purview account*

Explore Azure Purview

Now that you have successfully created a Purview account with the relevant access, you can begin exploring the various features of Purview including creating and registering data sources, managing credentials and access, creating scans, exploring the glossary, and browsing assets. Start by clicking "Open Purview Studio," shown in Figure 24-7, to launch Azure Purview.

Figure 24-7. *Click to open Purview from Azure Portal*

Create and Register Data Source

Creating and registering data sources is a key initial step when working in Azure Purview. Once you are in Purview Studio, click the sources icon, as shown in Figure 24-8, to add the data sources and register as assets within Purview.

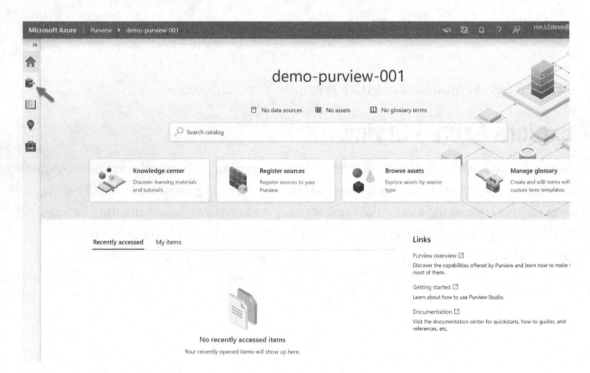

Figure 24-8. *Purview Studio homepage*

To add a new source, first create a collection by completing the following steps, which are illustrated in Figure 24-9:

1. Click sources.

2. Click New collection.

3. Give the new collection a name.

4. Click Finish.

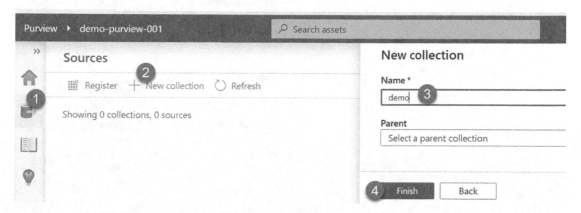

Figure 24-9. *Steps to create new Purview collection*

Next, you will need to register a source by searching and selecting from the list of registered sources shown in Figure 24-10. As Purview continues to mature, additional sources will be added with the hope being that it would also be able to accommodate multiple sources outside of Azure.

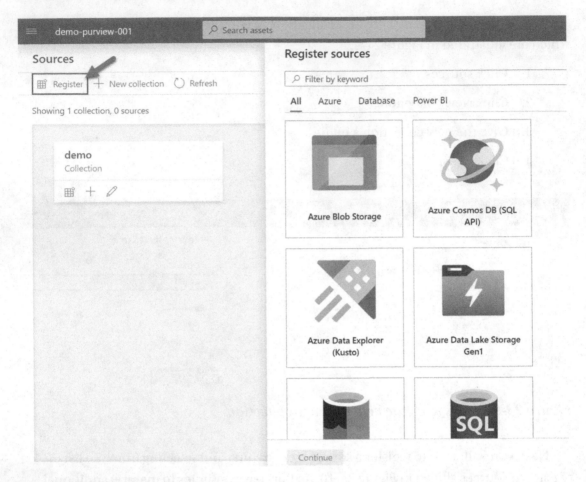

Figure 24-10. *Step to register sources*

Select Azure Data Lake Storage Gen2 as your source. Also, enter the details for the source, as shown in Figure 24-11.

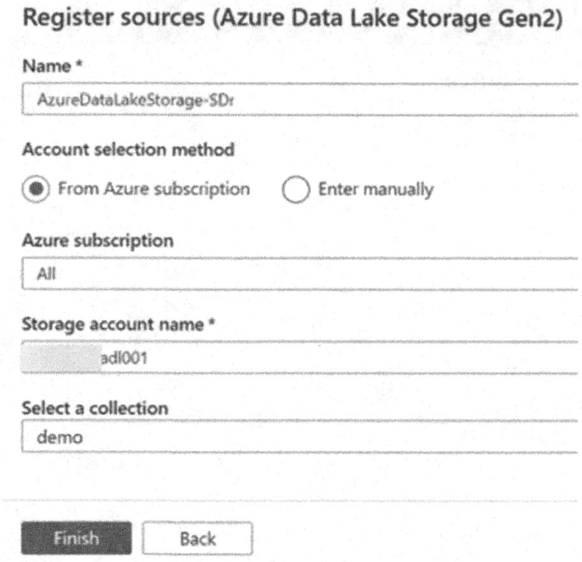

Register sources (Azure Data Lake Storage Gen2)

Name *

AzureDataLakeStorage-SDr

Account selection method

(●) From Azure subscription   ( ) Enter manually

Azure subscription

All

Storage account name *

adl001

Select a collection

demo

[Finish]   [Back]

Figure 24-11. *Register ADLS Gen2*

Notice in Figure 24-12 that the sources are added to the collection under which they are registered. A collection can have multiple sources, and there can be multiple collections in the canvas.

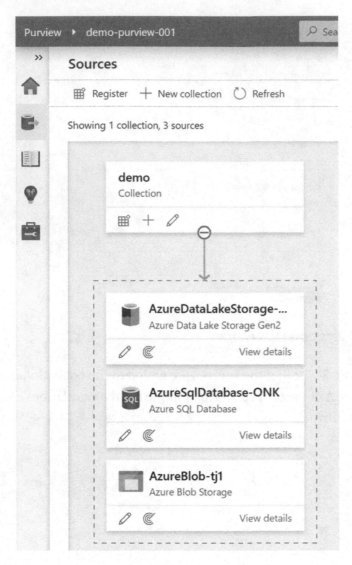

Figure 24-12. *Display of Purview collection of sources*

Manage Credentials and Access

Within Azure Purview, credentials are needed to quickly reuse and apply saved authentication information to your data source scans. Purview enforces the need to use Key Vault to store passwords and secrets as a minimum requirement to store credentials (https://docs.microsoft.com/en-us/azure/purview/manage-credentials).

Follow the steps illustrated in Figure 24-13 to add a new Key Vault credential. (1) Start by clicking the Management Center icon, (2) click Credentials, (3) click New, (4) name your Key Vault connection, (5) specify your Key Vault name, and (6) click Create.

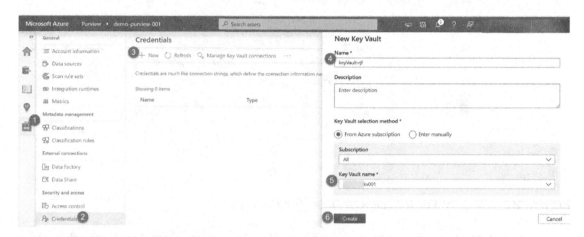

Figure 24-13. *Steps to register Purview credentials and Key Vault*

Purview will display the message shown in Figure 24-14 to confirm granting Purview access to Key Vault. Click Confirm to continue.

Figure 24-14. *Steps to grant access to Purview*

Figure 24-15 illustrates steps taken to provide the necessary GET access permissions to Purview from Azure Key Vault through Azure Portal. Simply navigate to the Key Vault account being used, select Access policies, and choose Add Access Policy.

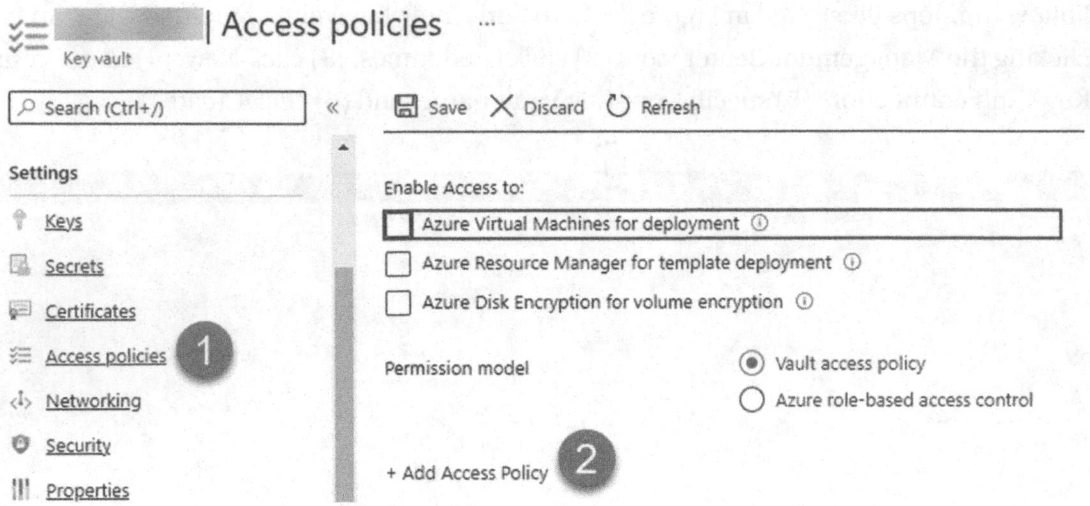

Figure 24-15. *Key Vault access policies for Purview*

Figure 24-16 shows the steps required to add an access policy. Select Get within Secret permissions, select your Purview account from the Select principal section, and click Add.

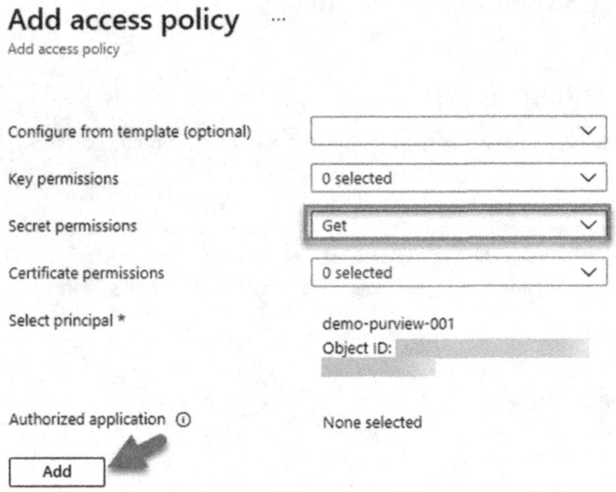

Figure 24-16. *Add access to Purview for Key Vault*

Once the connections are created and verified in Purview, they will appear in the "Manage Key Vault connections" section, as depicted in Figure 24-17.

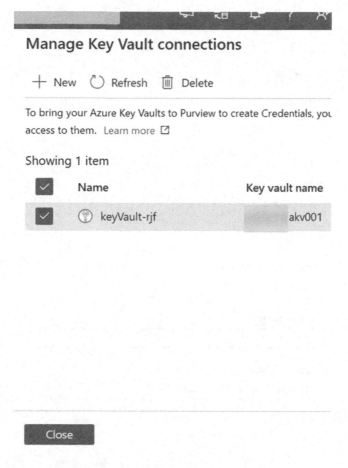

Figure 24-17. *Key Vault now visible in Purview*

The Key Vault connection can then be configured within the credential authentication UIs when adding new data sources and connections. For example, Figure 24-18 shows how to use Key Vault to extract the account key for the respective ADLS Gen2 account.

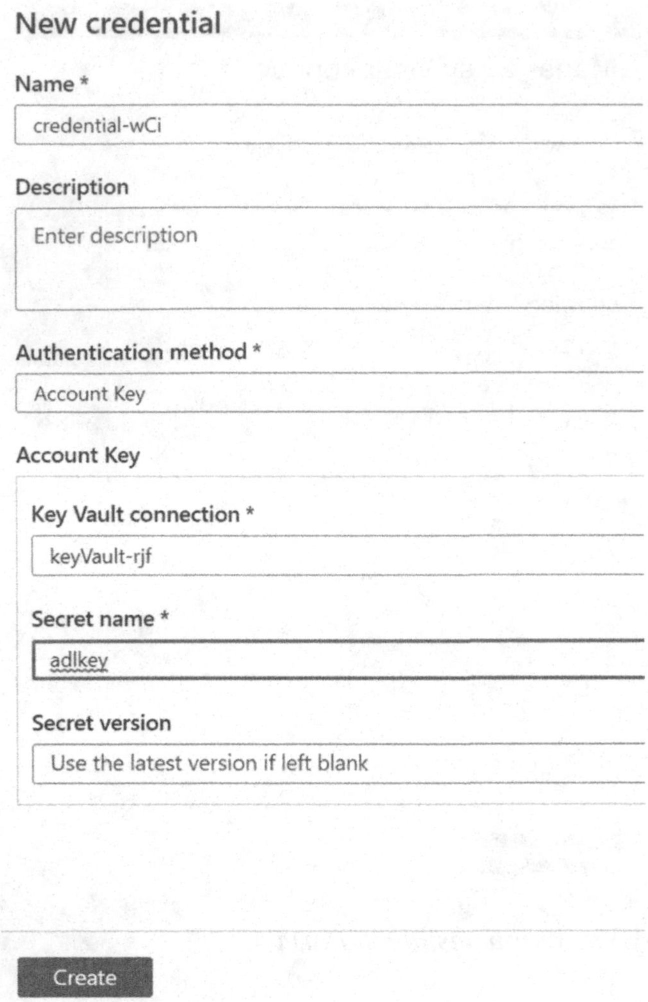

New credential

Name *

credential-wCi

Description

Enter description

Authentication method *

Account Key

Account Key

Key Vault connection *

keyVault-rjf

Secret name *

adlkey

Secret version

Use the latest version if left blank

Create

Figure 24-18. *Add credential for ADLS Gen2*

Create a Scan

Within the Azure Purview catalog, there is the capability to create scan rule sets to enable users to quickly scan data sources within the organization. A scan rule set is a container for grouping a set of scan rules together to easily associate them with a scan. To create a new scan for your sources, click the scan icon shown in Figure 24-19 and then populate the required credential information.

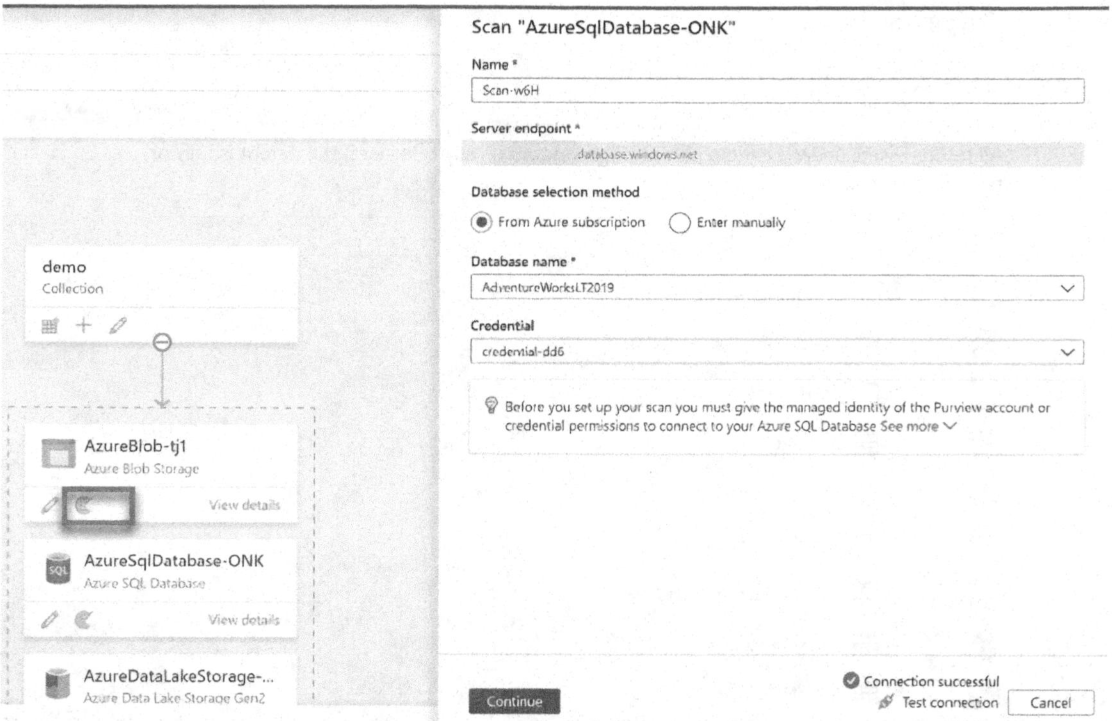

Figure 24-19. *Step to scan sources for Purview*

Additionally, Figure 24-20 shows how you can customize and alter the scope of the scan to either select or deselect objects within the source.

Scope your scan

↻ Refresh

All future assets under a certain parent will be automatically selected if the parent is fully or partially checked.

Search

☑ ∨ 📇 AzureSqlDatabase-ONK

☑   ▦ SalesLT.SalesOrderHeader

☑   ▦ dbo.BuildVersion

☑   ▦ dbo.ErrorLog

☑   ▦ SalesLT.Address

☑   ▦ SalesLT.Customer

☑   ▦ SalesLT.CustomerAddress

☑   ▦ SalesLT.Product

☑   ▦ SalesLT.ProductCategory

☑   ▦ SalesLT.ProductDescription

☑   ▦ SalesLT.ProductModel

☑   ▦ SalesLT.ProductModelProductDescription

☑   ▦ SalesLT.SalesOrderDetail

Continue   Back   Cancel

Figure 24-20. *Select tables as needed*

A scan rule set can utilize either a default rule set to include all supported system classification rules, as shown in Figure 24-21, or you can customize the scope of this rule set.

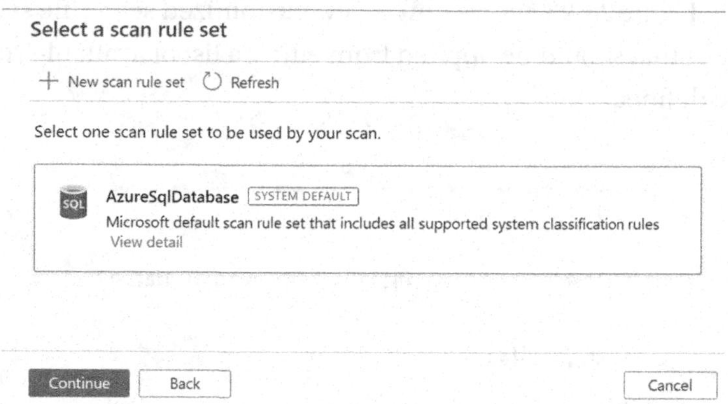

Select a scan rule set

＋ New scan rule set   ⟳ Refresh

Select one scan rule set to be used by your scan.

> **AzureSqlDatabase** [SYSTEM DEFAULT]
> Microsoft default scan rule set that includes all supported system classification rules
> View detail

Continue   Back   Cancel

Figure 24-21. *Select a scan rule set*

Figure 24-22 shows a sample list of classification system rules that are available.

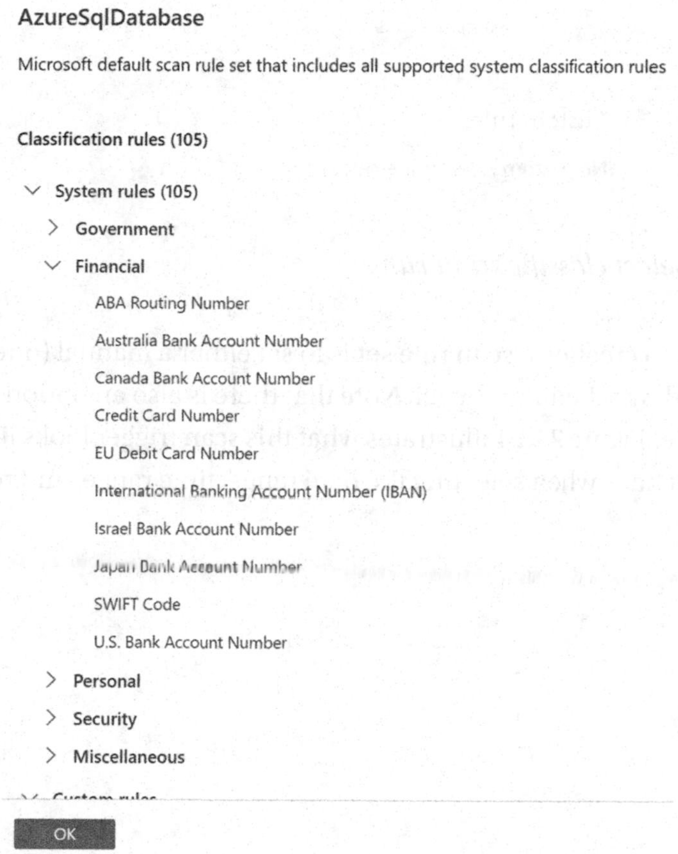

AzureSqlDatabase

Microsoft default scan rule set that includes all supported system classification rules

Classification rules (105)

∨ **System rules (105)**

  ＞ **Government**

  ∨ **Financial**

      ABA Routing Number

      Australia Bank Account Number

      Canada Bank Account Number

      Credit Card Number

      EU Debit Card Number

      International Banking Account Number (IBAN)

      Israel Bank Account Number

      Japan Bank Account Number

      SWIFT Code

      U.S. Bank Account Number

  ＞ **Personal**

  ＞ **Security**

  ＞ **Miscellaneous**

  ∨ Custom rules

OK

Figure 24-22. *Select multiple rules*

581

Alternatively, Figure 24-23 shows how a new customized scan rule set can be created to specify the rules that should be applied from either a list of available rules or custom rules that can be defined.

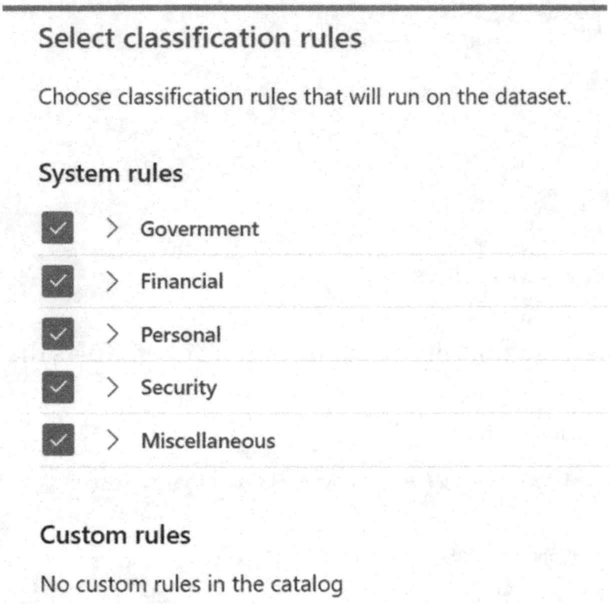

Select classification rules

Choose classification rules that will run on the dataset.

System rules

- ✓ 〉 Government
- ✓ 〉 Financial
- ✓ 〉 Personal
- ✓ 〉 Security
- ✓ 〉 Miscellaneous

Custom rules

No custom rules in the catalog

Figure 24-23. *Select classification rules*

The final step to creating a scan rule set is to set either a manual (one time) or recurring (defined by schedule) trigger. Note that there is also an option to set the recurring end date. Figure 24-24 illustrates what this scan trigger looks like. Notice the flexibility that you have when selecting the date range, time range, and recurrence of the scan trigger.

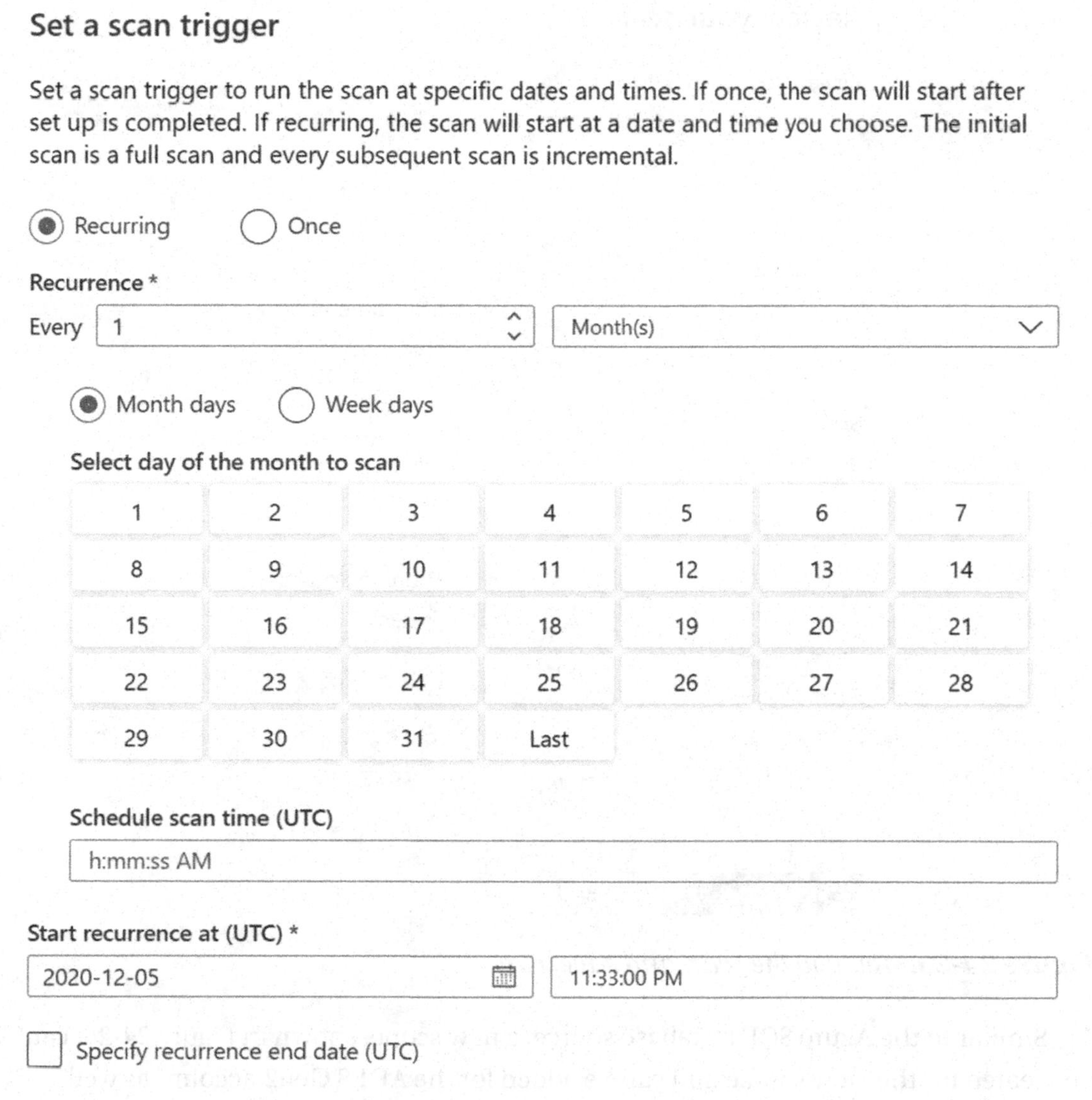

Set a scan trigger

Set a scan trigger to run the scan at specific dates and times. If once, the scan will start after set up is completed. If recurring, the scan will start at a date and time you choose. The initial scan is a full scan and every subsequent scan is incremental.

◉ Recurring   ◯ Once

Recurrence *

Every | 1 | ⌃⌄ | | Month(s) | ⌄ |

◉ Month days   ◯ Week days

Select day of the month to scan

1	2	3	4	5	6	7
8	9	10	11	12	13	14
15	16	17	18	19	20	21
22	23	24	25	26	27	28
29	30	31	Last			

Schedule scan time (UTC)

h:mm:ss AM

Start recurrence at (UTC) *

2020-12-05	🗓		11:33:00 PM

☐ Specify recurrence end date (UTC)

***Figure 24-24.** Trigger can be manual or recurring based on schedule*

Once you have configured the scan rule set, review it, and save and run it, as shown in Figure 24-25.

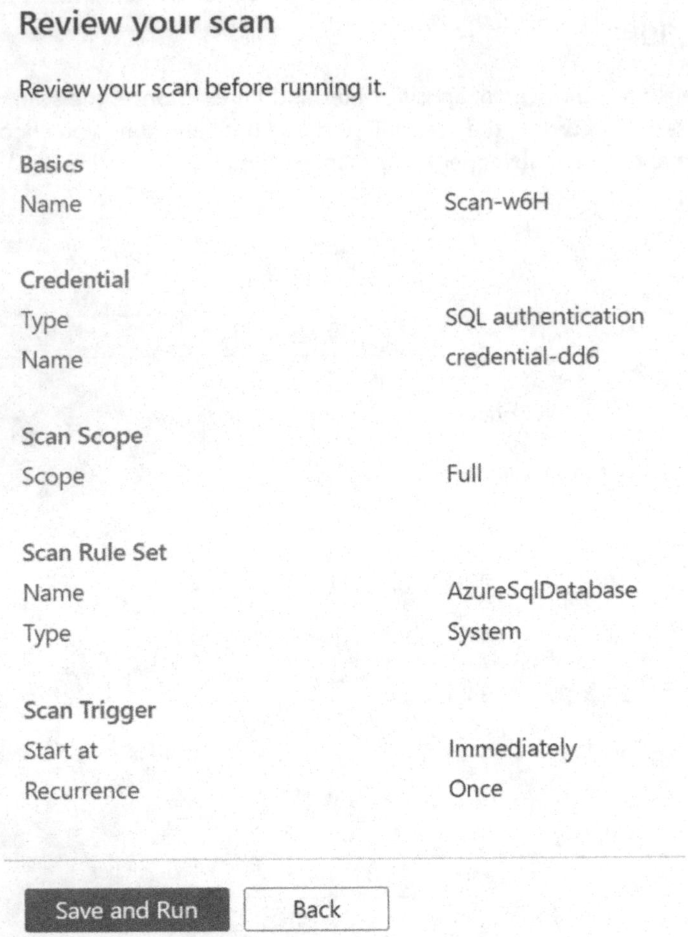

Review your scan

Review your scan before running it.

Basics

Name   Scan-w6H

Credential

Type   SQL authentication
Name   credential-dd6

Scan Scope

Scope   Full

Scan Rule Set

Name   AzureSqlDatabase
Type   System

Scan Trigger

Start at   Immediately
Recurrence   Once

[Save and Run]   [Back]

Figure 24-25. *Review the scan and save/run*

Similar to the Azure SQL Database source, a new scope, shown in Figure 24-26, can be created for this new source and can be added for the ADLS Gen2 account as well.

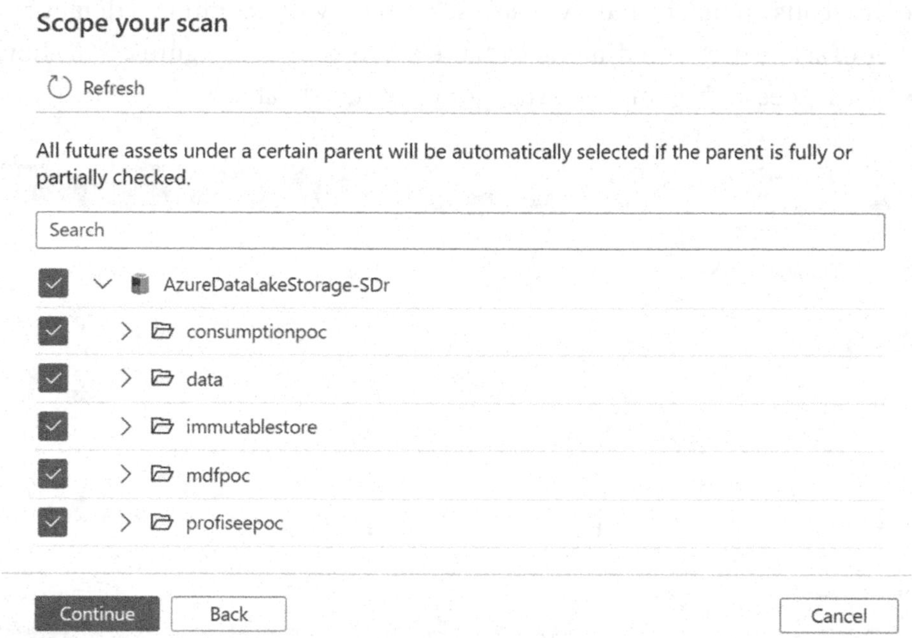

Scope your scan

◯ Refresh

All future assets under a certain parent will be automatically selected if the parent is fully or partially checked.

| Search |

☑ ⌄ ▮ AzureDataLakeStorage-SDr

☑   ⟩ 🗁 consumptionpoc

☑   ⟩ 🗁 data

☑   ⟩ 🗁 immutablestore

☑   ⟩ 🗁 mdfpoc

☑   ⟩ 🗁 profiseepoc

[Continue] [Back]   [Cancel]

Figure 24-26. *Scope the scan for ADLS Gen2*

You'll also need to create a new scan rule set shown in Figure 24-27 for this new ADLS Gen2 source. Select the system default scan rule set and click Continue.

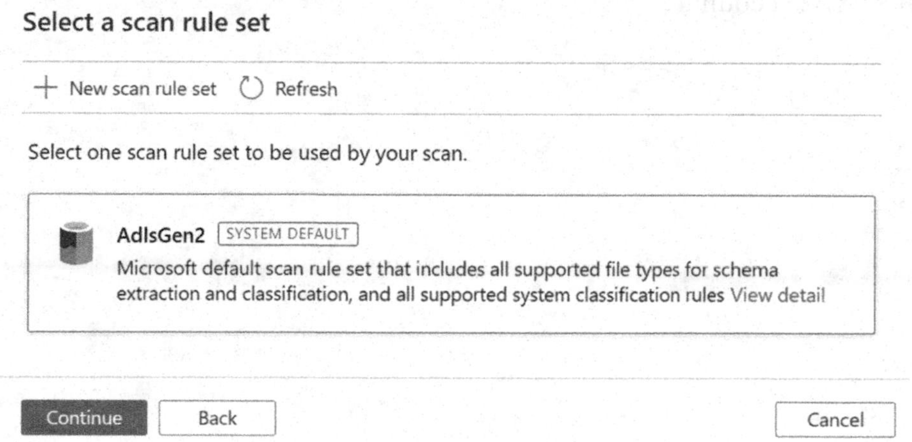

Select a scan rule set

＋ New scan rule set   ◯ Refresh

Select one scan rule set to be used by your scan.

> ▮ **AdlsGen2** [SYSTEM DEFAULT]
> Microsoft default scan rule set that includes all supported file types for schema
> extraction and classification, and all supported system classification rules View detail

[Continue] [Back]   [Cancel]

Figure 24-27. *Set the rules for ADLS Gen2*

Once the scans complete, the Overview section provides additional details related to the number of scans performed, along with the assets scanned. Figure 24-28 shows the overview of a successfully completed scan for your SQL database.

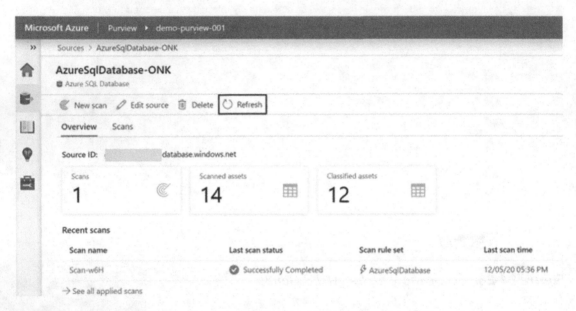

Figure 24-28. *Overview of successfully completed scan for SQL*

Additionally, Figure 24-29 shows the overview of a successfully completed scan for your ADLS Gen2 account.

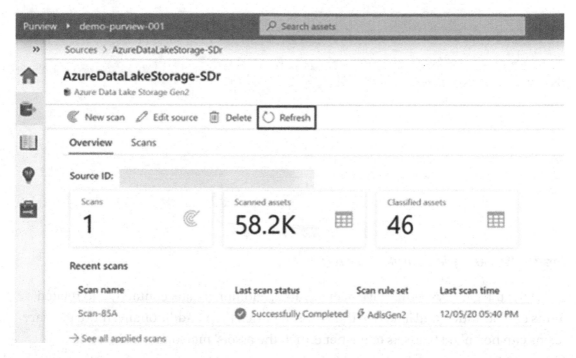

Figure 24-29. *Overview of successfully completed scan for ADLS Gen2*

Explore the Glossary

Purview's business glossary (`https://docs.microsoft.com/en-us/azure/purview/ concept-business-glossary`) allows users to define and manage glossary terms easily. Figure 24-30 illustrates the steps required to create glossary terms. (1) Navigate to the glossary by clicking the icon, (2) create a new term, (3) give your new term template a name, (4) add new attributes, and (5) click Create.

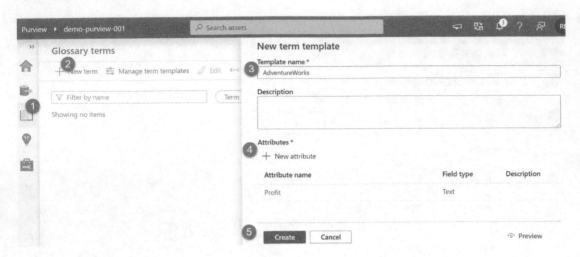

Figure 24-30. *Add new glossary terms*

Once the glossary terms have been created, additional data, contacts, and related items can be added and linked to the terms to enrich them. Additionally, these glossary terms can be linked to assets to further enrich the assets' metadata, as shown in Figure 24-31.

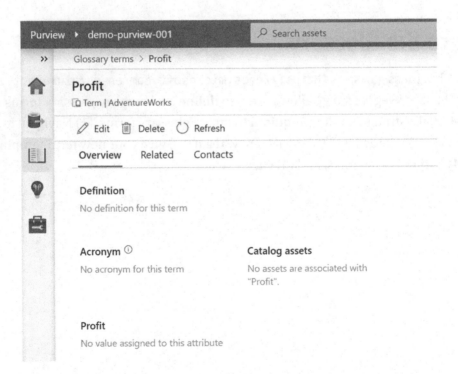

Figure 24-31. *Display of glossary*

From the illustration in Figure 24-32, notice that you can add new glossary terms using the UI.

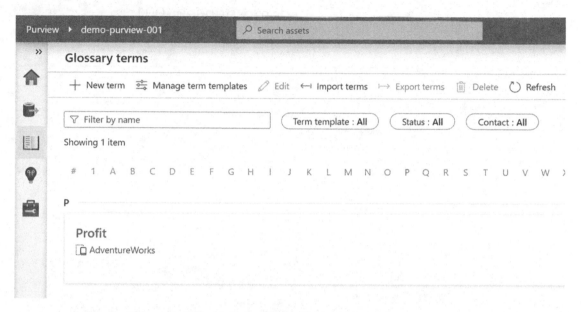

Figure 24-32. *Additional display of glossary*

Browse Assets

Finally, once assets have been registered in Purview, they can be accessed from the homepage by searching for them in the search box shown in Figure 24-33. Click into the search box and type the name of the asset you are interested in searching for.

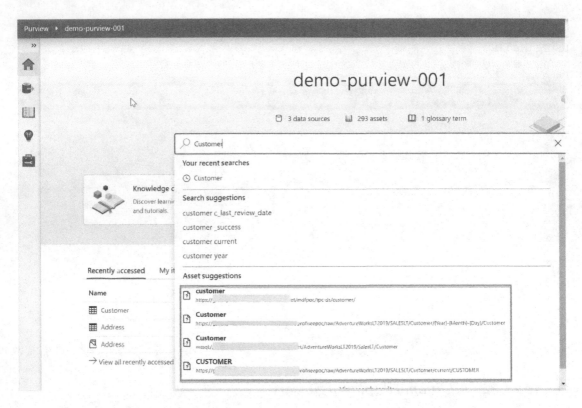

Figure 24-33. *Steps to search for assets in the Purview home screen page*

Alternatively, assets can be accessed from the "Browse assets" icon on the homepage shown in Figure 24-34.

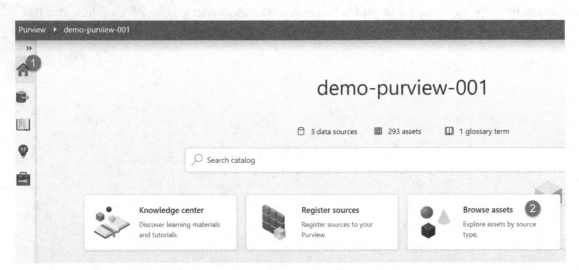

Figure 24-34. *Steps to browse assets in Purview*

The registered list of assets will also be available at a granular level, as shown in Figure 24-35.

Figure 24-35. *ADLS Gen2 assets are available in Purview*

Details along with lineage can be explored, edited, and enriched in these sections. Figure 24-36 shows the details of the Overview tab for ADLS Gen2.

Figure 24-36. *Overview of data in ADLS Gen2*

Figure 24-37 shows the details related to browsing the SQL database tables and schema.

Figure 24-37. *Steps to browse SQL tables in Purview*

Note that the Classifications section auto-captures the fields that adhere to the defined sensitivity and privacy rule sets, as shown in Figure 24-38.

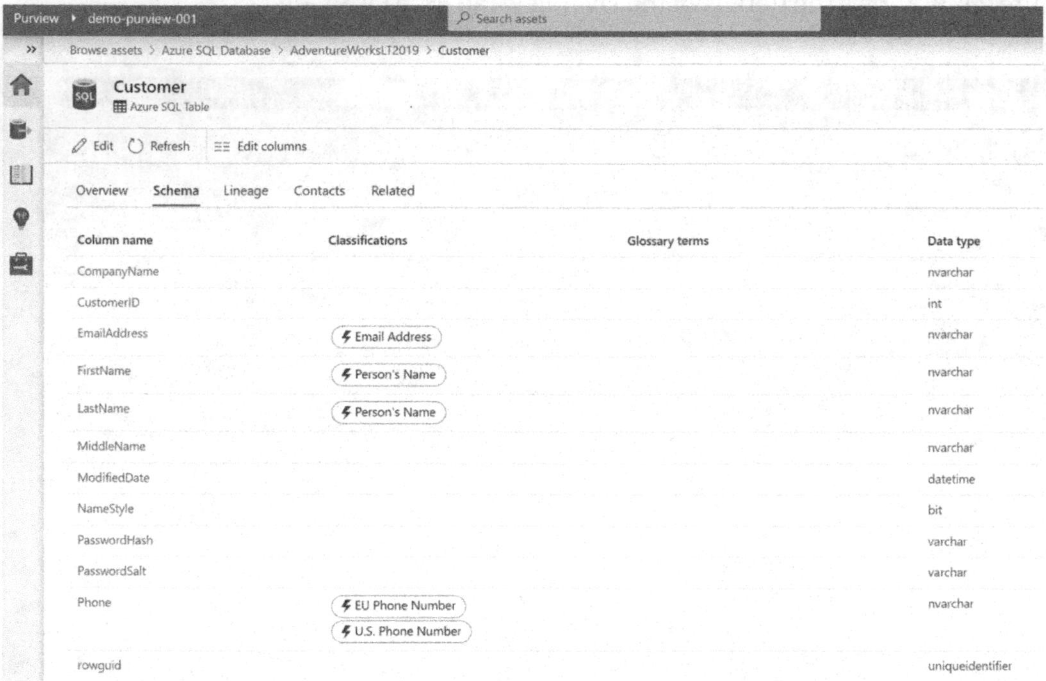

Figure 24-38. *Details that can be browsed in Purview about SQL tables*

Interestingly, within the Schema section shown in Figure 24-39, there is an option to change column names and data types of the tables, so this should be used sparingly and well governed from an access perspective.

Additionally, every field in the sample customer table can be linked to a glossary term, capture descriptions at the column level, and alter column-level classifications as needed, which help to further enrich the metadata and truly provide a detailed and unmatched data governance experience within Azure that has been missing in this space for quite some time.

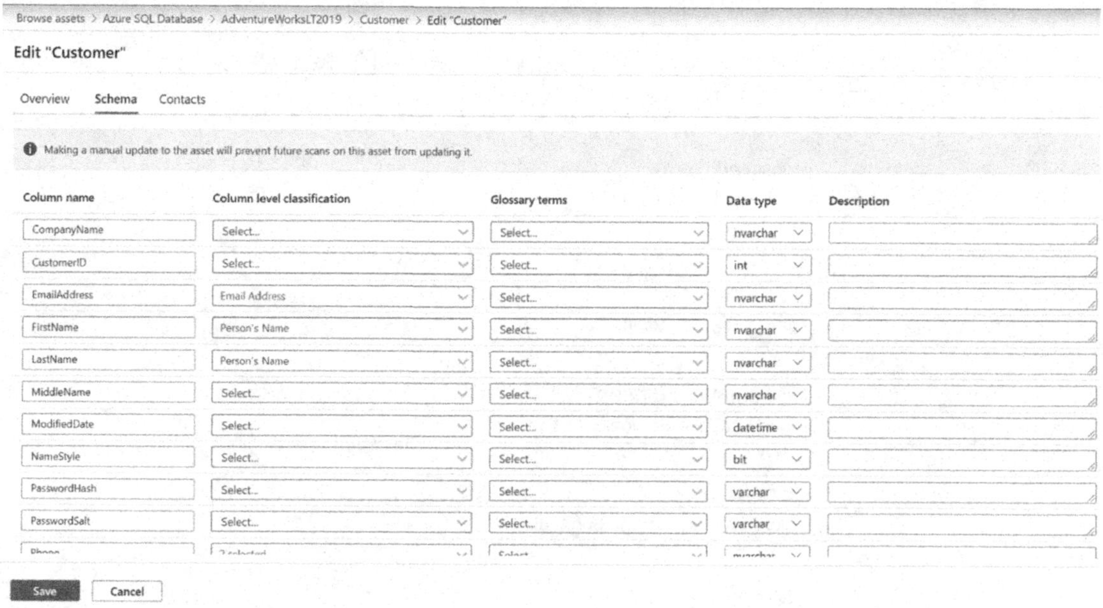

Figure 24-39. *Steps to edit SQL tables from Purview*

Figure 24-40 depicts more details about related assets in Purview. Notice the hierarchical view that helps with visually depicting and understanding the related items for your assets.

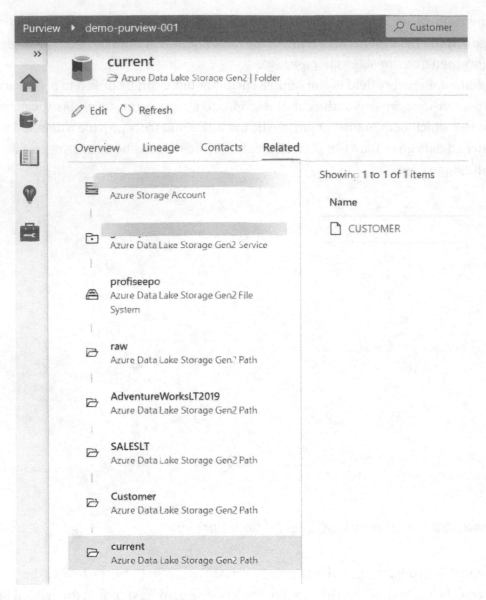

Figure 24-40. *Related assets in Purview*

By navigating to the Lineage tab shown in Figure 24-41, you are also able to view lineage of the registered assets in Purview. This is especially valuable when you are interested in learning about the end-to-end data flow process from an ETL perspective to further understand how data flows from source to sink.

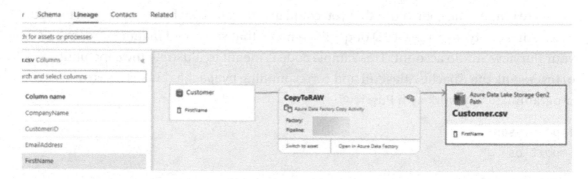

Figure 24-41. Lineage in Purview

Working with Purview Programmatically

You now have a good understanding of how to create a Purview account and work within Purview Studio to register assets, create scan rule sets and scheduled triggers, explore glossary terms, and view lineage all through the Purview Studio UI. Azure Purview also offers the capability of working with your Purview account programmatically, which we will briefly explore in this section.

Azure Purview supports integration via REST APIs for systems that Purview doesn't directly support. One such API is called Apache Atlas REST API. Atlas exposes a variety of REST endpoints to work with types, entities, lineage, and data discovery (`https://atlas.apache.org/api/v2/index.html`). To work with the Atlas API, you will first need to create a service principal application in your Azure Active Directory account in your subscription within Azure Portal. You will also need to give this newly created service principal application the "Purview Data Curator" role within the Access control (IAM) section of your Purview account. Once completed, you will need the `tenant_id`, `client_id`, and `client_secret` to begin working with Azure Purview programmatically.

At this point, you could connect to a new Databricks notebook and use the many samples available online or build your own to begin working with the Atlas API. Here is a sample GitHub repository that contains many samples that you could leverage to perform Create, Read, Update, and Delete operations on your Purview account using the Atlas API (`https://github.com/wjohnson/pyapacheatlas/tree/master/samples/CRUD`). You could also get even more creative by leveraging Azure Functions to integrate with both Databricks and your Purview account.

Here is a sample code block that you could add to your Databricks notebook to read an existing entity from the GUID or qualified name that you could manually get from your Purview Studio account. This sample code is meant to illustrate the capabilities of leveraging the Apache Atlas API and corresponding pyapacheatlas classes to programmatically work with Purview:

```python
import json
import os

# PyApacheAtlas packages
# Connect to Atlas via a Service Principal
from pyapacheatlas.auth import ServicePrincipalAuthentication
from pyapacheatlas.core import PurviewClient, AtlasEntity, AtlasProcess

if __name__ == "__main__":
    """
    This sample provides an example of reading an existing entity
    through the rest api / pyapacheatlas classes.
    You need either the Guid of the entity or the qualified name and
    type name.
    The schema of the response follows the /v2/entity/bulk GET operation
    even if you are requesting only one entity by Guid.
    https://atlas.apache.org/api/v2/json_AtlasEntitiesWithExtInfo.html
    The response of get_entity will be a dict that has an "entities" key
    that contains a list of the entities you requested.
    """

    # Authenticate against your Atlas server
    oauth = ServicePrincipalAuthentication(
        tenant_id=os.environ.get("TENANT_ID", "Enter-tenant-id"),
        client_id=os.environ.get("CLIENT_ID", "Enter-client-id"),
        client_secret=os.environ.get("CLIENT_SECRET", "Enter-Client-
        secret")
    )
```

```
client = PurviewClient(
    account_name = os.environ.get("PURVIEW_NAME", "Enter-Purview-
    name"),
    authentication=oauth
)

# When you know the GUID that you want to get
response = client.get_entity(guid="Enter-guid")
print(json.dumps(response, indent=2))

# When you need to find multiple Guids and they all are the same type
#entities = client.get_entity(
#    qualifiedName=["qualifiedname1", "qualifiedname2",
    "qualifiedname3"],
#    typeName="my_type"
#)

#for entity in entities.get("entities"):
#    print(json.dumps(entity, indent=2))
```

After running the preceding code, you'd be able to see the results about the respective GUID returned in the Databricks notebook, shown in Figure 24-42. Also consider exploring and integrating Spline (`https://absaoss.github.io/spline/blog.html`), which is a programmatic method of capturing and storing lineage information from internal Spark execution plans in a lightweight and easy-to-use manner.

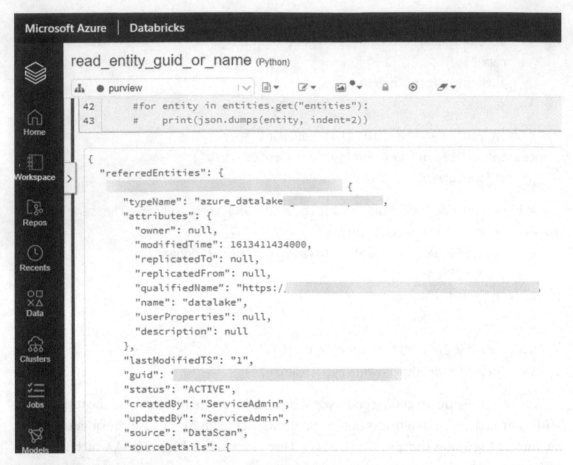

Figure 24-42. *Apache Atlas API read Purview account from Databricks*

Summary

In this chapter, I have explored how to get started with Azure Purview from setting up a new Purview account to managing credentials and access. I then demonstrated how to create and register data sources, create and trigger scans, and also browse assets. Finally, I talked about how to work with Azure Purview programmatically by using the Apache Atlas API.

Azure Purview is a promising modern cloud data governance product offering from Microsoft that offers a low-code, easy-to-use interface through Purview Studio. This ease of use offers the capability of empowering a variety of business and IT professionals to easily work with Purview. As this data governance product continues to mature, it will only grow in the features and sources that it supports.

Purview is also beginning to integrate well with other Azure products such as Synapse Analytics workspaces, which we covered in Chapter 23. Within Synapse Analytics workspaces, users can create new Data Factory pipelines, connect to a Purview account, and leverage the power of Purview to search for sources, sinks, and a variety of other assets in the Data Factory experience. This offers the capability of easily searching for and building ETL pipelines. Purview also integrates well with a variety of other Azure services such as Power BI and more, both programmatically and through its low-code studio interface. As it continues to mature, Azure Purview has the potential of being a full-fledged data governance product that could be compared to more mature cloud-based data governance tools on the market.

Index

A

© Ron C. L'Esteve 2021
R. C. L'Esteve, *The Definitive Guide to Azure Data Engineering*, https://doi.org/10.1007/978-1-4842-7182-7

N, O

P

Printed in the United States
by Baker & Taylor Publisher Services